Automation Technologies for Genome Characterization

Automation Technologies for Genome Characterization

Edited by

TONY J. BEUGELSDIJK
Los Alamos National Laboratory
Los Alamos, New Mexico

A Wiley-Interscience Publication

JOHN WILEY & SONS, INC.

New York • Chichester • Weinheim • Brisbane • Singapore • Toronto

Library of Congress Cataloging in Publication Data:
Automation technologies for genome characterization / edited by Tony
  J. Beugelsdijk.
       p.   cm. — (Wiley-Interscience series on laboratory
  automation)
     "A Wiley-Interscience publication."
     Includes index.
     ISBN 0-471-12806-6 (cloth : alk. paper)
     1. Gene mapping—Automation.   2. Gene mapping—Data processing.
  I. Beugelsdijk, Tony J., 1949-   . II. Series.
  QH445.2.A95   1997
  572.8′633′0285—dc21   96-37604

Printed in the United States of America

10 9 8 7 6 5 4 3 2 1

To the authors
for their contributions
and
to my wife Mary
for her love and support

Contributors

JOHN E. AGAPAKIS, PH.D., Acuity Imaging, Inc., 9 Townsend West, Nashua, NM 03063-1217

DAVID P. ALLISON, PH.D., Oak Ridge National Laboratory, P.O. Box 2008, MS-6123, Oak Ridge, Tennessee 37831-6123

DAVID R. BANCROFT, PH.D., Max-Planck Institute for Molecular Genetics, Abteilung Hans Lehrach, Ihnestrasse 73, D14195, Berlin-Dahlem, Germany

KENNETH L. BEATTIE, PH.D., Health Sciences Research Division, Oak Ridge National Laboratory, Oak Ridge, TN 37831-6123

BART BEEMAN, Micotechnology Center Lawrence Livermore National Laboratory, 7000 East Avenue, Mailstop L-222, Livermore, California 94550

BILL BENETT, Micotechnology Center Lawerence Livermore National Laboratory, 7000 East Avenue, Mailstop L-222, Livermore, California 94550

CHAU-WEN CHOU, Department of Chemistry and Biochemistry, Arizona State University, Tempe, AZ 85287-1604

BOBI K. DEN HARTOG, Los Alamos National Laboratory, P.O. Box 1663, Mailstop J580, Los Alamos, NM 87545

MITCHEL J. DOKTYCZ, PH.D., Health Sciences Research Division, Oak Ridge National Laboratory, Oak Ridge, TN 37831-6123

NORMAN J. DOVICHI, PH.D., Department of Chemistry, University of Alberta, Edmonton, Canada T6G 2G2

CHRIS FIELDS, PH.D., Molecular Informatics, Inc., 1800 Old Pecos Trail, Suite M, Santa Fe, NM 87505

H.R. (SKIP) GARNER, PH.D., University of Texas, Southwestern Medical Center at Dallas, 5323 Harry Hines Blvd., Dallas, Texas, 753-235-8591

DEAN HADLEY, Micotechnology Center Lawerence Livermore National Laboratory, 7000 East Avenue, Mailstop L-222, Livermore, California 94550

SCOTT P. HUNICKE-SMITH, PH.D., Stanford DNA Sequence and Technology Center and Stanford Department of Mechanical Engineering, 85 California Avenue, Palo Alto, CA 94304

SARATH KRISHNASWAMY, PH.D., Acuity Imaging, Inc., 9 Townsend West, Nashua, NM 03063-1217

PHOEBE LANDRE, Micotechnology Center Lawerence Livermore National Laboratory, 7000 East Avenue, Mailstop L-22, Livermore California 94550

HANS LEHRACH, PH.D., Director, Max-Planck Institute for Molecular Genetics, Abteilung Hans Lehrach, Ihnestrasse 73, D14195, Berlin-Dahlem, Germany

STACY LEHEW, Micotechnology Center Lawrence Livermore National Laboratory, 7000 East Avenue, Mailstop L-22, Livermore, California 94550

ELMAR MAIER, PH.D., Max-Planck Institute for Molecular Genetics, Abteilung Hans Lehrach, Ihnestrasse 73, D14195, Berlin-Dahlem, Germany

PATRICIA A. MEDVICK, PH.D., Los Alamos National Laboratory, P.O. Box 1663, Mailstop B295, Los Alamos, NM 87545

DEIRDRE MELDRUM, PH.D., Department of Electrical Engineering, University of Washington, Seattle, WA 98195

M. ALLEN NORTHRUP, PH.D., Lawrence Livermore National Laboratory, 7000 East Avenue, Mailstop L-222, Livermore, California 94550

MARTIN J. POLLARD, PH.D., Human Genome Center, Lawrence Berkeley National Laboratory, University of California, 1 Cyclotron Road, M/S 74-157, Berkeley, CA 94720

DAVID M. SCHIELTZ, PH.D., Department of Molecular Biotechnology, University of Washington, Seattle, WA 98195

THOMAS G. THUNDAT, PH.D., Oak Ridge National Laboratory, P.O. Box 2008, MS-6123, Oak Ridge, Tennessee 37831-6123

EMERSON TONGCO, Department of Electrical Engineering, University of Washington, Seattle, WA 98195

ROBERT J. WARMACK, PH.D., Oak Ridge National Laboratory, P.O. Box 2008, MS-6123, Oak Ridge, Tennessee 37831-6123

PETER WILLIAMS, PH.D., Department of Chemistry and Biochemistry, Arizona State University, Tempe, AZ 85287-1604

Contents

Preface

The discovery of human disease genes has historically been an arduous undertaking. Extensive and exhaustive studies of genetic inheritance and pedigrees in generations of families led to the discovery of the color blindness gene on chromosome Y in the early 1990s. As more biological tools became available, the pace of gene discovery increased. However, as recently as 1983, when the locus of the Huntington's disease gene was reported and the 1987 discovery of the muscular dystrophy gene, much of the biological laboratory practices were still rooted in intensively manual procedures. By October 1987, only about 1200 genes had been mapped to specific chromosomes or regions of chromosomes. Many of them like Huntington's and muscular dystrophy merited headline attention.

By the late 1980s genes were being discovered at a rate to merit only scant mention in the popular press. Yet, considering that the human genome has somewhere between 50,000 and 100,000 genes, even this "accelerated" pace is hardly breathtaking. It is especially disconcerting to see the enormous resources being applied to hundreds of individual gene discovery efforts when a concerted, large-scale effort at mapping and sequencing the human genome could efficiently address the gene discovery problem and free those resources for the study of gene function and expression. It was with this realization and the temporal juxtaposition of requisite technologies, that the Human Genome Project (HGP) was deemed possible and launched officially in 1990.

The HGP is now well under way. It is the largest, concerted effort in human genetics. The HGP traces its beginnings to the mid-1980s with a series of exploratory meetings on the feasibility of determining the physical map and ultimately the genetic sequence of the entire human genome. These meetings were attended with much excitement and high levels of energy as it was recognized that after decades of key advances in molecular biology, laboratory procedures were in place and well enough understood to begin a systematic and directed effort at this enormous undertaking. However, the scale of the project knew no precedent and far exceeded the then current experience of laboratorians.

With the Human Genome Project, the work of molecular biology expanded beyond the field in a major way. Scientists and engineers in other disciplines began to apply their crafts to the technological challenges presented by the HGP. Human genome centers were formed, funded by the United States Department of Energy and the National Institutes of Health. These centers were chartered with different tasks centered around goals of the Human Genome Project and the interests of their members.

However, the centers that built solid interdisciplinary programs have contributed the most toward the goals of the HGP. A strong presence in molecular biology with the fostering of a culture of interdisciplinary teaming marked their success. Moreover these centers were guided by a strategic vision of the leverage made possible by automation technologies.

Under the sponsorship of these centers and in other laboratories around the world, the latest advances in informatics, optical techniques, robotics, microfabrication technologies, and laboratory information management systems were applied to the challenges of the HGP. Many of the early critics of the genome factory approach have been silenced by the advances made by those organizations that effectively developed and deployed automated systems. The predominant single investigator model of biological research made room for the focused interdisciplinary team model. Successes such as those experienced at Centre de Etude Polymorphisme Humain (CEPH) in France and Human Genome Sciences (HGS) in the United States proved the value of large-scale automated assaults on the human genome. These organizations have made significant advances in mapping and sequencing knowledge, with heavy reliance on automation technologies, and have validated the automation paradigm. In addition to dramatically increasing the pace of gene discovery and furthering genomic knowledge, significant economic value has also been generated by these organizations. Contracts totaling hundreds of millions of dollars have been signed, guaranteeing access to their technology and databases in the highly competitive races for new drug discovery. Several bench scientists have become multimillionaires.

Much has been written about the science of the Human Genome Project. The science is enabled and accelerated by the technology that moves the project toward its goal. This book tells that story and brings together a crosscut of the various technologies in current use or being developed. These technologies were either not present or not extensively applied to DNA characterization at the start of the HGP. The technologies and methods were developed through the stimulus provided by the project. While many systems were initially met with skepticism, it is unthinkable for laboratories today to not rely on these machines, many of which are now commercially available. The authors' work has not only transformed the science of molecular biology but also transformed the "PI-centric" culture of biological science and validated the interdisciplinary team model of investigation.

It is impossible to present in this volume all of the technologies currently being developed or being used in the HGP. Such an undertaking is, necessarily and happily, always incomplete. Exciting new ideas are generated daily as the skills of other disciplines are brought to bear on the challenges of the HGP. Many of the work horse technologies that could have been featured in this book such as fluorescence-based sequencing and flow cytometry, to name two, have already been well documented elsewhere. A review of these techniques would have contributed little new information to an already large body of knowledge. Indeed, this suite of existing technologies form collectively the "shoulders of giants" that the HGP stands on. Omissions of such technologies from this work are not intended to diminish their importance or impact.

One of the early goals of the HGP was to develop detailed physical and genetic maps of each of the human chromosomes. Unlike DNA sequencing, there was a paucity of commercially available technology to meet the mapping task. The market for such technology is specialized and largely restricted to the large human genome centers, so no technology products were even under development. This situation presented an opportunity for the larger centers and laboratories to staff instrumentation development groups to build the automated systems required. While some of these early systems had limited success, many of them were later refined with the feedback of experimentalists and became very important in the successes of their sponsoring centers. Some have even achieved commercial status.

This book is organized into four sections. The first section describes laboratory automation activities at several major human genome centers and research laboratories. A consistent philosophy emerges for successful implementation of new automation technologies based on the experience of these authors. This section also includes a chapter on imaging technologies and the major role played by them in the image-rich biology laboratory. The second section discusses control system approaches. Both chapters in this section address the problem of integration of dissimilar systems. General features for control systems are distilled from the experience of these authors. The third section of this book describes some advanced and nontraditional technologies being applied to DNA characterization. Finally, this book covers some of the analysis, modeling, and database issues accompanying the characterization of genomes.

Dr. H. R. (Skip) Garner describes his experiences and success of the University of Texas Southwestern Medical Center's laboratory in their automation efforts. He outlines the requirements for assembling an interdisciplinary team and a successful strategy in approaching automation projects, and describes the cultural and nontechnical issues surrounding the acceptance of a new technology. The extremes wherein home-grown automation projects arise are those between "little utility and commercially valuable." While this gives birth to opportunity, it is also a limiting factor in widespread technology transfer. He underscores a common experience that automation is not a product but a continuous process of which only a portion is technological.

Dr. Martin J. Pollard of Lawrence Berkeley National Laboratory (LBNL) describes his labotatory's approach to automated systems design. The bottom-up approach used at LBNL was dictated in the early stages of the HGP by the need to establish new working relationships with experimentalists and the low funding levels for automation development. Building on early successes, LBNL has built several systems that have been of great utility and commercial interest.

Drs. Elmar Maier and David R. Bancroft of the Max-Planck Institute for Molecular Genetics present their organization's hybridization approach to the construction to physical maps and their development of successive generations of robots to meet the demand generated by their reference library database. They have developed and describe their systems for clone picking, high-density filter array generation, image analysis, and high-throughput polymerase chain reaction (PCR). Several commercial products trace their origins to their very innovative laboratory.

One of the key technologies incorporated into the above systems is machine vision. Drs. Sarath Krishnaswamy and John E. Agapakis of Acuity Imaging, Inc. describe machine vision and give an overview of the contributions imaging has made to machine guidance as well as data collection. They discuss the development of a system that uses machine vision in conjunction with a motion control device that enhances a previously "blind" process—that of colony picking. A once tedious and error-prone operation has become very accurate, efficient, and highly automated using vision technology.

One of the pervasive difficulties in building automated systems is the lack of interconnect standards and interfaces for both hardware and software modules. Each system thus becomes a unique design, and except for systems developed at a particular institution, exchange of systems or components is not possible. While standardization activities are underway, progress has been very slow. The two chapters in Part II of this book are devoted to attempts to develop a standardized software interface. Dr. Pat A. Medvick and Bobi K. Den Hartog of Los Alamos National Laboratory (LANL) describe a high-level software controller developed at LANL. Building upon and extending an existing automation effort at LANL, they describe the use of "Services" which, when linked together, become a program called the script of a method. Taking another approach, Scott Hunicke-Smith has developed the Graphical User Interface to Laboratory Equipment (GUILE) technology. GUILE was created to present a consistent, extensible, yet simple interface to devices and protocols. Both developments recognize the fundamental interconnection and integration problem facing developers of new automation systems and propose models for standardization. While still early to predict widespread commercial acceptance of these technologies, the ground-breaking work of these authors generates a certain momentum and features of their work may eventually emerge in future powerful development and integration tools.

Part III of this book treats several emerging and advanced technologies that show great promise.

In Chapter 7 Dr. Norm J. Dovichi discusses recent developments in capillary gel electrophoresis. The last decade has witnessed major improvements in gel technology, speed, and efficiency of separation, miniaturization, and detection. The early manual Sanger and Maxam-Gilbert sequencing protocols yielded little more than a few hundred bases of raw sequence per day. The use of fluorescent labels, parallel capillary arrays, and laser-based detection has led the way to dramatic increases in sequencing rates. Modern technologies have given an old separation technique a significant new life: Rates of several million bases of raw sequence per day are now imminent.

Drs. Dave P. Allison, Thomas G. Thundat, and Robert J. Warmack at Oak Ridge National Laboratory (ORNL) review the application of scanning microscopy techniques to mapping and sequencing DNA. Both atomic force microscopy (AFM) and scanning tunneling microscopy (STM) have been used successfully to image DNA strands. Current research focuses on increasing the image quality and developing techniques to read actual nucleotide sequences.

Miniaturization presents some of the most exciting possibilities for genome characterization. The marriage of semiconductor microfabrication manufacturing techniques with microchannel fluid devices and chemistry on solid supports brings the possibilities massively parallel and dramatic speeds we have come to associate with computers to DNA analysis. Dr. M. Allen Northrup and co-workers of Lawrence Livermore National Laboratory work at this molecular biology–semiconductor technology interface.

Drs. Mitchel J. Doktycz and Kenneth L. Beattie present another implementation of semiconductor technology in their work with "DNA-chips." These devices, or "genosensors," consist of individually addressable hybridization targets covalently linked to silicon wafers in high-density arrays. Massively parallel experimentation is achievable with these systems. Their application to DNA sequencing and DNA diagnostics is discussed.

Dr. Peter Williams and coworkers at Arizona State University give an overview of the application of mass spectrometry to DNA characterization. The recent development of electrospray ionization (ESI) and matrix-assisted laser desorption-ionization (MALDI) techniques have permitted the study of large biomolecules without excessive fragmentation. Mass spectrometry has an inherent speed advantage over gels and can be used where short, highly accurate sequences need to be determined. The techniques they describe form an important emerging suite of technologies complementary to gels for sequence determination.

The final section of this book describes work in the areas of simulation and data management. Simulation studies are very valuable in designing a system. Significant improvements can be made at this stage to process, system, and machine design. No automation effort is complete without consideration and design of the databases and the uses they serve. Many are designed after the fact and, if poorly designed, can negate the entire benefit of automated systems.

Dr. Dierdre Meldrum and Emerson Tongco of the University of Washington describe the use of Petri nets in simulation of the "genome factory." The analogy of agents, parts, operations by the agents on parts, and process conditions or states closely mirrors industrial manufacturing operations. The graphical representation of the Petri net model makes it relatively easy to simulate on a computer, to study the performance of the system without first reducing it to hardware, and also to translate the model to control code for the "genome factory."

Dr. Chris Fields of the National Center for Genome Resources addresses the information management issues of the HGP. The traditional style of information management typical of small laboratories is entirely inappropriate to production laboratories in which high throughput, efficient division of labor, and cost effectiveness are central concerns. Published results from genome laboratories are not complete works; they are starting points for further work to be carried out by others. The working and intermediate nature of these scientific results has raised the data management issues to the point of funding agency policy. Both the highly automated generation of large data sets and public expectations of the data impose demands for information systems beyond those of the small-scale biology laboratory. Dr. Fields presents a model for database design and engineering and argues eloquently for these systems to be designed during the conceptual phase of constructing genome factories.

These authors who have contributed to this book have formulated a vision of the future. This vision is based on weaving together the best technologies other discplines have to offer and the enormous advances these new technologies and methods have to offer to the science of biology. Through their creative energies they have made that vision a reality.

It is with great pleasure that I have worked with the authors who describe their work in this book. I have learned much from their contributions and anticipate that my experience will be shared by others.

Tony J. Beugelsdijk

Automation Technologies for Genome Characterization

LABORATORY AUTOMATION

1

Custom Hardware and Software for Genome Center Operations: From Robotic Control to Databases

H.R. (SKIP) GARNER

CONTENTS

Automation Technologies for Genome Characterization, Edited by Tony J. Beugelsdijk.
ISBN 0-471-12806-6   © 1997 John Wiley & Sons, Inc.

INTRODUCTION

Each center or investigator in the genome project takes a different approach to producing physical maps, sequence, reagents, and any other type of data or results [1, 2, 3, 4]. The particular approach, the scale of the research, and the particular predefined concepts of the investigators and their team determine how much is to be automated and how. Of course this can lead to tremendous successes for the automation component of a genome research project, namely the fluorescent automated sequencer, but it can also lead to systems being developed of little universal utility or even justification [5]. Some strategies, such as large-scale M13 shotgun sequencing, have a mix of automation that relies little on systems that have been integrated [6]. For some strategies, automation can play an important role, while in others, only a little automation or very modular automation of specific tasks may be sufficient.

In this chapter the bulk of the discussion will be on proven technologies, namely systems that have been developed and are deployed. These systems, which include both hardware and software, are in use by production biologists. Toward the end of the chapter, a discussion will be made of some new, yet unproved approaches being investigated. A further intent of this chapter is to provide insights, hints, and tool resources identified as having true utility. These were accumulated over a history of developing hardware and software, with both successes and failures, that may be of aid to others setting out to develop new systems on their own.

Before proceeding with details of development tools and specific automation systems, a few observations of the process follow. Some of these issues may be as important as the concept and execution of the concept on the final success of an automation project.

Scientific Demonstration versus Deployment

It should be emphasized that there is a large gap between a demonstration of an automation or software system and its mature form, such as can be used daily without incident by active biological experimenters, not the developers. Occasionally it is possible to make a new piece of hardware and its associated software work almost immediately, but that is rare. The time until it is truly useful is inversely proportional to its complexity. For simple modules like new software for the Beckman Biomek robots, this can be almost immediate, certainly within a day with sufficient testing. But even for the simplest systems, training of the user biologists is always necessary. To aid in the conversion of an instrument or software package from the development team to the biologist user, it is essential to make complete user manuals and to keep those manuals up to date as modifications are made. Manuals are valuable references, for most people do not read them even if they are written with painstaking clarity. It is therefore important to produce a simplified step-by-step procedural list that can be referred to regularly. The manual and list are greatly simplified if the system is simple and easy to use. Many users are uncomfortable using computers, so for some systems that involve interactions with software, the user interface is kept simple. Training may also be required on the general use of computers.

For more complex systems, like the Prepper, PhD. system discussed later, the development time was about one year, but it took another year of repeated attempts to infuse the system into production biology, until it became sufficiently robust for daily use by nonengineers. The canonical number for commercial instrument developers for the expense (time and money) in bringing a new instrument to market is 5:1. That is, five times the effort is required in bringing a system to salability than was originally expended in the development of the demonstration system.

Qualifications of the Development Team

The automation system development team must fully understand the task to be automated before attempting to start a project. In the lab, this means two things: (1) Operations research has concluded the need for the project and the proposed solution has a potential to work, and (2) most members of the development team, especially the team leader, has sufficient hands-on experience performing the task manually to understand the intricacies of the biology. A typical multidisciplinary team consists of two or more of the following: a biologist technician, physicist, engineer, computer programmer, or electronics/mechanical technician. Each member with a nonbiological background should obtain a minimum broad-base knowledge of biology methods and procedures by reading and performing the experiments in texts like *DNA Science,* a very good text [7]. They have also attended an in-house class, Genomics 101, that covers the specific application of biological methods to genomics and the Genome Sequence Sampling (GSS) strategy that a group is executing is also valuable [8]. For the biologists in the group, they have had to become familiar with the hardware, software, and tools of the physicists/engineers to effectively communicate details obtained from experience that the nonbiologists would have missed. The team leader and those most involved in a given project must first master the at-the-bench manual version of what they are trying to automate or improve. In the simple example, the construction of the Timing Resource Electrophoresis Controller (TREC) system, it was important for the team to make agarose gels, load them, electrophorese the samples, and record the results. This gave the builders of the systems a feeling for the ergonomics of the system to be built (how close vertically the gel boxes can be spaced) as well as functionality; the rails that the gel box shelves move on must move smoothly to prevent unwanted jostling of the samples, possibly washing them from the wells, and the parameters to be controlled by the software.

Prototyping Hardware and Software

There is an intermediate step between the manual performance of a task and a fully customized or integrated system. Following operations research, which determines that a particular bottleneck can be solved without creating disproportionate demands on other parts of an entire process, a decision to build a system is made. It is then possible in many cases to prototype a process on a general purpose laboratory pipetting robot, centrifuges, and analytical instrumentation. One system of choice for the pipettor robot is the Biomek 1000. This is the older model that can be programmed using Quick Basic and therefore can perform any rudimentary function within its operating envelope. The newer Biomek 2000 has yet to have a scripting language or basic routine library made available for it, so prototyping can be difficult [9]. Of course any of the other available general purpose machines can be used, but it is best to use one that the group has used. As an example of a prototype, the Biomek 1000 was first programmed to manipulate samples with a new 864 well plate, designed before attempting any custom hardware. This demonstrated that the 20-μl, 2-mm diameter wells could be reproducibly addressed by the pipetting system. It did indicate a need for a new type of tip, one that interfaced with the standard Beckman pipetting tool but had a smaller, more precise end like that on a standard P2 pipetting tip. Luckily a vendor was found who could work with us to make that tip. The vender, Robbins Scientific, was able to make it into a very successful, general purpose product [10]. In this case the prototyping was successful and had a broader outcome than originally planned. The pipetting to and from the 20-μl well showed that pipetting (aspirating or dispensing liquids from/ to the pipette tip) and vertical motion were required because the pipette tip was of a significant volume compared to that of the well. This lesson was important for the implementation of the pipetting steps to later be done on the custom pipetting station on the Genome Automation System (GAS) discussed later. In another case a gel-loading system was prototyped on the Biomek 1000 that worked so well that the original goals could be met without a dedicated system. A dedicated system may have had advantages over the Biomek 1000, since more gels could be loaded at once. Such a system could be integrated into a larger system that could incorporate the electrophoresis and documentation parts of the process. The Biomek 1000 implementation has shown to be so sufficiently fast and robust that it can keep up with our current demands, reducing the need for a dedicated system and freeing up developer time for other projects.

Acceptance by Production Biologists

The acceptance threshold of biological technicians to new approaches, methods, or hardware is significant. Biological technicians are resistive to giving up control of their experiment to a piece of equipment, and that threshold is even higher if the equipment is complex to operate or fails in the first initial trials. Therefore, in the laboratory, one strives for ease of use—startup and shutdown procedures, simple user interface—and robustness. Obtaining robustness is difficult and time-consuming. Equipment that works well and reliably is what technicians are accustomed to from the commercial market. It is therefore difficult, and can be very counterproductive, to introduce an early prototype into the lab. Instead, it is recommended that some select biologist technicians work the development group, become desensitized to the system development pains, and be involved in the first use of the systems. Further, as part of the process of infusing custom equipment or software into the lab, the development team must for a time become operators. They perform some of the production biology so that they and the equipment can gain experience to determine things like equipment drift, mean time between failures, and its true value to the production effort. After this time in the lab, it is necessary to develop a complete set of instructions that can be used to train the production biological technicians. Training is necessary before they can assume full responsibility for the operations of a system. For prototype machines the automation development group still maintains and services the machines to minimize downtime. The bottom line is that the value of a system to the production biology can be seen in two ways: (1) if it increases productivity or reduces cost and/or (2) if when the machines breakdown, there is a demand by the production technicians to get them back on line as soon as possible!

Exporting and Importing Technology

Because of the uniqueness of the mission of each genome center, the importing of new technology, especially prototype-level apparatus developed by another genome center, may not be practical or desirable. However, there exist occasions when something of universal, or pseudouniversal value need to be exported or imported. There are many ways of doing this: The best is to commercialize a piece of equipment so that it can be developed completely and supported in all its new sites. This has been past practice; the Applied Biosystems Incorporated (ABI) sequencer family is an example. Commercialization was used to export plasticware, software, and spectrophotometer accessories from the San Diego and UT Southwestern Genome Centers, and that has worked well. There is a problem, however, when the profitability falls for a given device or is marginally profitable. Genome-specific equipment may not have a wide enough audience to maintain it on the market. Another method for exporting or codeveloping is to identify a need for an item that would be universal and profitable, and then to work with a vender that can make and sell it. The Robbins Scientific example, to develop a new type of tip for the Beckman Biomek robot family, was cited. Other groups have taken such a tact. That worked in our case because a company like Robbins Scientific has a niche market and is able to make new injection molds for biological products. Of course the genome labs know when something is or will be valuable and this is especially true if a genome center can guarantee purchases for the product to be made, namely showing that there is a sales base.

The most challenging to import/export are custom automation systems and genome task-specific accessories or software. To transfer technology successfully, several conditions must exist: (1) sufficient documentation of the device, (2) a stable and rugged device or software, (3) expertise by the importing organization to replicate or support the item, (4) a solid communication link established between each group, (5) enough upper-level support and resources to complete the task, and (6) a true desire by both the importer and exporter to get the technology transferred. Experience has shown that having all these conditions present together is rare. As a consequence not much technology has been transferred. The Lawrence Berkeley Laboratory (LBL) colony picker and the Staden Software package are examples [11]. To assist others in evaluating what is seen in the national and international forums, several groups put technology information as well a genome biology data on their World Wide Web Home Pages [12]. From the UT Southwestern Web Page (http://mcdermott.swmed.edu/) detailed plans, control software, and photos of systems can be downloaded. However, the web page is insufficient to construct anything except the most simple devices. When the UT Southwestern Automation group desired to replicate the Stanford Oligo Synthesizer System, a trip to the exporting site was required [13]. It should be mentioned that there is another barrier to successful importing of technology. If a technology falls into the category that it is commercially viable, or potentially so, the exporters or the exporters organization may be reluctant to release all the information required to replicate the device for licensing, patent or legal. This may require a group to first look for another technology that will perform at adequate levels and to then design their own system from scratch. Patent and legal issues led the UT Southwestern group to also study the LBL oligo synthesizer design and then construct a similar, but upgraded version. There are two conclusions to be drawn: First, direct importing may only be possible for systems that lie in the middle ground, between little utility and commercially valuable, and second, groups and the funding agents may sometimes be forced into the situation where re-inventing is required. Re-inventing of a technology may be required for a number of reasons (legal, different development team expertice mix, unavailability of some unique hardware or software, ability for the receiving group to construct or support an exact copy of a system), and since the funding agencies have usually paid for the initial research, and the receiving group probably similarly funded, the agencies could be paying more than once for a given technology. This is not necessarily bad, for the missions and approaches of a given group may be sufficiently different that a complete remake of a system may be required. But in some cases where the exact technology has been perfected but is not available to the wider community because of legal, publication, or commercialization reasons, there may be needless duplication of effort and cost.

Operations Research

The application of operations research techniques, which is the systematic study of complex problems to better understand them and improve them, is especially valuable in attempts to automate a process. Each step of a process is studied to determine what improvements can be made, how to best scale up the step, how to distribute development resources, and what steps to take to guarantee balance, monitoring and planning for the creation of new bottlenecks, guaranteeing that the output and format of one step matches the input requirements of the next step. This is of particular importance in scale-up when the organizational paradigm changes drastically. Most biological processes are executed in a linear and serial manner, which for a small or focused experiment will yield a detailed set of data about a small experimental sample set. Operations research will focus on how to parallelize certain steps, the logistics of each subprocess, and where holds or pauses can occur without adversely affecting the results. As an example, given 4 technicians, each can operate independently and process 20 fingerprinting samples each day, including culturing, prepping, pouring agarose gels, digesting, running the samples on an agarose gel, and data analysis. A more efficient approach would be to process the samples out of sequence. One technician would pour many gels for everyone and store them until needed, another would do the digestions, and another would load and run the gels, and the last would analyze data. The problem here is that while this may be more efficient, technicians can loose the feel that they control their data and can also become bored with handling only a subprocess. Operations research also involves a balance between process efficiency and technician efficiency to obtain the best overall efficiency.

Scale-up Decisions

A decision to scale up can be very disruptive to the instantaneous throughput of a laboratory or process. The changes necessary to make a meaningful increase in productivity, a factor of 3 to 10 per person, often require a complete shift in operational paradigm, as just mentioned. And no scale-up should be attempted without much experience in how the process is performed. The decision to make a major change comes when it is realized that just adding additional effort, without any assumed productivity enhancement, will not achieve the targeted goals when the proper planning has been done and when all the equipment and new methods are tested and in place ready to infuse into the production effort. A decision to automate may also be driven by the desire to investigate how much can be done with automation—by an automation experiment. However, again success is judged only by the rate of the biology being completed and the cost.

Productivity Monitoring

By monitoring the throughput of the center at its various stages, we have been able to predict equipment failure, identify bottlenecks in the processes and mix ups in the sample or data handling, and determine the best task and technician match. Database software in use in the laboratory as an electronic notebook can be depended on for the productivity/process monitoring as well as experiment planning and coordination. Described below is a suite of databases written in the database shell, 4th Dimension, that perform these tasks in the Southwestern genome center [14]. In the first example, the average read length tracked over hundreds to thousands of samples has shown in one occasion a continuous downward trend in the average read length of sequencing from cosmids. After noting this trend, the process was inspected for technical drift by the technicians. After running and checking standards, eventually it was traced to one ABI 373A sequencer. The problem was a failure of the optical filter system in the machine. As a second example, in fingerprinting cosmids by enzyme digestion in mass (~400 cosmids/day in two enzymes with membrane transfer and end probing), each step of the process is monitored using a database. In this experiment the database indicated where additional technician help was needed and when technicians had to be reassigned to make sure that each step in the complex pipeline is not waiting on a previous step.

GSS Strategy

The UT Southwestern Genome Science and Technology Center (GESTEC) is pursuing a strategy called Genome Sequence Sampling [1, 15] that is a prelude to complete, contiguous genomic sequencing of human chromosome 11. It is this strategy, and all the methods and equipment that it encompasses, that are being automated. In addition the center conducts automation, informatics, and instrumentation research of a higher-risk, longer-payoff nature. As a starting point the GSS approach utilizes the low resolution Yeast Artificial Chromosome (YAC) based maps which have been constructed by the center and others [16, 17]. Efforts are continuing to increase the density of Sequence Tagged Sites (STSs) on the low-resolution map. The highest-quality and most efficient YACs, on which to base higher-resolution maps, can be chosen. From a minimum tiling path set of primarily nonchimeric YACs, cosmid libraries can be probed via hybridization or the YAC can be subcloned into cosmids to produce a set of cosmids with high depth that can be assembled by standard, but automated, enzyme-digested fragment size analysis. Each approach has its advantages and disadvantages—biologically and in its automation. From this contig, sequence can be produced in basically two paths, either directly from the T3 and T7 vector ends of the cosmids or by shotgunning from a selected number of minimum tiling path cosmids. In the former a very dense set of sequence samples are produced, approximately 350 bp every 2 kb for a ~20x deep library. This is a prelude to complete sequencing that can be done via a short parallel walk using primers designed from the sampled sequence. Also long-range Polymerase Chain Reaction (PCR) can be done to recover any DNA segment less than ~30–40 kb between any large gaps, and standard PCR can be used to recover DNA from the typically <2 kb gap. In the latter, once the cosmids are mapped, the minimum tiling path cosmids can be subcloned efficiently into M13, which can then be sequenced using the well-proven shotgun approaches (see Figs. 1.1 and 1.2).

Robotic Hardware and Control

BIOMEK 1000   Fixtures have been built that adapt agarose gel casting trays (for gel electrophoresis) to the Biomek robot (see Fig. 1.3). Along with specialized software written in Basic, and the robotic control Basic library, Biomek QB, technicians can now load 40 to 120 lane gels (including marker lanes) automatically in about ten minutes and then prerun the gels to drive the samples into the gel prior to moving

GSS fingerprinting and contig
building

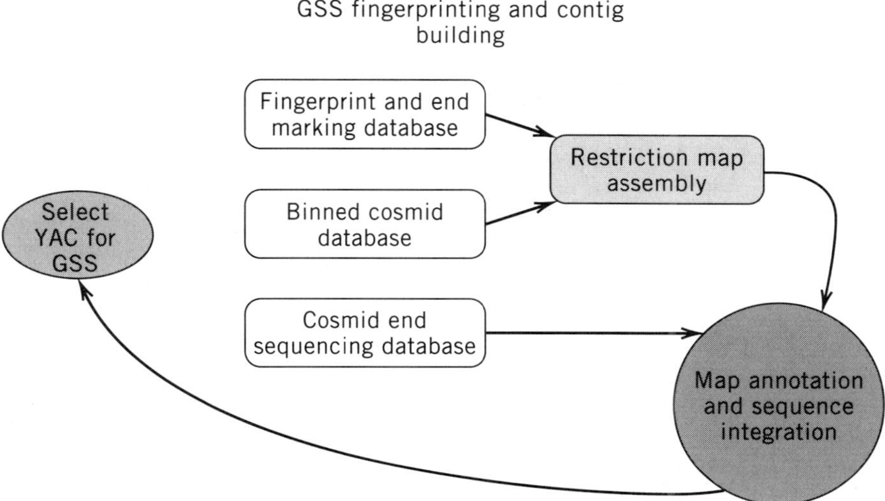

Fig. 1.1 GSS Strategy, a progression from low-resolution YAC based STS content maps to high-resolution cosmid based maps with sequence sampling to complete genomic sequencing is carried out using YAC and cosmid libraries.

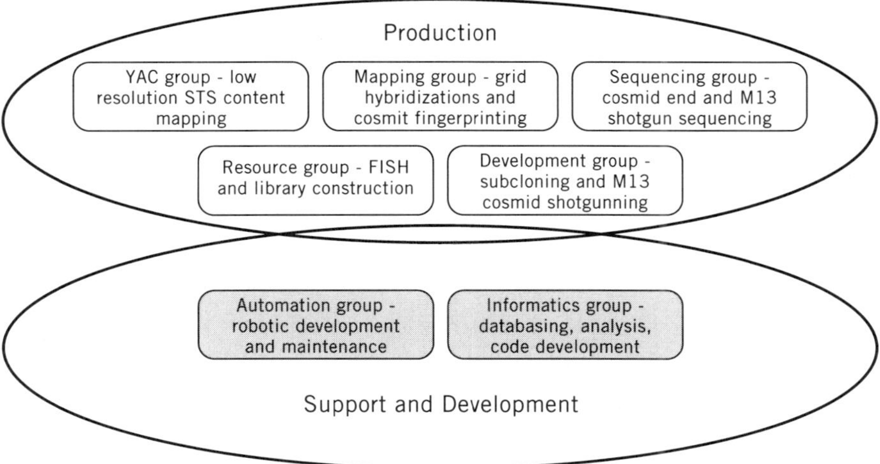

Fig. 1.2 Organization of the research groups was done to optimize the throughput of the various production biology groups. Production modules do cosmid end sequencing, fingerprinting, and binning (grid hybridization using mapped YACs).

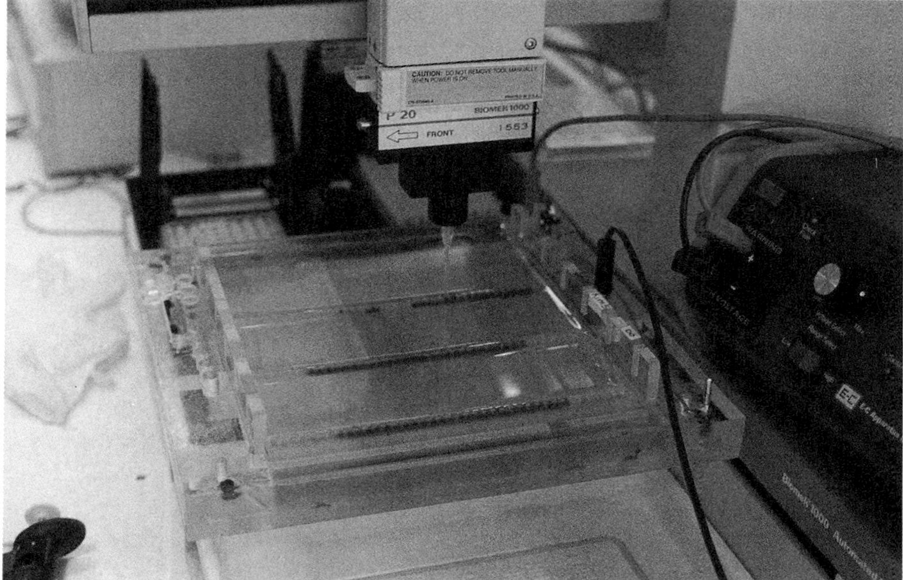

Fig. 1.3 Beckman Biomek 1000 gel-loading fixture in place on the tablet of the robot. This fixture holds the gel-casting tray, has provisions for microfuge tubes to hold markers, and is electrically connected to power supplies for gel pre-running. The fixture consumes two spaces, leaving room for tips and a sample plate on the tablet.

the gel by hand, to the standard gel boxes that operate manually or under TREC control. Upstream software that assembles and then incubates enzyme digestion of samples in various plate formats for restriction fragment mapping has been written. Special fixtures to hold enzyme reaction mixtures have also been constructed for efficiency and to conserve operational space on the Biomek tablet. Air-heated incubation towers that fit over the Biomek side loader have also been built for culturing or enzyme digestion.

Using the versatility of the Biomek QB routines, it is possible to use any plasticware format (96, 384, and 864) and to address any custom fixture like the gel-loading fixture above. The Biomek robots are installed in controlled environments with UV sterilization. This has allowed for prefilling of 384 well plates used on a commercial colony picker (Hybaid) that is not capable of dispensing uniformly at this low volume (50 μL) [18].

An 864-pin and 384-pin replicator tool and associated software were developed for the Biomek 1000 SL. Pin sets for the tool are available for transfer of 0.5- to 10-nL volumes. New software for the Biomek to efficiently transfer existing libraries from the 96-well to 864-well format was also developed. With these tools, libraries can be replicated or stamped out on membranes. For larger-format membranes (6000 clones), a Hybaid colony picker/transfer spotter robot is used.

BIOMEK 2000 The Biomek 2000 new generation machine, with its enhanced precision tools (transfer volumes down to 0.5 μL), is being used to set up sequencing reactions. With its larger work space (12 plates) it is possible to set up 96 dye primer reactions at a time or 4×96 dye terminator reactions. Two Biomek 2000s are sufficient to set up the cycle sequencing reactions for eight sequencers run at least twice a day (36 lanes). In 1996 the new scripting language for the Biomek 2000s should become fully capable and be of more utility. Plans are to download from the Sequencing Unit–Mapping Unit (SUMU) DatAssistant information about the quality and quantity of DNA template, which will allow the Biomek 2000 to set up the reactions using varying amount of DNA and increase the success rate of sequencing. The current success rate (number of samples with greater than 300 bp read) is 30% for dye primer chemistry and 71% for dye terminator chemistry [19].

PREPPER PhD   The centrifuge-based DNA prep system, Prepper, PhD., can process a variety of different samples (cosmids, plasmids, YACs, and M13) in 1- to 2-ml deep well plates using a number of different protocols. This machine produces DNA for use in automated sequencing, fingerprint mapping, and STS content mapping (YAC and cosmid library screening) [20, 21]. Operations are completely done by the machine, and take from two to five hours to process 2×96 samples, depending on the type of prep and parameters, such as the number of ethanol precipitations. Consumables costs are about $0.05 per sample using this machine. There are plans to upgraded to 4×96 samples per run. Prior to making this machine, a sensitivity study of the various parameters involved in DNA preparations was conducted, in particular, the Alkaline Lysis method [22]. The most significant result was that pelleting can be done at 1/10th the centrifugal force typically used, which allows us to process samples in a deep well plate (see Figs. 1.4 and 1.5).

The quality and quantity of DNA produced by this machine make it useful for dye primer and dye terminator sequencing chemistry directly from the machine output without further cleanup steps. For cosmids, a 6-mL culture is grown in 15-mL snap cap tubes overnight, centrifuged, resuspended in 200 μL, and transferred to 96-well deep well plates. The yield from the 6-mL culture is sufficient for two enzyme digestions (fingerprinting) and two sequencing reactions (each end of the cosmid).

A Macintosh IIvx computer, two stepper motor boards, and a 96-channel input/output board are the heart of the control system. A Macintosh computer was chosen because of this group's experience programming it and the availability of stepper motor controller boards, and because it is the system of choice for the biological technicians that use it. There are 5 stepper motors in the system, 8 pneumatics, and 16 fluid control valves.

The systems mounted above the centrifuge include an 8-channel pipettor, an 8-channel aspirator, eight 8-channel fluids dispensers, a pellet fan drying tool, a rotor alignment system, a plate gripper for moving the plates in and out of the centrifuge, and a variety of sensors to monitor system status and safety. Key to the ruggedness of the machine was the inclusion of a large number of status sensors. These sensors monitor pipette tip seating, tip ejection, plate seating, pneumatics position, and important linear drive positions (centrifuge closed, centrifuge open). These sensors, coupled with the increased amount of structural considerations in the third-generation machine, have made it very reliable. Two third-generation machines and one second-generation machine, operating daily in the lab, have produced an average of greater than 500 preps

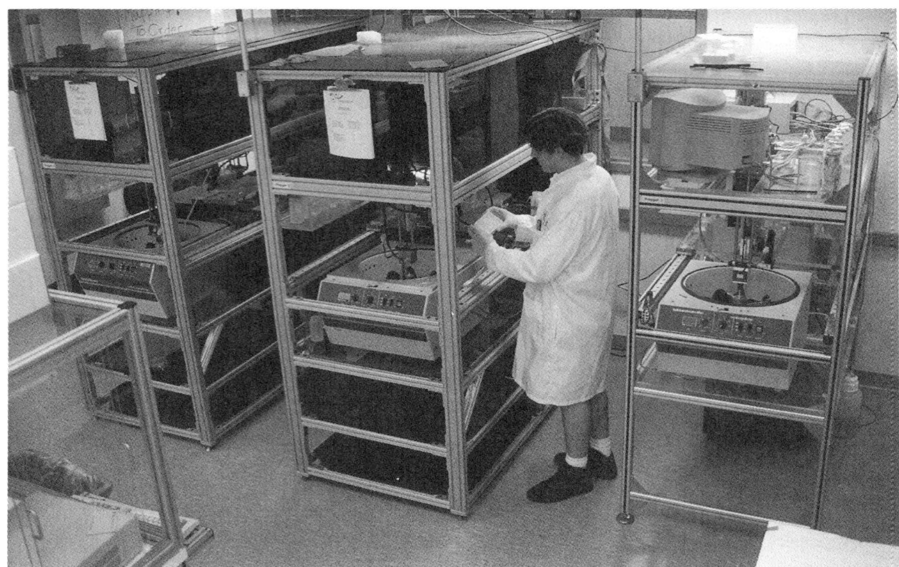

Fig. 1.4 Version 3.0 of the Prepper, PhD. along side of version 2.0. The newest version was "made more robust" by the addition of additional status checks and sensors.

Fig. 1.5 Prepper, PhD. v3.0 operating tablet with new expanded work space to enable the processing of 2 × 96 samples at a time.

per day over a recent three-month period, with no loss of samples and a one machine-day failure (caused by the failure of a component in the commercial centrifuge).

The Think C 6.0 compiler and debugger was used for this project. The interface is user friendly, and all systems parameters can be modified from that interface. Various protocols are written using any standard editor in an English-like sentence structure. This file is parsed by the software into a series of actions that are carried out by the hardware. There are approximately 100,000 lines of code, the first 60,000 of which were generated using AppMaker, a software generation tool [23, 24].

GAS The Genome Automation System (GAS) performs PCR-based library screening for STS content mapping of YACs for the expansion of the low-resolution map of chromosome 11. The Genome Automation System consists of a computer controlled central linear hub to which modules, each with their own functionality, are attached and can work independently. The modules are centrifugation, fluidics, thermal cycling (done off-line at up to six 864 well plates at a time), detection and incubation. The central linear hub consists of a 3-meter-long *XYZ* gantry that can move plates, tips, and so on, from module to module for processing. All the control and interfacing electronics (stepper motor controls, pneumatics, ac/dc switching) are operated by a Macintosh Quadra 950 computer. Figure 1.6 shows the GAS, with the modules cases removed for visibility. Figure 1.7 shows the GUI.

The intelligence that controls the hardware to execute a given protocol is the software. The software for the GAS is written in Think C. The current version of the control software is over 150,000 lines of code. This is increasing daily as new modules of hardware are brought on-line, as the implementation of new protocols are written, as ancillary components of the control system such as sample tracking or activity scheduling are completed, or as bugs are found and repaired. The Think C compiler in use is version 6.0. We also use the code generator, AppMaker to speed code writing, simplify the code and aid in adhering to the Macintosh programming conventions. The code is modular, thus allowing for easy expansion and maintenance [25].

The user interface is an icon-based Graphic User Interface (GUI) in which various windows, menus, and controls appear to the user as needed. The GUI is complete and functional; however, as additional equipment is brought on-line or as a consequence of experience during operations, the GUI is adjusted to maximize its utility and maintain its user-friendly nature.

With properly selected Sequence Tagged Site (STS) primers generated by end sequencing of random cosmid clones, PCR product detection assay can be automated. Detection of a double-stranded product by an Ethidium Bromide stain added directly to the PCR mixture, before or after thermal cycling without significant background, can be done (see Figs. 1.8 and 1.9). This method functions by detecting the change in the amount of double-stranded product and is not foolproof, but with well-designed primers the false positive and negative rates are very low. Primer dimers and nonspecific amplification are the usual causes for false positives. Putative positives detected by this method are subsequently loaded into agarose gels for verification of product size. This greatly reduces the amount of gels to be run, and for the depth of the YAC libraries, the number of positives are typically 3–5%, so only 1/20th the amount of gels need

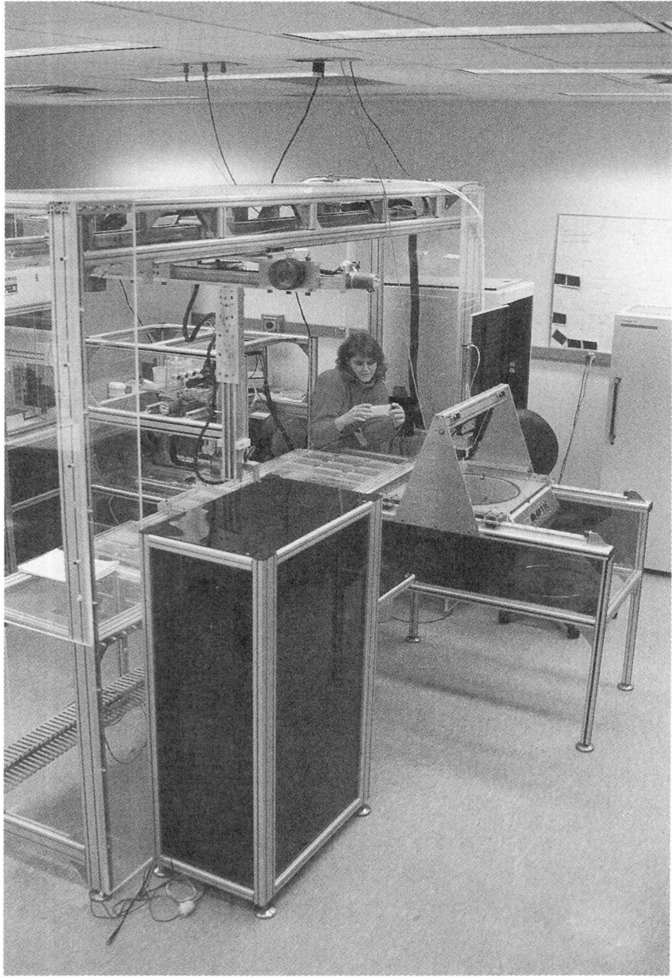

Fig. 1.6 Photo of the GAS in operational mode. The GAS consists of a central linear hub around which are modules dedicated to fluidics, centrifugation, sample detection, thermal cycling, incubation, and refrigeration.

be run. The autoloading of these positives is implemented on a Biomek 1000. Plastic gel trays that have a microwell plate footprint for loading on the GAS fluidics station are being constructed, and a new modular station is in design so that the entire operation of gel verification (gel loading, electrophoresis, and detection) can be done in an integrated fashion.

During the development of the high-throughput methods implemented on the GAS several observations were made applicable to both PCR and cycle sequencing [25, 26]. It should be noted that this process involves the screening of libraries using primers

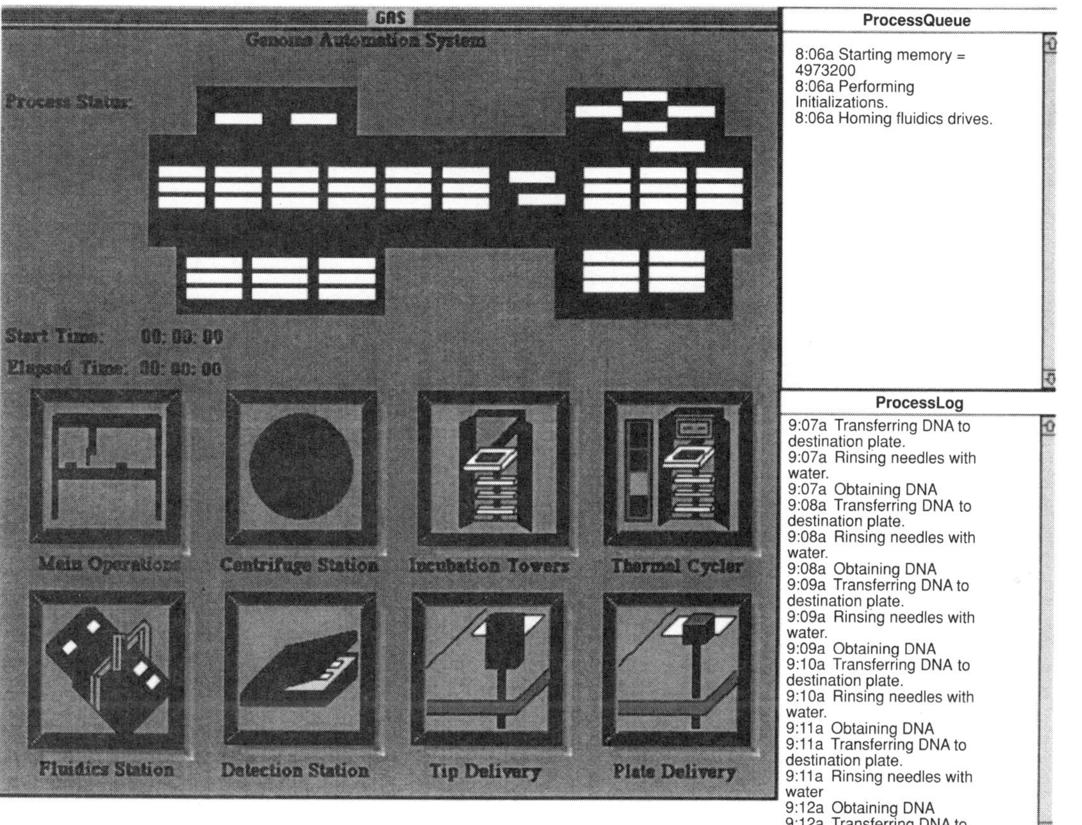

Fig. 1.7 The user interface contains a menu, icon buttons, and a status window.

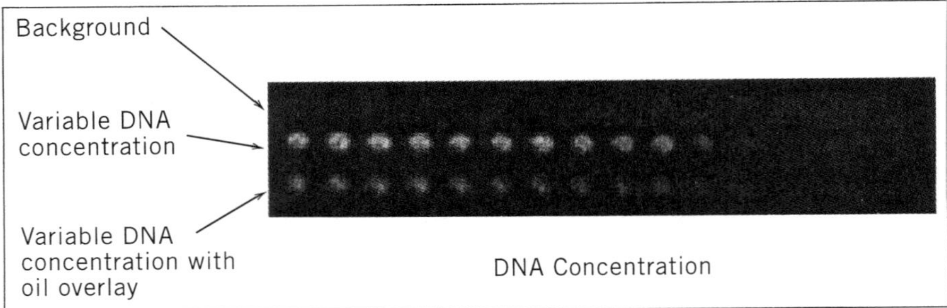

Fig. 1.8 *Dilution series for DNA Ethidium Bromide complexes within an 864-well plate under UV illumination.*

that have been extensively tested for the production of a single product; they do so robustly and using only a single round of amplification (<35 cycles).

1. Thermal cycle time and ramp rate are set by thermal transport diffusion times through plastics and the plasticware geometry. To obtain uniform amplification, long cycle times (~12 minutes per cycle) for some plasticware like our 864 well plate are required. The efficacy of Taq polymerase is such that amplification can be done without significant loss of fidelity or gain, thus increasing the sensitivity of screening experiments, provided that the denaturization time spent above 90°C is minimized.

2. Although there are many brands of polymerases, few performance differences among them have been noted. Batch-to-batch variation in polymerases/buffers has been observed.

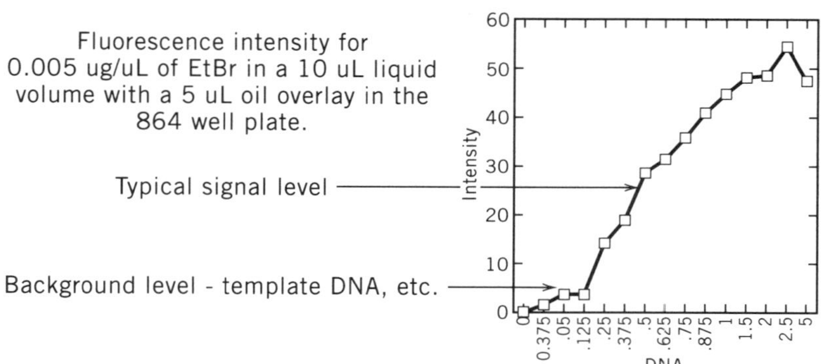

Fig. 1.9 *Calibration curve for absolute intensity derived from the dilution series in Fig. 1.8.*

3. Some polymerase formulations include Triton-X in the buffer which fluoresces heavily for the combination of excitation and emission wavelengths (300 and 500 nm) used to detect bound Ethidium Bromide.

4. Higher concentrations of polymerase make screening results more reproducible.

5. Experiments using an air oven confirm that a post cycling holding temperature of 25°C (room temperature) is adequate for sustaining the products and reaction such that it can be detected hours or days following thermal cycling.

6. Assembled reactions can be stored at 4°C for months (three months was the longest time studied) prior to thermal cycling, with no noticeable change in the amount or quality of the product or any other aspect of the PCR analysis.

7. Special precautions such as aerosol filter tips are not required to eliminate cross-contamination for the parameter regime used for screening. Decontamination of pipettors by alcohol or repeated rinsing with water (usually >3 cycles) is sufficient. Alcohol was found to be a large inhibitor to amplification, so complete drying is very important.

8. All the leading brands of thermal cyclers were evaluated with a calibrated thermal couple and chart recorder, and there were found (a) wide variations in the actual temperatures reached for given programmed temperatures, (b) variations from device-to-device within a brand of cyclers were observed, (c) temperature accuracy not based on price, and (d) variations in overshoot conditions.

9. High-quality/purity DNA templates are not always required for reliable amplification. For library screening, one can make very simple, crude DNA preparations prior to amplification with regular success.

TREC   The Timing Resource Electrophoresis Controller is a hardware/software system that is used to control and monitor the 16 gel electrophoresis boxes in our mapping laboratory. This is expandable to 96. The same binary I/O card used in the Prepper, PhD. and the GAS is used as the link between a set of solid state AC relays and the Macintosh computer. Software written using AppMaker and then compiled with the Think C compiler is used to control and monitor the status of the gel power supplies. The status of each of the electrophoresis runs are displayed on the monitor. An icon representing the gel box is green if the box is ready for use, red if in use, and blue after completion of the run. Each icon also has a countdown timer so that technicians know how much time is remaining. The duration of each electrophoresis run is independently controllable (see Figs. 1.10 and 1.11).

This system was built to allow for external monitoring of the status of the lab over the network, and it can also control any AC powered system, especially electrophoresis power supplies. It offers ease of use and is less expensive than discrete timers for each electrophoresis power supply.

The entire system, including gel boxes, was installed in a custom rack fixture to minimize the laboratory space required.

TOOLS   A 864-well microwell plate was designed for multifaceted utility to meet many of the needs of human genome project researchers: such as library storage, PCR,

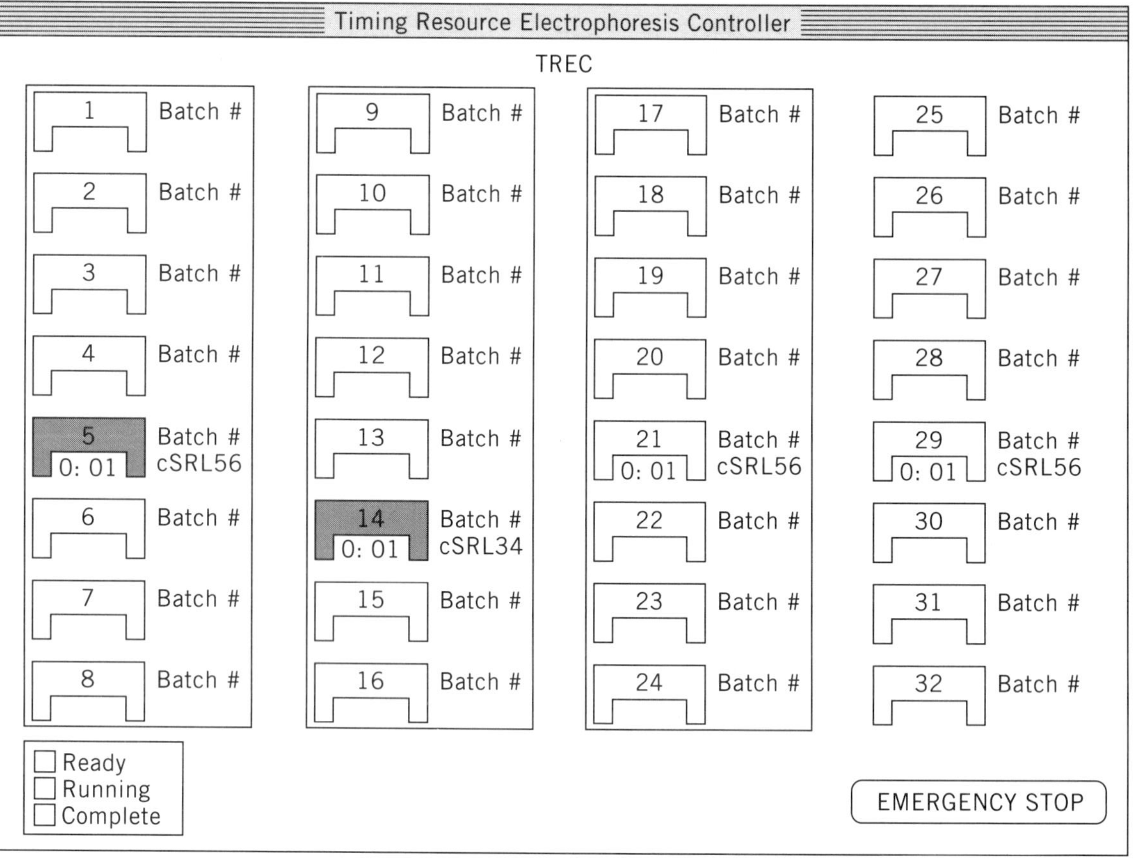

Fig. 1.10 The TREC graphics user interface. Each icon represents a gel box and its associated power supply. The status of the individual gel box is given by the color of the icon. Controls such as the duration of each gel box are adjustable from the menu.

Fig. 1.11 The TREC system within its mounting frame. The TREC hardware consists of a Macintosh computer with internal I/O board, a solid state relay set in a box adjacent to the Macintosh, power supplies, and gel boxes.

or *in situ* diagnostics. The design criteria included (1) a sample volume of 20 μL, (2) a well width-to-depth ratio such that inexpensive "yellow" pipette tips, Beckman Biomek tips (or equivalent) or Ranin miniature tips may be used to reduce consumables costs, (3) a 3-mm well separation to provide maximum compatibility with existing tools such as ganged pipettors, (4) optically polished surfaces for in-the-well sample diagnosis, (5) a bar code site for robotic plate recognition, (6) alignment wells for precision targeting by automated systems, (7) a well-wall angle and well-bottom geometry designed to accommodate high-accuracy, low-volume tip-touch pipetting, and (8) production using various plastics to obtain a good match with the desired use. This plate is now commercially available and is used by the GAS and Biomek robots [26].

Methods for culturing bacteria and yeast within the 20-μL wells of the new plate have been developed. This was essential if the new format is to house cosmid or YAC libraries. Central to this effort is the control of culture evaporation, aeration, and agitation. After identifying appropriate sealing materials (standard book tape or Beckman metallized plate sealers), cultures can be grown to cell densities (within a factor of two) typical in larger volumes without well-to-well contamination.

Insights and Lessons Learned

It should be realized from the onset that automation systems, both hardware and software, are never finished. Most understand this about software, for it is customary for every software package to have a new version go to market every year or so. Similarly, for any hardware tied to software, there are revisions. These upgrades involve software improvements, bug fixes, expanded capabilities, improved method incorporation, and hardware renovations. These are necessary for the type of high-throughput prototype systems being developed for the genome project because the technology level must still evolve to complete the project on time and on budget. But it means that there is rarely a stable platform, and therefore continuous effort, to maintain and support the systems.

The importance of process feedback control and monitoring should not be minimized. Experience on several systems, some in third or fourth generation, has shown that errors and malfunctions that can affect productivity, sample integrity, safety, and operator moral contribute strongly to system value and acceptance. Although it is possible to a priori guess at possible error conditions that need to be monitored, most of the time they are discovered after a system is pushed into production. It is usually simple to install new logic into the control programs for feedback sensors, but it is important to provide hooks in the hardware, especially the computer interface for additional sensors to be added after the system has started into operation.

There is a differing in opinions as to whether automation and robotics motion systems should be open or closed loop. Closed loop usually means that encoders are included with the motion system to measure the position of a given component at any time. Open-loop systems have no encoders and depend on systems precisely and reproducibly achieving their planned target position. While closed-loop systems are the most robust, they are also more complex and significantly more expensive. With care in calculating trajectories, acceleration profiles, speeds, and mass loading of motion systems, it is possible to design safety margins into the open-loop motion driver selection. Most commercial systems are open loop, the Beckman Biomek robots, for example. The prototype designs discussed here are all open loop. The best choice for making open-loop systems work well is to use microstepping motors that are over specification (torque), and motion drives (linear, circular, etc.) that are also over specification and have little hysteresis or backlash. To achieve most reproducible motion, it is best to always approach a given point from the same direction, even if it means that it is necessary to overshoot a target point and then return.

The changes in available hardware and software affect the evolution and maintenance of automation systems. Although rarely, when hardware becomes unavailable, it is sometimes necessary to find backup sources. Software, however, is very transitory. Compilers, associated libraries, drivers (for given hardware), code generators, and operating systems are continually evolving, making it necessary to upgrade if development is to continue. Experience has shown that sometimes it is important to freeze a system, that is, not to upgrade the operating system, compilers, or development environments for some systems that need slight adjustment. This is because of the tremendous effort required to "port" a large operational code to a new version of a compiler that result in the code being inoperable for some time. For developers, there should be a compelling reason for upgrading, not simply because a new compiler or other software has come out. Sometimes software upgrades are forced because of new hardware, for example, the introduction of the PowerPC Macintosh computers to replace the 68xxx systems. This has required new compilers and other software. Because these systems are still relatively new (one year old), the compilers, operating systems, and nuances are not stable and well understood. Changing to a recently introduced hardware platform can cause the learning curve to be reset, again requiring a large investment in time to master the software development tasks.

LABORATORY NOTEBOOKS, EXPERIMENT SCHEDULING, AND PLANNING

SUMU

SUMU DatAssistant is a laboratory sample-tracking, experiment-planning, and results-reporting software tool written in the database language 4th Dimension [14]. SUMU currently exists in two versions, CosmidSUMU and M13SUMU. We use another 4th Dimension database called Genome Notebook, originally written by Steve Clark to track the production and testing of STSs to assist in the production of low-resolution, and YAC-based maps of Chromosome 11 [27]. The first portions of SUMU to be developed and put to work were in support of the sequencing effort, thus the two varieties. SUMU stands for Sequencing Unit–Mapping Unit, but also means to "clarify" in Japanese. The reasons for the diversion of the SUMU DatAssistant into two varieties rather than one larger variety were (1) the types of data and types of organization of M13 shotgun and cosmid end sequencing are very different, (2) smaller modules are easier to support, and (3) accessing a smaller subset of data is faster. However, common to each is the underlying principles, much of the user interface and the basic structure of the relational database.

The SUMU DatAssistant was designed as an aid to the production biologists and not only as a database. It serves as a database, holding data, and has provisions for accessing the data. Often databases are viewed by the technicians as something they put data into and never see again, and view them only as a distraction from doing biology. For most data entry this can be overcome by insisting that only data in the database indicate a biologist's progress. Data in a written notebook are not accessible to many beyond the researcher. SUMU DatAssistant was designed to overcome this resistance by providing something of value to the biological technicians. It helps them organize the complex "batching" of the various processes involved in mapping and sequencing and also outputs ABI 373A and 377 sequencer "sample sheet" files, thus nullifying the need for manual entry of these required files by the technicians. These features have made the SUMU DatAssistant very useful and used. In addition to creating sample sheets, this relational database interfaces with laboratory equipment, specifically a plate-reading UV spectrophotometer to select only high-quality (DNA yield and purity) samples for automated sequencing or mapping, should this be desired.

CosmidSUMU functions as follows: (1) A mouse-clickable data entry sheet, with 96 buttons designed to match the 96-well microwell plate format, is used to enter cosmids to be sequenced as determined by UV spectrophotometric assays or from score sheets generated in the analysis of the fingerprinting of cosmids in the mapping phase [28]. This also keeps track of the cosmids that have been successfully fingerprinted, namely those that have not failed for a host of reasons—dead clone, bad prep, partial digestion, deleted clone, and the like. Cosmids are processed in 96-well plate format in the culturing, prepping, and fingerprinting stages of the analysis process. (2) A status of each cosmid entered is kept as to whether it has completed cycle sequencing, loaded on the sequencing gel, and ultimately produced usable sequence. This list can be queried at any time to determine the sample-processing backlog. (3) Cosmid names from the list are then batched for dye primer or dye terminator chemistry cycle sequencing reaction assembly, which is done manually. There are provisions for outputting a file to be downloaded to Biomek 2000 pipetting robots that will, in the future, select cosmids from 96-well plates according to the list in the output file. Samples are batched in sets of 24, so the total of dye primer reactions fit within the 96-well format of the Perkin Elmer 9600 thermal cyclers [29]. CosmidSUMU will automatically instruct the technicians to include standards, and will stock DNAs with known sequences for the fixed primer sets used to cosmid end sequencing—T3 and T7 primers. These standards are included with every sequencing run to diagnose problems associated with either the cycle sequencing, gel loading, or reading. CosmidSUMU batch for dye terminator chemistry, which use one tube per reaction, can be done as well. (4) After thermal cycling, the samples are again rebatched in sets of 36 to match the number of lanes in the ABI sequencers. An ABI "sample sheet" is automatically produced [30]. In file form it is moved to the appropriate sequencer. A hard copy is also printed so that the technician is sure of the sample loading order and has a permanent record to which comments and notes can be attached (see Figs. 1.12, 1.13, and 1.14).

M13SUMU is very similar to CosmidSUMU but, because of the nature of the problem, is much simpler. The major difference is that M13 templates, which are cultured and extracted in batches of 96, fall into two classes. A new M13 subclone library from a cosmid either does or does not work. If, after a test sequencing run, a few M13 clones from a new subclone library (the libraries are 600 to 1,000 in size) are found to be good, M13SUMU is given the names of the plates of M13 templates that are ready to sequence. M13SUMU then can proceed to assemble batches for cycle sequencing, gel loading, and running as was done in CosmidSUMU. The internal

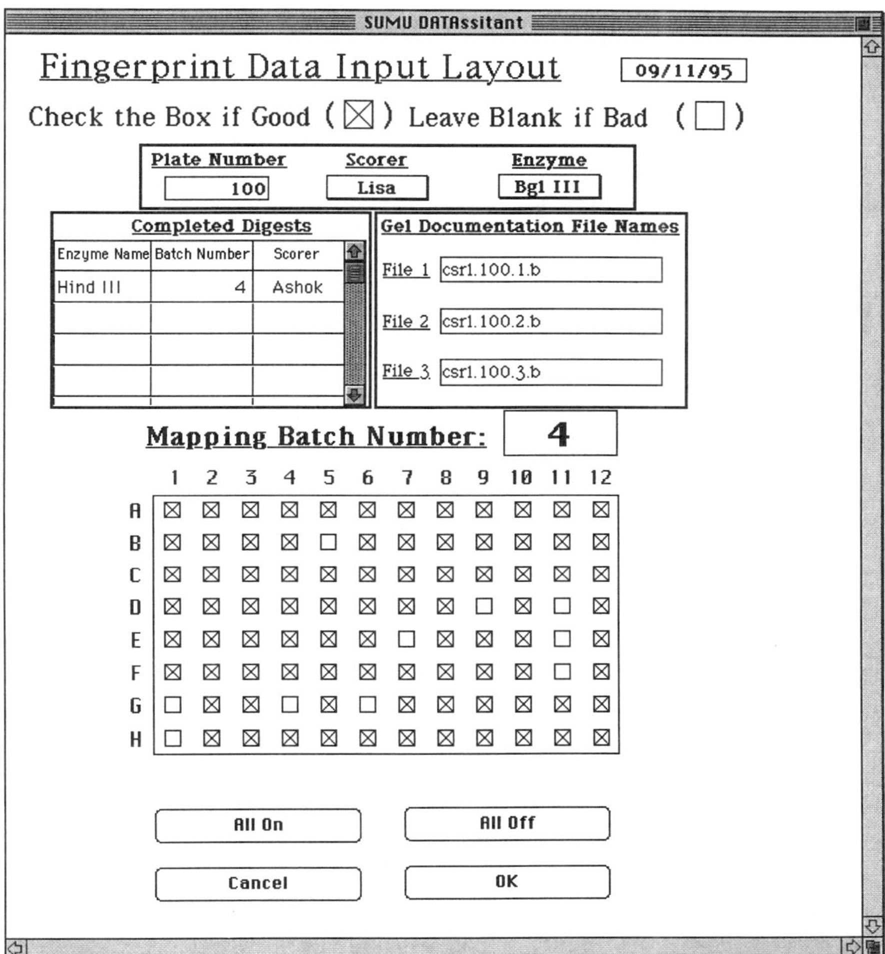

Fig. 1.12 CosmidSUMU data input sheet made in an array of 96 for easy entry. This can also be filled in automatically from a UV plate reader or in the future other imaging software.

logic of the program reflects the process: Since M13 templates are always available in guaranteed sets of 96, they can be handled knowing that all 96 are to be processed.

Another valuable and widely used tool for archiving, analyzing, and tracking data progress is the Excel spreadsheet. For each of the production groups doing sequencing, fingerprinting, or PCR-based STS screening, an Excel spreadsheet keeps track of the immediate results and the plans for the group. Excel spreadsheets are especially useful because they are cross-platform, there is universal knowledge of spreadsheets, and data can be exported and imported from many sources, including 4th Dimension databases.

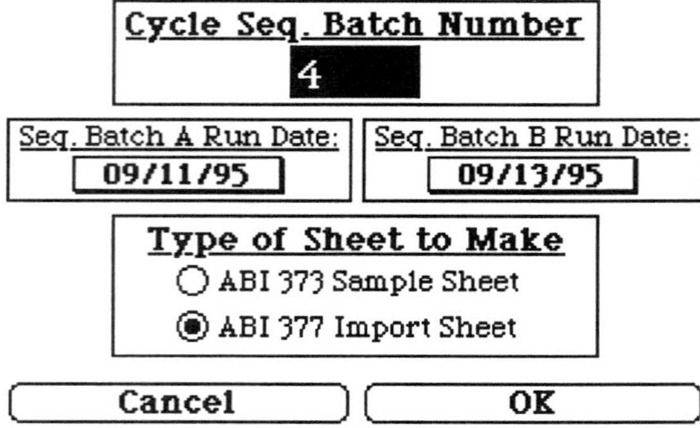

Fig. 1.13   Window setup for CosmidSUMU "sample sheet" production.

With the great amount and large variety of data and the large number of people accessing and inputting data, data distribution into many different spreadsheet storage media for later incorporation into a unified database is a valid approach. The laboratory computers, linked by the Macintosh network, house the various spreadsheets, 4th Dimension databases, and files of data (e.g., raw sequence files) within "shared folders." The reliability and user friendliness of this network approach, which automatically takes care of issues like multiple users accessing the data at once and read only permission for certain users, has driven the group to use this type of data handling. Ultimately, after certain experimental procedures become sufficiently well defined (the sequencing process, the fingerprinting process), dedicated databases and experiment schedulers can be written, a.k.a. CosmidSUMU. Until that time, and to be on-line as fast as possible, networked Excel spreadsheets work well [31]. With the recent release of Excel 5.0 which includes the ability to incorporate Visual Basic functionality into the spreadsheet, it is anticipated that more databaselike features (GUIs, logic) can be built into the spreadsheets.

Tools

The 4th Dimension database is a relational database development shell (ACI US, Inc.). It is similar to many other relational database systems, Oracle, and so on, in that it has all the power and capabilities necessary for a large job. It was selected as the database of choice for the project because it was Macintosh based. The Macintosh is the most comfortable platform for the users, biological technicians and those who write grants and papers. It has two further advantages: It is far less expensive than other database systems of similar performance, and it allows for the easy import and export of data with other databases, spreadsheets, graphics programs, and word processors. The learn-

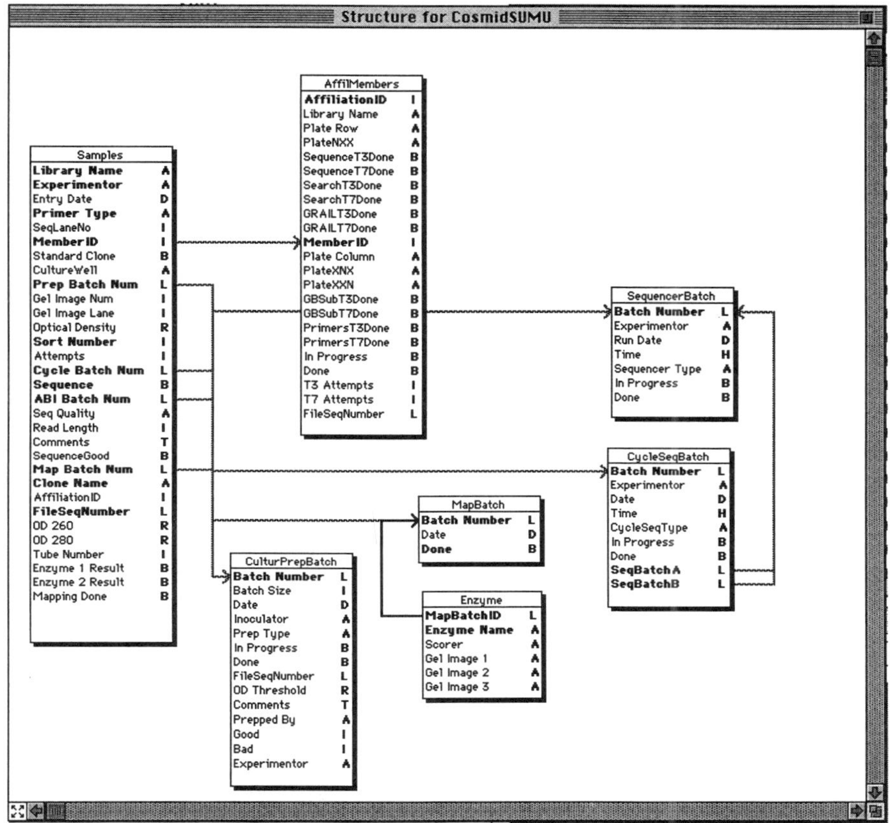

Fig. 1.14 The structure for the CosmidSUMU database.

ing curve on starting out to develop one's first 4th Dimension database is not steep. It is possible to have something working in a couple of weeks, even if the programmer has not used the database before. However, like all powerful databases, as the sophistication and complexity of the problem grows, the hidden details and nuances do take time and experience to master. The 4th Dimension is fully networkable: It also includes a simple structure display, full multi-user password protected capabilities, an easy to use GUI design tool set, and a low-power, adequate drawing tool set.

Insights and Lessons Learned

Perhaps the most important lesson learned in developing our databases is to make them usable and valuable to the users. "Usable" means that (1) the GUI and other features are simple, obvious, and follow Macintosh standards (so the general layout and functionality resembles other Macintosh software); (2) the database contains all the features required and not a lot of extra baggage that slows it down; and (3) the database can be easily adjusted, expanded, or reconfigured to respond to changes in the experimental strategy, experimental methods or suggestions from the users. "Valuable" means that it provides a necessary service for all that are involved. It was possible to create data assistants rather than simple databases. Most biological technicians that generate data but do not analyze or assemble much of it view databases as a time-consuming black hole. The class of data assistants created for the lab to do sample tracking and experiment planning not only serve as data repositories but also as software tools that help the lab technicians to do more. They avoid data re-entry and, more important, create output that becomes input for other required software in the lab.

The "keep-it-simple" concept is very applicable here. Developers have avoided large, all-encompassing databases because of the nature of the GSS strategy, and the cost in time to develop and debug such a system. Although ultimately there are tools to integrate the data, it is not necessary for every component of the data, or every person in the lab, to have access to all the data. It is therefore possible to modularize the databases for very specific purposes, which makes them simple and quick to develop.

DATA INPUT AND ANALYSIS

There are three types of data that move around within the production and development labs: biological information, sample-tracking and experiment-planning information, and design information. Biological information consists of sequence files, enzyme digestion data, hybridization data, Fluorescent In situ Hybridization (FISH) data (image files and spreadsheet results files), and STS content mapping data. Sample-tracking and experiment-planning information are lists of samples to be FISHed, probed against a gridded clone library, or fingerprinted, and individual or plates of clones in an experiment cycle (e.g., sequencing). Development data include raw data images, computer-aided design (CAD) files, parts lists, and design proposals. As described earlier, each of these data is generally held in relational databases (4th Dimension), spreadsheets (Excel), word-processing documents (Word) or simply as files within the hierarchy of the Macintosh or Sun Sparcstation file systems. The efficient movement of these data into, among, and out of these various programs is important.

The Macintosh applications are defined by developer standards that are very rigorous. It is for this reason that most programs are compatible and can freely exchange information via cutting and pasting from a clipboard buffer. These standards make programming a Macintosh computer more complicated, but there are tools designed to aid the development process. The ease in exchange of information among programs along with its simplicity in use contributes to the popularity among researchers and technicians in the lab.

Both Macintosh and Sun Sparcstation servers are used for the analysis and archival storage of data. They also serve as repositories of our applications (programs) backups and updates, as our World Wide Web (WWW) site servers and as SLIP line sites for remote phone based logon, which allow experimenters to access all the data from home or while traveling. The movement of raw data to and from these servers is done both manually and automatically. As new projects are started or as unique situations arise, some manual movement of data is accomplished over the local ethernet network. Automatic data archiving and movement is done using Retrospect Remote [32]. This commercial product is very powerful and can be controlled by scripts to wake up and move data at any time, and it also interacts directly with 4- and 8-mm digital audio tape (DAT) drives. This program is currently set to automatically retrieve data at midnight from the various distributed lab computers. These data are mainly sequence ".tra" and ".seq" files, the restriction fingerprint image, and called fragment files [30]. The data are then available to any user directly from the servers without requiring knowledge of exactly which computer generated the original data. After completion of the retrieval, the backup portion of the program initiates. Nightly, there are backups of all new files from all computers (servers and remote). Weekly, there are full backups of all the computer systems. Currently this is about 10 Gbytes.

DATA OUTPUT

The funding agencies measure the progress of each research laboratory by the amount of sequence and map information deposited in one of the world's databases. Most biological results are deposited in either GenBank or Genome Data Base (GDB) [33, 34]. For each of these databases, there exist tools for "automated" submission. Although these tools are of utility, they do not aid in the most time-consuming aspect of the job—sequence or map annotation. The GSS strategy generates a considerable amount of sequence information, but it mainly consists of 300 to 500 bp of data in the form of ordered markers and random sequence. This makes it very difficult to annotate completely. Also output includes whole cosmid sequences (~40 kb) by shotgun or a directed strategy. For this type of data, the effort to annotate is a small fraction of the labor required to generate the data in the first place, mainly because the annotation is for a larger, quantized block of sequence. The rules for sequence deposition in GenBank are defined well enough so that some annotation can be automated. For deposition of mapping or gene information to GDB, the variation in the type of information and the rigidity of the information is necessarily less stringent. For example, a Sequence Tagged Site (STS) is a location in the genome mapped by a sequence signpost that can be recovered via PCR amplification [35]. The question arises, and has yet to be clarified, whether an STS is really an STS if only the sequence is known. This is true especially if PCR primers have not been made and the ability to uniquely amplify only that fragment (in the human genome and not another possible contaminating genome) has not been demonstrated in the lab. A second example are data generated by groups that do STS content mapping, but the ordering of clones using STSs or by screening, the generation of a greater than minimum tiling path set of clones anchored by STSs, can be confusing. It is possible to end the screening process at any tier in the analysis process: Some efforts screen the library down to the level of the exact clone; other efforts screen only down to the level of the plate pool. At the plate pool level a clone is known to exist, but the data are being localized only down to one or more possibilities out of typically a 96-well pool. At UT Southwestern, in the two examples just given, a more stringent criterion is used. STSs are verified by primer synthesis and PCR, and libraries are screened to the clone level. Both of these criteria add additional work and make automation more difficult because of the amount of decision points made by humans within the processes.

In addition to GenBank and GDB, data are released in several other ways—via journal publications, collaborators, the World Wide Web (WWW), and other Internet tools such as CU-SeeMe video transmission [36]. The WWW has provided an important new tool for immediate publication and visualization of data and is used extensively by many groups. Almost every group conducting genome research has a Home Page on the WWW. The WWW home page is easy to access, simple to implement, and easy to maintain. But like any other media, the labor is in collecting and formatting the data in a useful form. With collaborators, data may be in an intermediate state of development, not complete and not ready for formal publication. The WWW provides password protection for certain Home Pages, which facilitates the exchange of confidential information among collaborators.

To effectively use the WWW and the Home Page as a medium of information exchange, there is a minimum tool set required by both the developers (publishers) and those wishing to access the information:

- *Web browser.* There are a few options to maneuver and search on the Web. The program of choice is Netscape [37]. This program is mature and has all the tools necessary to easily navigate, search, and retrieve information. In conjunction with a Hyper Text Markup Language (HTML) editor, Netscape can also display pages in development.

- *HTML Editor.* The control language used to format and establish the execution logic is called a Hyper Text Markup Language, HTML [38]. The HTML Editor, written by Rick Giles, provides all the necessary tools for standard web page composition and compilation. Information is available at

<http://dragon.acadiau.ca:1667/~giles/HTML_Editor/Documentation.html>.

HTML is a very simple language to use. Composition is English-like, and even standard word processors can be used to edit HTML code.

- *GIF Converter.* Written by Kevin Mitchell, this shareware is used to convert various image file formats into a compressed standard Graphics Interchange Format, GIF form [39]. Most programs used on personal computers output images in PICT or TIFF format. These programs include drawing, drafting, CAD, 3-D rendering, and video analysis software such as NIH Image. GIF Converter is used to edit and save the images in a form that can be associated with the HTML code, which is compressed to minimize the bandwidth necessary to move the data around the Internet.

- *Stuffit Deluxe.* File compression and decompression tools are required for those publishing on the Web or maintaining a FTP site, and are also required by those that download this compressed information. The current tool of choice is Stuffit Delux, a commercial product that can work in both modes and encode and decode information in a variety of popular formats [40].

- *JPEGView.* For large image files and for other formats like movies, data are usually downloaded directly from the Web and must be viewed by other tools. JPEGView works well to view these large files or to play movies [41].

It is also possible to publish live video and perform video conferencing directly over the Internet. The CU-SeeMe software package is used to both transmit and receive data over the Internet. Rather inexpensive (~$100) black-and-white cameras can be attached to personal computers such as the Macintosh in order to acquire video data [42]. The microphones and speakers that accompany every new computer are used to attach sound to the video. On the Internet there are servers, called "Reflector" sites, that can be used to route video CU-SeeMe data through or to socialize. In the laboratory CU-SeeMe cameras are used to monitor processes and robotic systems. With these cameras and the CU-SeeMe software, it is possible to monitor from one's personal computer up to eight active cameras at a time, with a video refresh rate of about one frame per second.

AUTOMATION DEVELOPMENT LAB OUTFITTING

In developing custom fixtures or complete robotic systems, it is essential to have access to information and facilities to construct the hardware and software. Information includes suppliers of the many and various hardware, where one can get hints to speed development, and which are the best hardware and software products for a given job. Some of this only comes with experience, but some can be obtained by establishing and maintaining a network of associates within the genome and biological automation community. Included in Tables 1.1 and 1.2 are sources for common equipment used in the automation development lab and for collections of product listings published annually by professional magazines. This information is kept up to date on the UT Southwestern Genome Center World Wide Web Home Page at <http://mcdermott.swmed.edu/>.

Some of the basic equipment required within an automation development group includes shop tools (milling machine, lathe, drill press, table saw, etc.), electronics tools (oscilloscope, function generator, ROM programmer, in-circuit emulator, etc.), and computer tools (CAD station, off-line programming station, compilers, code generators, debuggers, Internet access for help, complete networking of laboratory computers). It is also essential to have a stockroom of parts to construct new test fixtures and spare parts for custom equipment. This stockroom should include electronics parts as well as bulk material such as extruded aluminum, plexiglas, and other plastics that can be machined. It is also important to establish a good relationship with the local campus, corporate, or private machine shops. In general, the experts in these shops can aid with basic design, design for machinability, and can make parts faster and better than a

TABLE 1.1 Equipment Sources

Macintosh stepper motor interfaces	nuLogic	617-444-7680
Computer input/output control boards, accessories	National Instruments	800-433-3488
Video frame grabber	Data Translation	800-525-8528
Macintosh A/D, D/A converters, software	GW Instruments	617-625-4096
	Metrabyte	508-880-3000
Precision electronics	Keithly Instruments	216-248-0400
	HP	800-452-4844
	Tektronix	800-835-6100
Extruded aluminum structure, frames	80/20	219-478-8020
	Avantech	214-596-3399
864-well plates, thermal controllers, software	Helix/General Atomics	800-424-3549
Miscellaneous hardware, tools	McMaster Carr	310-692-5911
	Tooling Solutions	800-400-6890
Miscellaneous electronic components	DigiKey	800-344-4539
Macintosh software	Macwarehouse	800-696-1727
	SciTech	800-622-3345
Macintosh C compiler	Symantic Think C	800-441-7234
	MPW C	800-SOS-APPL
Best Macintosh C compiler	Metrowerks Code Warrior	514-747-5999
GUI designer, code generator	AppMaker	508-369-8175
Database	4th Dimension	408-252-4444
Optics, Miscellaneous scientific stuff	Edmund Scientific	800-573-6250
	Newport	800-222-6440
	Melles Griot	800-326-4363
Pneumatic actuators	Tol-o-matic	800-328-2174
Linear motion drives, miscellaneous equipment	Velmex	800-642-6446
	Techno	516-328-3970
	Servo Systems	800-922-1103
Microbeads	Polysciences	800-523-2575
	Duke Scientific	800-334-3883
Microinjectors, biological electronics	World Precision Instruments	813-371-1003
	Stoeling	708-860-9775
Miscellaneous mechanical parts, screws, tools	Small Parts	800-423-9009
	Ted Pella	800-237-3526
Plastic injection molding	Prototype Model & Mold	619-455-0371
Plastic material	Port Plastics	818-333-7678
Plasticware, glassware, array pipettors	Robbins Scientific	800-752-8585
	Adams and Chittendon Scie. Glass	510-524-9551
Custom turnkey robotics, thermal cyclers	Intelligent Automation	617-354-3830
Sequencers, thermal cyclers	ABI/Perkin Elmer	415-570-6667
Optical detectors	Hamamatsu	908-231-0960
Photo sensors	STI	800-221-7060

TABLE 1.1 *(Continued)*

Robotic Vision	Acuity Imaging	508-667-7900
Robotic actuators, pneumatics, valves, fittings	Phd.	800-624-8511
	Upchurch	360-679-2528
	Lee Valve	800-533-7584
Biomek, centrifuges	Beckman	800-742-2345
Gel boxes	CBS Scientific	800-243-4959
chip probe cards	Cerprobe	214-644-9590
ITO device fabrication, sputter coating	Thin Film Devices	714-630-7127
UV, photography	Fotodyne	800-362-3686
fluorophores, chemicals	Molecular Probes	503-344-3007
	Oncor	800-776-6267
	Glen Research	703-437-6191
General Labware	Baxter	800-234-5227
	Scientific Products	800-234-5227
	Cole Parmer	800-323-4340
	Fisher	800-766-7000
	PGC Scientific	800-424-3300
	VWR	800-932-5000
	Hoefer	415-282-2307
	BioRad	800-234-5227
	Stratagene	800-424-5444
Incubators, Hoods, Instrumentation	Forma	800-872-7133
	Sciemetrics	800-231-2065
	Precision	312-227-2660
	Techware	800-325-3010
Scanning Probe Microscopes	Digital Instruments	805-899-3380
Gasses and equipment	Altair Gasses	800-660-2066

nonprofessional machinist. The local machine tools are for simple jobs, to modify the shop-made items, and to make some items fast. It is also important to establish a relationship with the machinists so that instructions can be communicated correctly. This, in general, translates to high-quality drawings. It is important for automation developers to have access to, and use, CAD drawing tools to make standard blueprint-type drawings, assembly drawings, and 3-D renderings.

FUTURE TECHNOLOGIES

Biochips

In development is an optoelectronic device to control and detect DNA hybridization for rapid, accurate, large-scale genome analysis and sequencing. The GSS strategy and the

TABLE 1.2 Sources of Information:
Buyers Guides

American Chemical Society	202-872-4600
American Laboratory	203-926-9300
Photonics	413-499-0514
Physics Today	301-209-3040

construction of low-resolution STS-based maps can be accelerated using miniaturized technologies to perform a large number of screenings, rapidly and inexpensively. Further, to associate the sequence and mapping information with medical conditions, large-scale genotyping efforts are made possible using chip-based technologies.

For purposes of hybridization analysis of large collections of clones in parallel, a variety of approaches have been invented using DNA clone arrays on solid supports, membranes, or solid substrates. These include arrayed libraries for multiplex- or hybridization-based fingerprinting, reverse dot-blots for DNA diagnostics, and large-scale clone analysis. The value of array-type hybridization techniques are (1) the parallel analysis of multiple samples, (2) salability to large clone sets, and (3) the possibility for automated analysis using imaging systems. Disadvantages include the necessity of subjecting all of the samples to the same biochemical conditions during the hybridization [43, 44, 45].

Others have investigated the possibility of miniaturization and automation of array-type techniques, with the eventual possibility of directly combining biochemistry with electronic detection instrumentation [43]. Fodor developed a technique for combinatorial synthesis of peptide and DNA oligonucleotides directly on the surface of a glass support using photochemical masking [44]. A second approach, suggested by Beattie, is the use of electronic, rather than optical, detection methods for DNA hybridization [45]. This effort has resulted in the Genosenor consortium project, a DNA chip designed to directly detect single- and double-stranded DNA on the surface of the chip, through relaxation detection.

A second-generation electronic device has been constructed that uses microarray-based electrophoretic transport of DNA interfaced directly to a Charged Coupled Device (CCD) camera to form an Optoelectronic Hybridization Microsensor (OHM biochip). This biochip is a marriage of several existing DNA chip technologies and has superior performance with its combined, independently controllable, bioelectronic array and onboard/direct, high-efficiency, optical detection. The approach uses microlithography to produce a variety of array types to which oligonucleotide probes can be attached and probed. This is directly interfaced to a computer and is controllable by the computer and other external instrumentation. The properties of the OHM biochip and support technology are (1) an electric field optimized OHM biochip using the transparent conducting material Indium Tin Oxide (ITO) in an 8×8 array of $50\text{-}\mu^2$ electrodes—this biochip has been designed and fabricated using standard mask and sputter technology; (2) instrumentation and mounting devices for controlling the chip and reading results using external CCD cameras interfaced to a Macintosh computer is being constructed; (3) test protocols and methods for rapid, reproducible and stable fixation of oligonucleotide probes to the chip surface, for attaching high-sensitivity fluorescent dyes to test DNA targets, and for sample preparation; and (4) a computer-controlled chip "programming" device built to control the electronic deposition of a large array of hybridization/nucleic acid probes on the chip. The experiments and applications of the first chip are (1) specific oligonucleotide probes for human genetic diagnostic testing, genotyping, using binary hybridizations with hexamer strings, and probes for chip-based human genetic linkage analysis or library screening, and (2) rapid fragment size analysis and mismatch "electric" melting curve analysis (see Figs. 1.15, 1.16, and 1.17).

The first biochips were obtained from collaborators at Nanogen, Inc. (San Diego, CA) who are developing these chips as clinical genetic-testing and screening devices [46]. The newest UT Southwestern design differs in that the electrodes are ITO instead of metallic, and the basic mask design has been altered to optimize the electric field within the solution being analyzed. The Nanogen biochip has an array of individual, electronical controllable pads on which hybridization occur. The principle advantages

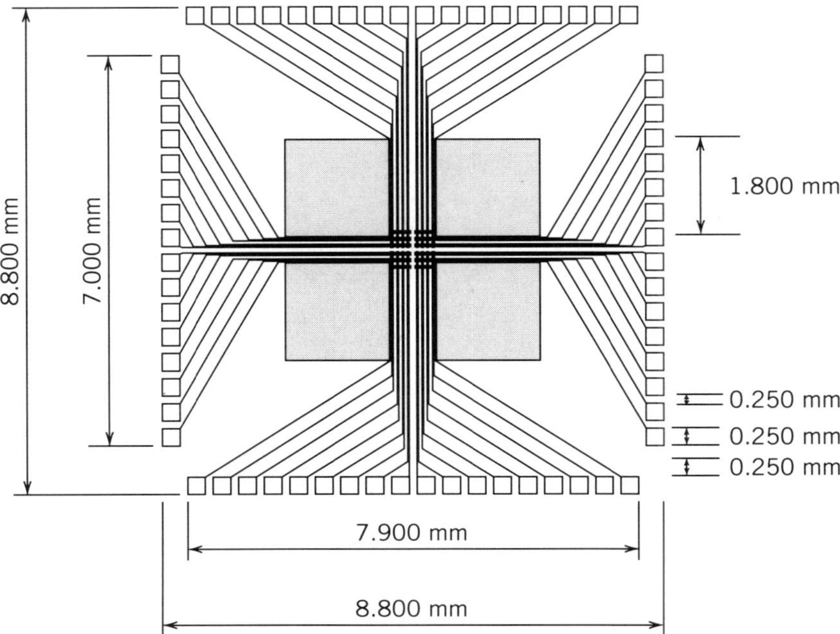

Fig. 1.15 Design of the 8 × 8 ITO biochip showing the central 50 × 50 micron pads and the leads to the external probe card electronics contacts.

Fig. 1.16 Photo of the 8 × 8 ITO biochip.

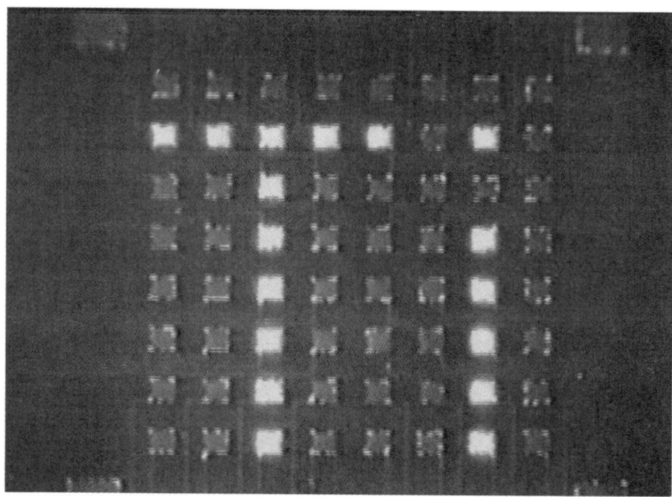

Fig. 1.17 Illustration of DNA intercalated with Ethidium Bromide that has been electrically migrated to the pads of the biochip.

of this approach are (1) each hybridization pad can be uniquely controlled for stringency, and (2) a simple oligo probe attachment method uses externally synthesized oligos. Its principle disadvantages are (1) low light collection efficiency that limits sensitivity, and (2) not yet perfected surface chemistry that mainly affects reproducibility.

Sequencer Development (Sample Preparation)

To support the evolution of the genome project from mapping to sequencing, the automation group and collaborators have focused on identifying the bottlenecks in the sequencing process. At the current production sequencing level of about 400 sequence reads per day, there is a sufficient level of activity to identify what is needed to make significant incremental improvements to the sequencing process. This has led to efforts to design, prototype, and eventually construct full working devices for integrated sample preparation, and for increased parallelization of gel-based sequencing methods.

Within the GSS strategy, the process of obtaining end sequence from an individual clone stored in a frozen library is labor intensive, repetitious, and subject to errors. For this and similar strategies, where sample preparation can be parallelized, it is possible to construct a fully integrated machine to perform the necessary tasks. A fully integrated sample preparation system has been designed to carry out these tasks. It is called the Sequence Automation System (SAS) and is based on the integrated system GAS, which was used to automate the PCR based library screening tasks. The SAS will automate the upstream steps required for DNA sequencing, including inoculation, culturing, DNA extraction and purification, cycle sequence reaction assembly, thermal cycling, reaction pooling, and gel, capillary, or biochip loading. This system, like its predecessor, the Genome Automation System, is designed to process up to 2560 samples per day, enough for a very large genome center with sequencing emphasis. This system is designed to substantially reduce the amount of hands-on human labor now utilized on repetitive tasks, which are prone to human-induced reproducibility variations and lead to technician burnout.

This new system will automate, in one united device, the entire process of inoculation, microculturing, DNA prep, sample inspection, cycle sequencing, and sample loading for the feeding of these new sequencers on a scale appropriate for genome sequencing centers. The design objectives of the machine are (1) supply sequencer reader ready templates at a rate of 2560 usable samples per day (sufficient for 50 sequencers) from cosmid, plasmid, and M13 reagents directly into the sequence reader, and (2) development of robot-adjusted biological and mechanical techniques required for volume manipulation down to less than 0.1 μL, and sample loading targeting down to sub-50-μ accuracy. This machine has been designed to be modular. In that way, as new sequence reading devices (thin gels, capillary arrays, biochips) become capable of significant sequence production, then dedicated modules designed to load those formats can be constructed and integrated into the full system. In essence the steps leading up to sample loading are very similar, but it is anticipated that as more DNA template efficient reading systems become available, the plasticware format can be adjusted for further reduced volumes and increased sample number densities (see Fig. 1.18).

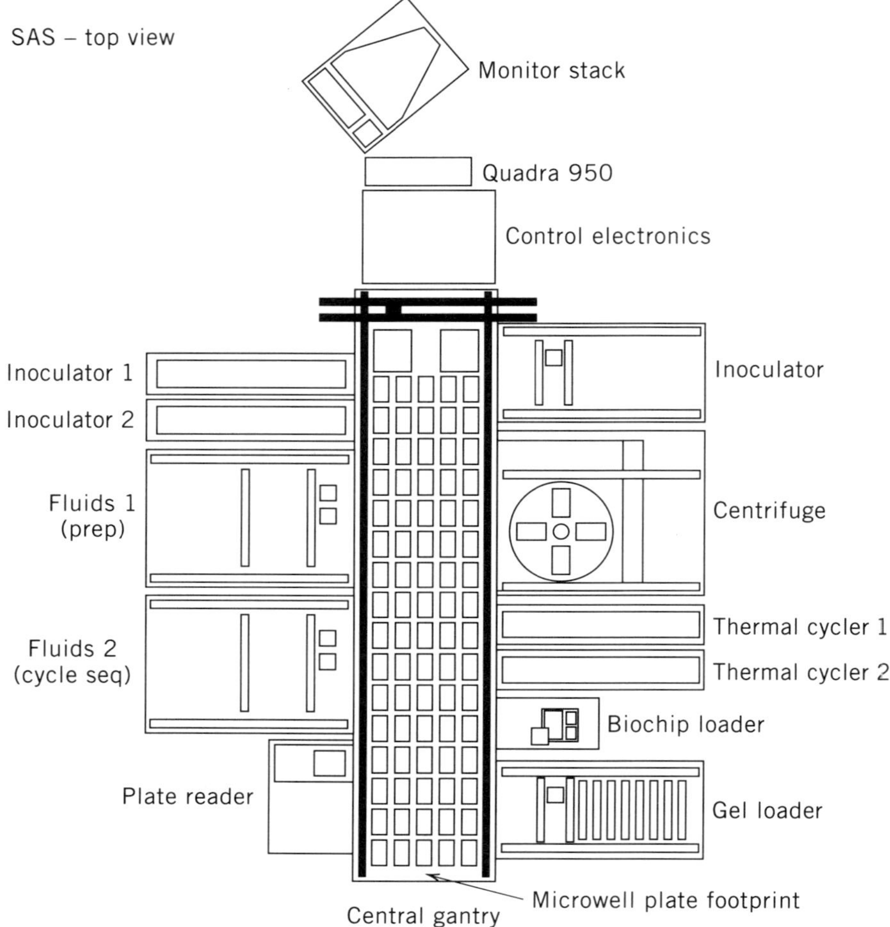

Fig. 1.18   CAD illustration of the design of the Sequence Automation System.

Sequencer Development (Advanced Gel Readers)

The multidisciplinary collaborative teams at UT Southwestern Medical Center and Science Applications International Corporation (SAIC, San Diego, CA) are jointly developing a new, gel-based, highly parallelized, sequence reader [47]. The Advanced Imaging Spectrograph System, for Slab-Gel DNA Sequencing, was conceived to respond to the needs for higher throughput sequencing and genotyping. The concept is based on the integration of an imaging spectrophotometer that records the entire emission spectra for each pixel across a sequencing gel (or capillary array) using a proven spectrophotometer system. A prototype Advanced Imaging Spectrograph System, Astral Sequencer, is being designed.

ACKNOWLEDGMENTS

The work summarized here reflects the efforts of a large number of people within the genome center at the University of Texas Southwestern Medical Center. In particular, Glen Evans, MD, Ph.D., the director of the center, has provided important insights and support for the work. Others who have contributed significantly to this work include Shane Probst, Kim Regan, Travis Ward, Stafford Brignac, Simon Rayner, and Greg Miller. These people are doing the hands-on engineering, programming, maintenance, development, and operation of the many automation systems within the center. In addition Ali Dibiri (SAIC, San Diego) and Mike Heller (Nanogen, San Diego) have been important collaborators.

REFERENCES

1. M. Smith, A.L. Holmsen, M. Peterson, Y. Wei, G.A. Evans (1994). Genomic sequence sample analysis: A rapid strategy for sequence based mapping of complex genomes. *Nature Genet.* 7:40–47.

2. M. Olson, L. Hood, C. Cantor, D. Bpydyrom (1989). A common language for physical mapping of the human genome. *Science* 245:1434–1435.

3. R.W. David (1995). An integrated system for genomic DNA sequencing, presented at the automation in mapping and sequencing conference. Berkeley, CA, November 5–8, 1995.

4. E.S. Lander, M.S. Waterman (1988). Genomic mapping by fingerprinting random clones: A mathematical analysis. *Genomics* 2:231–235.

5. T. Hunkapiller, R. Kaiser, B. Koop, L. Hood (1991). Large-scale and automated DNA sequence determination. *Science* 254:59–67.

6. R. Waterston, C. Martin, M. Craxton, C. Huynh, A. Coulson, L. Hillier, R. Durbin, P. Green, R. Shownkeen, N. Halloran, M. Metzstein, T. Hawkins, R. Wilson, M. Berks, Z. Du, K. Thomas, J. Thierry-Mieg, J. Suzston (1992). A survey of expressed genes in *Caenorhabditis elegans. Nature Genet.* 1:114–123.

7. D. Micklos, G. Freyer (1990). *DNA Science, A First Course in Recombinant DNA Technology.* Cold Spring Harbor Press, Cold Spring Harbor, NY.

8. G.A. Evans (1991). Recent advances in genetics. *Epilepsy Res. Suppl.* 4:189–198.

9. Beckman Instruments, product literature.

10. Robbins Scientific, product literature.

11. J.M. Jarlevic (1993). Automation of a large-scale directed sequencing strategy. DOE Contractor—Grantee Workshop III, Santa Fe, NM, February 7–10.

12. L.D. Stein (1995). *How to Set up and Maintain a World Wide Web Site.* Addison-Wesley, Reading, MA.

13. Ron W. Davis, private communication.

14. ACI Technologies, 4th Dimension.

15. K. Kupfer, M.W. Smith, J. Quackenbush, G.A. Evans (1995). Physical mapping of complex genomes by sampled sequencing: A theoretical analysis. *Genomics* 27:90–100.

16. E.D. Green, M.V. Olsen (1990). Chromosomal region of the cystic fibrosis gene on yeast artificial chromosomes: A model for human genome mapping. *Science* 250:94–98.

17. J. Quackenbush, J.M. Bailis, J.V. Khristich, K. Diggle, Y. Marchuck, J. Tobin, S.P. Clark, A. Rodkins, S. Marcano, A.C. Churukian, J.S. Hutchinson, Y.H. Wie, M.W. Smith, L. Selleri, G.A. Evans (1995). An STS content map of human chromosome 11: Localization of 910 YAC clones and 109 islands. *Genomics* 29:512–525.

18. Hybaid Ltd., colony picker, product literature.

19. Dave Burbee, private communication.

20. H.R. Garner, B. Armstrong, D. Kramarsky (1993). Dr. Prepper—An automated DNA extraction and purification system. *Sci. Comp. Auto.* 9:61–68.

21. H.R. Garner, B. Armstrong, D. Kramarsky (1992). High-throughput DNA prep system. *GATA* 9:127–133.

22. B. Armstrong, H.R. Garner (1992). Analysis of protocol variations on DNA yield. *GATA* 9:134–139.

23. Symantic Corporation, Think C.

24. Bowers Development Corporation, AppMaker.

25. H.R. Garner (1994). Automating the PCR process. In *The Polymerase Chain Reaction*, K. Mullis, F. Ferre, R. Gibbs eds. Chapter 16. Birkhauser, Boston.

26. H.R. Garner, D.M. Lininger, B. Armstrong (1993). High-throughput PCR. *BioTechniq.* 14:112–115.

27. S. Clark, G. Evans, H.R. Garner (1993). Informatics and automation used in the physical mapping of the genome. In *Biocomputing: Informatics and Genome Projects*, D. Smith (UCSD) ed. Chapter 2. Academic Press, San Diego.

28. Molecular Devices Corporation, Spectramax.

29. Perkin Elmer Corporation, product literature.

30. Applied Biosystems Incorporated.

31. Microsoft Corporation, Excel.

32. DANTZ Development Corporation, Retrospect Remote.

33. C. Burks, M.J. Cinkosky, W.M. Fischer, P. Gilna, J.E. Hayden, G.M. Keen, M. Kelly, D. Kristofferson, J. Lawrence (1992). GENBANK. *Nuc. Acid. Res.* 20:2065–2069.

34. P.L. Pearson (1991). The Genome Data Base (GDB)—A human gene mapping repository. *Nuc. Acid. Res.* 19(suppl.):2237–2239.

35. E.D. Green, M.V. Olsen (1990). Systematic screening of yeast artificial chromosome libraries using the polymerase chain reaction. *PNAS* 87:1213–1217.

36. CU-SeeMe, IP Video Conferencing, Cornell University.

37. Netscape Communications Corp., Netscape Navigator.

38. I.S. Graham, "HTML Sourcebook, A Complete Guide to HTML." John Wiley & Sons, Inc., New York 1995.

39. GIF Converter, written by K.A. Mitchell.

40. Stuffit Delux, Aladdin Systems, Inc.

41. JPEG view, written by Aaron Giles.

42. $99 camera, available from Connectix Corporation.

43. Biosensors reach out. Special issue of *IEEE Engineering in Medicine and Biology Magazine*, June 1994.

44. S. Fodor, J. Read, M. Pirrung, L. Stryer, A. Tsai Lu, D. Solas (1991). Light-directed, spatially addressable parallel chemical synthesis. *Science* 254:767–773.

45. M. Eggers, M. Hogan, R.K. Reich, J. Lamture, D. Ehrlich, M. Hollis, B. Kosicki, T. Powdrill, K. Beattie, S. Smith, R. Varma, R. Gangadharan, A. Mallik, B. Burke, D. Wallace (1994). A microchip for quantitative detection of molecules utilizing luminescent and radioisotope reporter groups. *BioTechniq.* 17:516–524.

46. G.A. Evans, E. Tu, W. Butler, R. Sosnowski, D. Montgomery, M. Heller (1994). Electronically controlled DNA hybridization on semiconductor devices: A miniaturized format for rapid medical diagnostics and DNA sequencing. A Decade of PCR—In Celebration of Ten Years of Amplification Conference, September 14–14, 1994, Cold Spring Harbor Laboratory, NY.

47. Ali Dabiri, private communication.

2

Automation Strategies: A Modular Approach

MARTIN J. POLLARD

CONTENTS

Automation Technologies for Genome Characterization, Edited by Tony J. Beugelsdijk.
ISBN 0-471-12806-6　© 1997 John Wiley & Sons, Inc.

INTRODUCTION

The goals of the Human Genome Program (HGP) are to develop the genetic and physical maps and determine the DNA sequence of the human genome and the genomes of several model organisms [1]. It is an international research project involving public and private laboratories primarily in the United States and Europe. This project arose from the discovery of recombinant DNA techniques during the early 1970s and the development of the Maxim-Gilbert and Sanger methods of sequencing DNA. Of critical importance was the development of automated sequencing machines. Simultaneously computer technology was reaching into the every laboratory in the form of automated instrumentation, personal computers, and workstations. Recombinant DNA techniques provided the biological tools necessary for a conceptually simple means of determining detailed DNA sequence information. However, the task of actually sequencing the human genome is enormous, even considering the use of automated sequencing machines. It was recognized early on that this project would require the application of instrumentation technology and computing at an unprecedented scale in the field of biology.

The HGP has changed not only the way we think about biology but how we do biology in the laboratory. A primary difficulty in the task of genomic sequencing originates from the fact that the size of the DNA that can be analyzed with today's technology is extremely small relative to the size of the genome itself. In admittedly oversimplified terms, genomic research entails breaking down the genome into small pieces that can be analyzed, determining the srequence of each piece, and then collating and reassembling the information to construct the finished sequence map of the original genome. The enormous number of samples that must be physically handled, and then whose information content must be processed, has stimulated collaboration with engineers and computer scientists to develop new ways to do biological research.

STRATEGY IMPLEMENTATION

The large-scale nature of genomic research requires strategic decisions as to how to apply engineering and computer technology to acquire accurate mapping and sequencing information in a cost-effective manner. There exists several variants of sequencing protocols that differ in their emphasis on the use of hardware automation relative to computer analysis. For example, in the area of DNA sequencing, the competing strategies have been random versus directed sequencing. Both strategies depend on the standard automated sequencing machines. Hardware automation is used to select and prepare samples for sequencing. Computer analysis is used to analyze the individual sample sequence information and assemble it into contiguous finished genomic sequence.

Random strategies are designed to minimize the complexity of laboratory procedures and automation and to emphasize computer analysis and assembly of large numbers of randomly distributed sample sequences. The automation is designed to prepare large numbers of samples for sequencing in large numbers of sequencing machines. These sequence data result in a relatively large and complex computer assembly problem. In contrast, directed strategies put relatively greater emphasis on the use of automation to implement more complex laboratory procedures. Additional data are generated that must be tracked throughout the process. Some of these data are mapping information which is used to select clones to be sequenced and reject clones from regions that have already been sequenced. This results in the processing of fewer samples and the use of fewer expensive sequencing machines, producing a significantly simpler sequence data assembly job.

The choice of a mapping or sequencing strategy will have a profound influence on the personnel and resource needs of a genome center. There are undoubtedly many reasons for choosing a particular strategy. This decision must in part be determined by the availability of resources in computer science and engineering at a particular institution. There has been very little tradition of collaboration between biology and the fields of mechanical and electrical engineering. This is evident in the existence of relatively few genome centers that have strong engineering programs in these areas. Access to engineering resources will provide an additional dimension of flexibility in the development of optimum genome strategies. In the area of genomic sequencing, there has been an evolution toward strategies comprising elements of both directed and random methods. This type of dynamic strategy development requires sufficient flexibility to implement new automation hardware to modified random strategies and alteration of existing highly automated directed strategies.

The remainder of this discussion will describe some of the principles of engineering as they are applied to genome research. This will hopefully provide a systematic means for evaluating current and future large-scale automation strategies as well as basis for evaluating individual instrument designs. I will focus on hardware automation for processing samples to subsequently be sequenced in automated sequencing machines. I will not discuss computer automation or informatics related to sequence assembly.

GLOBAL AUTOMATION STRATEGIES

The expectation early in the HGP was that it would be necessary to invent revolutionary new instruments to prepare and sequence samples at a rate appropriate to complete the sequencing of the human genome at a cost of $0.50 per base over a period of 15 years. While a number innovative research programs are underway that are directed toward this goal, none has contributed to the success of the existing multimegabase per year sequencing centers. These centers have relied on the application of conventional engineering technology to the implementation of effective production instrumentation. Revolutionary technology advancement remains an important goal to help complete the sequencing of the human genome and to make available the tools of rapid sequencing to smaller laboratories in universities, industry, and in the clinical health field.

Existing technology has been used to develop production automation that has been applied to the genetic and physical mapping of model organisms and human, and in the discovery of expressed sequence tag sites. Similarly existing technology has helped to make possible the large-scale sequencing of the genomes of model organisms (*S. cerevisiae, C. elegans, Drosophila*). Human sequencing is still in the early stages of scale-up and will build on the technology and experience of the sequencing efforts for model organisms. The sequencing of the human genome will present new technological challenges. Strategy development for human genome sequencing should reflect the still dynamic nature of the HGP.

The paradigm often invoked for genome automation is one of the "genome factory." Factory automation has been in use and has been very successful in the manufacturing field for much of this century. In recent years, with advances in automation, robotics, and computer technology, the modern factory automation environment is relatively adaptable to changes in product lines. Scientific research, on the other hand, has not been in an environment suitable for factorylike automation except in the fields of environmental sampling and analytical chemistry. The sample handling and analysis requirements for the HGP present another area amenable to laboratory automation. The implementation of an actual "genome factory" and the appropriate management structure will be a distinctly different environment than that of a research or academic institution. The sequencing effort at LBNL is driven by the metrics: cost per basepair, throughput, cycle time per contig, and errors per megabase.

A BOTTOM-UP MODULAR APPROACH

There are two contrasting approaches to laboratory automation that can be characterized as being either top-down or bottom-up. A bottom-up approach is one in which technology is implemented in relatively small discrete modules to address bottlenecks in the current state of the project at any given time. The emphasis with this approach is to get technology into the laboratory as soon as possible in the areas that need it the most to realize an increase in productivity. A top-down approach would be to examine the entire experimental process and design a highly efficient and highly integrated hardware and software solution to all aspects of the problem to be solved. This is a long-term view that depends on a well-defined and stable "problem" to be engineered. Top-down systems are relatively inflexible to changes in experimental protocols. Systems of this design generally come on-line all at once after a considerable period of design, fabrication, implementation, and testing. The Lawrence Berkeley National Laboratory (LBNL) Human Genome Center (HGC) decided to adopt a modular bottom-up approach to the planning and application of automation technology to *Drosophila* and human sequencing.

The bottom-up approach practiced at LBNL recognized that in the early stages of the HGP the experimental environment both within the lab and on a worldwide basis would be quite fluid. The biological technology that led to the conceptually simple notions of how to sequence the genome was also undergoing rapid change. The period from the mid-1980s to the present has been a period of experimentation involving a variety of large-scale mapping and sequencing strategies. Because of its ability to adapt to a changing experimental environment, a bottom-up modular approach is often preferred.

This early period of experimentation with biological protocols was also an opportunity to develop the working relationships between biologists and engineers that would be required for the implementation of more massive automation systems in the future. Early introduction of well-defined small-scale engineering technology into the laboratory led to manageable projects to develop these working relationships. Finally, and perhaps most realistically, a modular approach is an initially less expensive means to introduce engineering technology into the laboratory during the funding ramp-up of the project.

These advantages of a bottom-up development and design philosophy must be weighed against the disadvantages of this approach. The primary disadvantage is the difficulty of building a well-integrated collection of automation modules throughout all aspects of the experiment over a long period of time. There will always be pressure to utilize new and better technology as it becomes available. Modules designed and built during the early stages of the project may constrain or influence later technology development. Alternatively modules developed early in the project will have to be rebuilt to keep up with current technology. As more modules are developed, they will begin to connect together and the interfacing of modules will become a serious concern. It is important in the adoption of a bottom-up design philosophy to understand the difficulties with this approach. The disadvantages can be partly mitigated by a continuous dynamic analysis of the project throughout implementation of the automation while maintaining the distinctive features and advantages of a bottom-up approach, which are early introduction of individual automation modules and continuous flexibility to adapt to changing protocols and new technology.

A modular approach is applicable to a wide variety of institutional environments. It is often the default approach for institutions unable to make the large initial investment of a top-down system. A number of genome centers, including the LBNL HGC, have focused on particular biological strategies to genomic sequencing. Focusing on a particular sequencing strategy may appear to lend itself to a top-down automation design. However, the availability and productivity of automation technology has driven the evolution of new biological strategies and has contributed to a changing laboratory environment. A modular approach is also advantageous in laboratories that may have a number of biology groups pursuing different experimental programs in genomic research. These types of laboratories push automation development into areas of common interest among biology research groups with different goals. Finally, a modular approach is convenient for laboratories that choose to go to outside contractors for engineering expertise and equipment. It requires laboratories to define discrete instruments for purchase.

LABORATORY AUTOMATION AND ROBOTICS

Decisions regarding global automation strategies as well as designs for particular instruments should be based on a realistic expectation of what benefits can be achieved by automation. The principles learned and applied to factory automation can also be applied to genome laboratory automation. An understanding of the types of processes that are automatable and a reasonable expectation of the performance that can be achieved is essential for designing a high-quality, efficient, and cost-effective strategy for mapping and/or sequencing.

The following is a list of benefits attributed to factory automation that are frequently cited throughout the business and the engineering world. It is worth considering their implications to genome research. This list is not ordered with respect to importance. It is should also be realized that any global or particular automation design does not have to meet all of these criteria. It only has to realize an improvement in the production sequencing metrics of cost per basepair, throughput, cycle time per contig, and errors per megabase.

Speed/High Throughput

Automated instruments can often perform a task much faster than manual methods to achieve higher throughput. Throughput is the number of samples processed per unit time. One area where this is clearly the case is in the use of imaging technology. The imaging of fluorescent samples in polyacrilamide and agarose gels, coupled with the initial computer analysis of these images, is considerably faster than manual methods. Similarly film exposures to radioactive probes can be digitally scanned and analyzed.

Automated instruments can also perform mechanical tasks faster. The machine may actually move faster than a person performing the same operation. The important elements to consider in order to achieve an increase in the speed of a mechanical operation are the speed of the drive elements, the weight of the samples (total load), and the distance to be traveled. Mechanical speed will increase for light loads, moved over short distances, very fast. Average overall speeds can be increased with machines that operate at a steady pace for long periods of time.

Throughput can be increased by operating on many samples at once or doing multiple tasks at one time—multiplexing. An example of multiplexing is the thermal cycling of samples in an air oven. While the cycling time of an air oven is relatively slow, the use of a dozen 864-well microtiter plates results in a very high sample throughput.

Repetition/Boredom

Tasks that are highly repetitive and boring are good candidates for automation. Repetitive tasks are often a poor use of skilled personnel and often lead to process errors.

Cost Effectiveness

Automated machines can reduce material and personnel costs. Training for complex tasks can be reduced, and the people operating machines may be able to do more tasks or process more samples. Machines can also improve cost by reducing the amount of consumables such as reagents and plasticware. In some cases a custom instrument may be more cost effective than a modified commercial instrument. The actual cost of any instrument must be amortized over the life of the instrument.

Better-Quality Results/Can't Do It Any Other Way

Automated machinery can often do tasks that cannot practically or efficiently be done by manual methods. The manipulation of objects on a small scale is an example. Accurately moving a probe (pipet tip or a colony-picking needle) to a small target (a well in a 384-well plate or a bacteria colony) is best done by an automated instrument. Another example is the placement of a gridding tool for very high-density gridding of samples on filters.

Complexity

Automated machines can often do complex tasks provided that the task can be broken down to simple discrete operations. These operations can be chained together into a single complex task performed by a machine. A machine can perform the overall task with more consistency than can be done manually. Predictable results from consistent machine operation can be used to plan and optimize subsequent steps in the laboratory. Some tasks are complex enough to require a significant learning period for new employees to perform them adequately. A high turnover in personnel for this task will result in constant training and low productivity. It may be simpler and faster to teach operators to use a machine. DNA sample preparation for sequencing is an example of a complex operation that can be automated.

Standardization

Automation at some level requires materials and format specifications in order to process the samples. Standardization on, for example, the 96-well microtiter plate format provides a known reference format upon which everything else should conform. Standardization can enforce a common input and output for each module within an integrated system. It can also make it possible for instruments to be transferred or commercialized for wide distribution.

SAFETY

Safety considerations are a major reason for the introduction of automation in industry. Automated machines and robots can be used to perform tasks in environments that are hazardous to humans. The handling of toxic chemicals and radioactive materials are good examples of where automation can provide essential or desired health and safety protection. However, experience has shown that the use of radioactive and other chemicals commonly used in biology laboratories can be done safely with proper training. The use of radioactive and toxic chemicals is generally a waste disposal problem rather than a safety problem in the genome laboratory. Often, rather than solving safety problems, automated machines themselves can be dangerous, and the safety issues become one of protecting laboratory personnel from the machines.

FIXED VERSUS FLEXIBLE (ROBOTIC) AUTOMATION

The hardware configuration of an automated system can be classified as either fixed or flexible. Both top-down and bottom-up automation strategies can use fixed and/or flexible automation.

The term "fixed automation" is used to describe specialized machines that have been designed to perform a particular task or set of tasks very efficiently to achieve very high throughput. Fixed automation cannot easily be reconfigured to perform a substantially different task. This is an appropriate automation strategy for an unchanging process that demands high throughput. The entire cost of the machine must be justified by the throughput of the process over the period of time during which the fixed process remains valid.

A flexible or robotic instrument is defined as a reprogrammable, multifunctional manipulator designed to move material, parts, tools, or specialized devices through various programmed motions for the performance of a variety of tasks (Robot Institute of America, 1979). The cost of the robot is justified by the longer lifetime achieved by adaptability to changing processes. Robots generally take the form of a coordinate system frame and an arm, and a end effector (i.e., gripper) that moves within the work envelope. A stated advantage of flexible automation versus fixed automation is the speed of implementation of new and different protocols. Robots can be purchased "off the shelf." The remaining work is then limited to design and fabrication of peripheral components and programming of the final system. Since few parts are custom manufactured, there is less mechanical debugging. Finally, the "reaction time" to implement a new protocol on a robot platform is shorter than the time to design and build a custom automated machine.

Flexible automation tends to be slow because the single arm and gripper performs tasks one at a time. These robots are used to best advantage when servicing multiple custom applications modules located within the work envelope, thereby providing a multiplex capability to the whole system.

There are various classification schemes to describe robots, but the categories of robots generally seen in the biology laboratory are the following:

- An articulated arm operating in generally a Cartesian (XYZ) coordinate system. The Hewlett Packard/Sagian ORCA and the CRS T265 are examples of this type of robot which is commonly found in genome laboratories. Both of these robots move on a linear rail that can be as long as two or more meters and therefore have a very large work envelope. They have very dexterous jointed arms and grippers that can tilt and rotate giving 6 degrees of freedom.

- A cylindrical coordinate system robot can access tools and materials by moving a gripper radially and circularly around a vertical center axis as well as vertically on the center axis. The robots produced by Zymark, Adept Technology, and CRS Robotics operate in this type of coordinate system. These robots access side stations that are located radially around the center of the robot.

- Cantilever and Gantry robots operate in a Cartesian coordinate system but they are built from linear rails oriented along the *XYZ* axes. These robots generally do not have articulated arms, though wrists with additional degrees of freedom can be attached to the gripper mechanism. The Beckman Biomek 1000 and 2000 Workstations and the Hamilton Microlab Workstation are designed with this format.

The advantages of having a capable automation group (mechanical technicians, engineers, and programmers) to implement flexible automation cannot be understated. Commercial robots are more often than not deficient in either applications hardware and/or software. The off-the-shelf advantages of commercial robots are overstated relative to the ease in which a custom gantry robot system can be assembled by a competent mechanical engineer. Finally, the reconfigurability feature is seldom utilized in a production environment. Most robot systems are configured for one application and are seldom used for anything else. It is not uncommon to see two identical robot systems each configured to a particular application. This "actual" use of robot platforms should be weighed against the high-throughput advantages of a fixed automation platform.

LARGE VERSUS SMALL MACHINES

Important considerations in modular instrument design are space, access, and reliability. Automated instruments and robots come in a wide range of sizes depending on their functions and applications. Instruments can be as small as a microtiter plate filler (9″w × 12″d × 10″h) to a Beckman Biomek Workstation (3′w × 2′d × 30″h) to machines that will fill a large room. The size of the instrument will influence how it is used on a daily basis.

Large machines can be characterized as having large work envelopes to provide access to many microtiter plates, filters, or specialized automation module side stations. These instruments occupy a significant amount of laboratory space and are often located some distance from the general laboratory area. They must be designed to operate unattended for significant periods of time so that there will be less frequent operator interaction for an instrument that is located a long distance from the main laboratory. Large instruments are generally more complex and expensive to meet the criteria of handling large numbers of samples and to achieve reliable operation. They will take a long time (1–2 years) to design, fabricate, and test; therefore the resulting throughput and productivity goals should be set very high to justify the investment of time and money. Machines of this type are generally only built to implement very stable processes. The largest machines are often found in top-down strategies and in mature bottom-up strategies.

Small machines can occupy a small amount of lab space, and therefore it is more practicable to locate them in proximity to the main laboratory activity. This means that operators can interact with them more frequently if necessary. Often they will perform limited tasks and are therefore less complex to design, fabricate, and they can be implemented fairly quickly (1–12 months). This also means that they are easily duplicated. There may be an advantage to having multiple small instruments versus a single large instrument. A single large instrument that is down for maintenance or repairs for any significant amount of time can seriously affect productivity.

ROBOT CAPABILITIES

An essential element in designing automation modules is to understand the type of things that robots and automated machinery can do in the laboratory. Much of the automation developed to date is devoted to materials handling and processing, although a few instruments, typically imaging systems, generate information to be used in later stages of the process. This reflects what normally happens at the laboratory bench. Any task that is proposed for automation must be reduced to its elemental robot compatible components. The automation solution may look very different from the manual procedure but will achieve the same result. An appreciation of these capabilities is essential for developing biological protocols that are automatable.

The following is a list of the type of practicable robot capabilities that can be used as building-blocks in order to automate benchtop protocols:

Materials Handling

Robots can move objects from one place to another place. This is often done with grippers configured with custom-designed fingers to accommodate the item being moved and the location into which the item is moved. Materials can be grasped with suction cups, pushed and pulled, moved with conveyor belts or shuttle mechanisms.

Sophisticated robots have the capability of changing grippers and tools. The typical actions are to move microtiter plates from one location to another and to move pipetting or dispensing tools to microtiter plates or reservoirs. Colony-picking machine needles move bacteria from a colony to a microtiter plate by adhesion of the cells to a picking needle.

The major performance specifications for materials handling and motion control using a robot are as follows:

Repeatability. The ability to return to the same point.

Accuracy. The ability to go a specified distance from a known reference point.

Velocity. How fast the robot can move from one location to another.

Acceleration. The speed with which a load can be brought from rest to a particular Velocity.

Load. The amount of mass that can be lifted, pushed, or pulled.

Dexterity. The ability to trace complex paths.

Tool interchangability. The ability to attach and operate a variety of tools.

Liquid Handling

ASPIRATE/DISPENSE   Liquid delivery is usually performed using robotically compatible pipettors (to aspirate and dispense using fixed and removable tips) and dispensers. These tools can either attach to the robot arm and gripper or reside as a fixed module within the robot work envelope.

Pipetting tools can be used to mix liquids and, in combination with movement, transfer liquids from one location to another. Pipettors can be found in two configurations. Syringe pump pipettors have a remote motorized syringe station connected by tubing to a dispensing nozzle. The second configuration is similar to manual pipettors where the entire plunger mechanism is in a compact enclosure attached to the gripper. This configuration of the pipetting tools used in the Beckman Biomek robots. Pipettors and dispensers are usually configured with 1, 4, 8, or 12 liquid delivery channels. Pipettors can operate in a volume ranging from 1000 to 0.5 μl, though this entire range is generally not available in a single tool.

A dispenser delivers liquid from a large reservoir. This is done using pumps or by pressuring the reservoir and releasing fixed volumes, opening a valve for a calibrated period of time. An interesting dispenser research area is the use of ink-jet dispenser mechanisms to dispense submicron volumes of liquid. Liquid delivery at volumes less than 2 ml can be difficult to accomplish due to the surface tension of the liquid. Careful selection and design of pipet tips and dispenser nozzles is required.

ADHESION   Liquid transfer can also be achieved by simple adhesion of liquid to a pin as is the case in 96/384 pin replicating tools used to copy bacteria libraries by inoculating microtiter plate wells from a source microtiter plate.

PUMPING   Machines can actively pump liquids from one reservoir to another or to be used as a thermal transfer medium for heating and cooling.

Heating and Cooling

Heating and cooling sidestations can be found in various formats. Temperature-controlled heating and/or cooling blocks transfer heat by conduction through a metal block in close contact with a thin-walled microtiter plate. Heat is also transferred by convection using circulating water baths or air ovens. Heating by quartz lamps or hot air-heating elements is used to sterilize picking needles in colony-picking machines.

Image Acquisition

Image acquisition is accomplished with a digital camera or scanning device. It is used to record fluorescent images, such as ethidium-stained DNA in an agarose gels, or to locate objects, as is done in a colony picker or pick and place robot. The typical colony-picker camera is a room temperature black-and-white CCD camera with a 480 × 640 pixel detector. Cooled digital cameras (for high sensitivity to weak fluorescent signals) and high-resolution cameras (larger pixel array detectors) are very expensive and are less commonly used. Color cameras are also available, but these typically reduce the pixel resolution to a third of that of a black-and-white camera. Additional hardware requirements are a camera lens, a digital frame grabber card, and an image analysis software package.

Data Analysis

The computers that control a robot can process data as well. An agarose gel imaging station can acquire the image, calculate and plot calibration curves, and calculate the length of the DNA samples recorded in the gel. Computers can also be used to track bar code data and interface with LIMS or other information/materials tracking software.

Sensing

Robots and automated machines can sense a variety of physical parameters. Camera vision images can be processed to guide a robot to an object (and verify its presence and correct placement). Other sensors can measure temperature, pressure, humidity, proximity (i.e., is an object located near the sensor), color, liquid level, force, and a variety of chemical parameters.

COMPUTERS AND SOFTWARE

Typically both flexible and fixed automation modules are controlled by stand-alone personal computers, usually either an IBM-style PC or a Macintosh computer. Most laboratory-scale automation does not require powerful computers. The computationally intensive features of motion control and robot kinematics are built into custom input/output boards, amplifiers, and other stand-alone instrumentation modules. Much of the computer power of a PC is used to execute graphics applications such as a graphical Windows user interface or digital image display and analysis software. Robot systems designed for industrial use may require a more powerful custom computer and electronic control equipment system.

The actual operation of automated systems is through simple software interfaces intended to be used by nonspecialized personnel. A good user interface is one that is intuitive and simple to use. The software interface should be well laid out visually on the screen, should not produce unexpected and confusing messages or machine responses, should remember previous user parameters, and should operate with a minimum of user input. Ideally the user interfaces for all modules should have a similar "look and feel" to minimize training on each instrument. Custom software interfaces should be designed to accommodate the users. It is extremely important to realize that a user can declare a functional instrument unacceptable simply based on experience with a poorly designed software interface.

Small commercial liquid handling robots such as those produced by Beckman, Hamilton, and Tecan are design to be "programmed" by novice users. Users can typically learn to program these systems in a single day. Implementing a protocol usually involves entering simple hardware parameters and the chaining together of a library of preprogrammed robot actions into a complete protocol. Most robot systems, both commercial or custom built, required expert dedicated personnel to program protocols and maintain the software. The types of software skills necessary to program automation modules range from learning the complex and powerful custom macro languages of a commercial articulated robot, programming in high-level languages such as Visual Basic or low-level languages such as C or assembly code, and finally programming in custom low-level command languages that operate specific input/output boards, motor driver boards, and so on.

Software platforms should be chosen with the following criteria in mind: the ability to develop a well-designed user interface, control the equipment, and be easily maintained by software personnel. It is very important that more than one person be familiar with any software platform that is implemented. At LBNL, we have chosen Visual Basic as the standard software package used to write motion control applications because it provides a powerful Windows-based graphical user interface, is simple for anyone to program and maintain, and can be linked to other Windows compatible software (i.e., the C programming language).

Buy or Build?

In the course of instrumentation module development and implementation, the question of whether to buy or to build an instrument will invariably be encountered. If a commercial instrument meets the stated goals, then it is the best to purchase the instrument. This has the obvious advantages of a machine that has been extensively tested, which may have a historical track record of laboratory use, and has the technical support and expertise of the manufacturer.

If no commercial instrument meets the stated goals, then it still may be possible to purchase a commercial instrument and modify its hardware and/or software capabilities. The manufacturer may or may not cooperate in a modification of its instrument. Many manufacturers maintain research groups that are often willing to work with customers on special projects, usually with the expectation that they can develop something new to add to the product line. They may recommend a third-party engineering consultant or direct you to customers who may have solved similar problems.

Building a large-scale instrument should be considered a major undertaking, requiring proper resources in personnel and laboratory and machine shop space. Requesting a custom instrument from an engineering group is not the same as purchasing an instrument from a commercial manufacturer. It requires not only a financial commitment but an intellectual and time commitment from the biology management to be an integral participant in the project. This means close interaction between engineers and the biology management during the design phase and with biology users during the installation and beta-testing phase.

A custom instrumentation project will generally require one or more engineers (mechanical, electrical engineering, and computer science) as well as mechanical and electrical technicians and machine shop personnel. It will also require facilities and space to fabricate parts and assemble the instruments separate from the final installation point. Depending on the size and complexity of the project, it can take from 2 to 12 months to design and fabricate an instrument, and 2 to 6 months to debug the instrument and integrate it into production.

Many laboratories do not have the proper engineering personnel and facilities for instrumentation development and thus must seek assistance from outside engineering firms. If the instrument is to be built by an outside engineering firm, then the commercial aspects of the project should be discussed early in the project. One option is to contract for a one-time special project to be built by an engineering firm for a particular laboratory. It is also possible to collaborate with a commercial company to develop an instrument with the intention of commercializing the instrument for sale in a wider market.

The build or buy question touches on the issue of technology transfer. Technology transfer is only going to be successful for instrument designs that are easily adaptable to new environments. Both custom and commercial instruments will have to be designed for a range of performance, size, and cost for a given customer base. However, the demands of genome research is for efficient high-throughput (specialized) instrumentation. It is not surprising that there are many unique instruments in genome centers around the world that do not have wide applicability in other genome laboratories or in the commercial sector. This also illustrates that strong engineering capabilities are necessary to take full advantage of commercial instruments.

AUTOMATION MODULES

A modular approach in the application of engineering solutions to automation at the LBNL HGC and elsewhere can best be illustrated by a list and description of some of the custom instruments that have been produced. These are instruments that exist in the laboratory and are in use today. This list contains instruments that range from modifications of commercial instruments to entirely custom fabrications. The following list, description, and notes on instrumentation are devoted primarily but not exclusively to the instruments pursued by the LBNL HGC. In some cases similar instruments have simultaneously been developed elsewhere. The descriptions also list some of the primary automation benefits achieved by each machine. This is an illustrative list and is not intended to be a survey of all of the instruments developed worldwide.

INSTRUMENTATION MODULES

Imaging

There are a number of custom and commercial imaging instruments dedicated to recording radioactive or fluorescent sample signals or scanning photographic films. Imaging of hybridization filters is common in many laboratories. The imaging instrumentation at LBNL (Fig. 2.1) is dedicated to imaging ethidium bromide stained agarose gels. It consists of UV illumination of the gels, a cooled CCD camera for image capture, and commercial and custom image analysis software. There is also an essential imaging component to colony-picking machines.

AUTOMATION BENEFITS   Digital image capture and analysis are much faster than manual methods and are cost effective in performing repetitive work. Moreover, image instrumentation can perform complex computational analysis of the data.

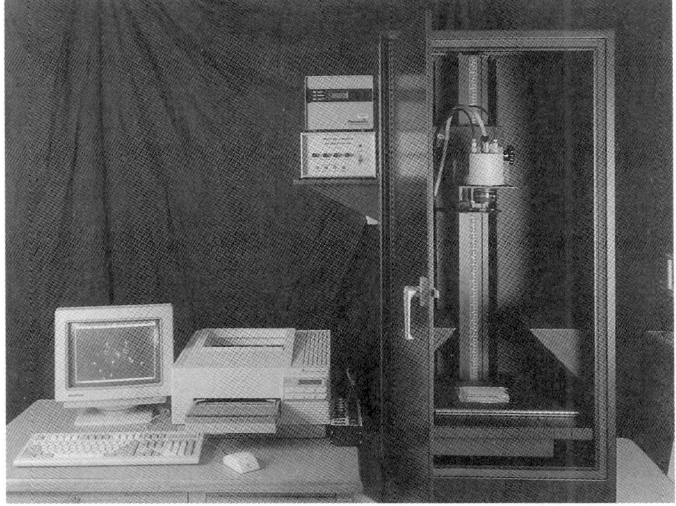

Fig. 2.1   *Agarose gel-imaging station with cooled CCD camera and data analysis computer.*

Colony Picking

This LBNL custom colony picker consists of a digital imaging station used to determine the location of the colonies in 10×10 cmn agar plates and a separate colony-picking machine (Fig. 2.2). The machine consists of 12 picking needles mounted on a carousel above two *XY* tables. The *XY* tables move the source colony plate and the destination microtiter plate underneath the picking needles. The needles pass through a sterilizing station with each cycle. The digital imaging of the plates, the colony picking, the placing of the colonies into the deep-well microtiter plate, and the needle sterilization all take place simultaneously. The machine requires that the user continuously image and feed plates to the colony picker. The entire system picks colonies at a rate of about 1200 colonies per hour. The typical use of the colony-picking machine is for low-volume continuous colony picking that consists of multiple (20–30) experiments of 96 colonies every couple of days. Another mode of use, more prevalent at other laboratories, is to pick large libraries (25,000–100,000 colonies) on an infrequent basis.

AUTOMATION BENEFITS   The machine performs extremely repetitious work, and it picks colonies faster, on average, than manual methods. Picking colonies from 10×10 cm agar plates into deep-well microtiter plates is unique to this machine and is particularly well suited to the colony-picking experiments done at the LBNL HGC.

OTHER IMPLEMENTATIONS   Colony picking has been implemented in a large number of laboratories, including the Sanger Centre, the Max Plank Institute for

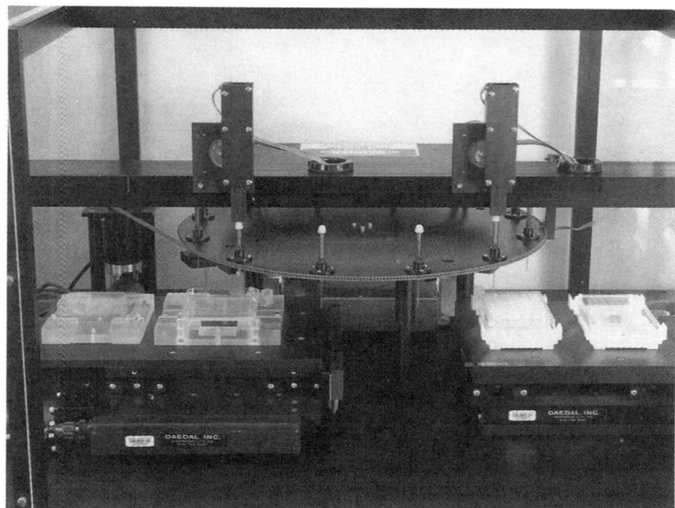

Fig. 2.2 Bacterial colony-picking machine. This machine picks bacterial colonies at a rate of 1200 colonies per hour. The digital image station is not shown.

Molecular Biology, and Stanford University (plaque picking). Commercial instruments have been developed by PBA Technology, Genetix Ltd., and BioRobotics in England.

Liquid-Handling Robots

Commercial liquid-handling robots produced by Beckman, Hamilton, and Tecan are common in genome laboratories. They are accurate, easy to program, and have relatively high throughput if configured with 8-channel pipetting tools. Automated pipetting at LBNL is implemented on the Beckman Biomek 1000 and the Hewlett Packard/Sagian ORCA robot. The Biomek 1000 has proved to be an accurate and reliable low-volume (1–20 ml) multichannel pipetting machine. Virtually all of our programming on the Biomek 1000 is done in Microsoft QuickBasic coupled with a custom low-level software package (BiomekQB). With the ability to completely control the Biomek 1000 through our custom software, we were able to develop custom hardware modules to implement PCR reaction setup, sequencing reaction setup, sequencing reaction pooling, and agarose gel-loading applications.

The Hewlett Packard/Sagian ORCA is used for pooling and library replication applications that require a large number of source and destination microtiter plates, and therefore a large work envelope. We developed a disposable tip 12-channel pipetting head that is used extensively for row, column, and plate pooling protocols. A separate fixed tip 16-channel manifold is used for plate filling applications.

AUTOMATION BENEFITS   Both the Biomek and the ORCA perform repetitive work, implement complex pooling protocols, and are cost effective as unattended machines that allow personnel to perform other tasks.

Thermal Cycling

Water-based thermal cycling has been implemented primarily as robotically controlled baskets of sealed microtiter plates immersed into temperature-controlled reservoirs. While the individual temperature cycles are fairly long, the overall throughput is very high in some of these machines which can run many 384-well microtiter plate samples. Circulating temperature-controlled air ovens can also thermally cycle samples in a batch mode. Numerous commercial thermal cycling instruments are on the market based on resistance heating and water or Peltier cooling. The commercial machines typically cycle one microtiter plate at a time.

AUTOMATION BENEFITS   Custom thermal cycling instruments have higher throughputs than commercial instruments which can generally cycle only one plate at a time.

DNA Synthesis

LBNL developed a custom 12-channel DNA synthesizer (Fig. 2.3) that can produce 20 nm of 21-mer DNA at about $0.10/per base in 2.5 hours. The instrument was

Fig. 2.3   12-channel DNA synthesizer. The machine synrthesizes 12 different 20-mers in 2.5 hours for approximately $0.10 per base.

designed to have dedicated liquid delivery lines for each reagent to minimize reagent loss that occurs in commercial instruments when reagent manifolds must be flushed between each step. The instrument is manually loaded with the correct CPG beads in each channel, all of the dispensing steps and chemistry are automated, and the final deprotection and capping steps are done manually. The instrument has been designed for scale-up in multiples of 12 channels, but the LBNL HGC has not needed additional capacity in a single machine.

AUTOMATION BENEFITS   This instrument is extremely cost effective. It has significantly higher throughput than commercial instruments.

OTHER IMPLEMENTATIONS   Stanford University and Intelligent Automation Systems (Cambridge, MA) have developed 96 channel DNA synthesizers. A 192 channel DNA synthesizer has been developed at the University of Texas Southwestern Medical Center (Dallas, TX). The ABI 3948 is a 48 channel oligonucleotide Synthesizer. There are a number of single channel DNA synthesizers.

Microtiter Plate Pooling/Microtiter Plate Filling/Library Replication

These applications have been implemented on a Hewlett Packard/Sagian ORCA articulated robot platform (Fig. 2.4). A substantial amount of peripheral hardware was developed around the robot, including 96/384 pin replicating tools, microtiter plate locating and stacking plasticware, a 12-channel syringe pump pipettor, and a multipin

Fig. 2.4 The Hewlett Parkard/Sagian ORCA robot is used for pooling, plate filling, and library replication applications.

tool ultrasonic bath and heater sterilizer. Applications programming was done using the software package provided with the ORCA robot.

AUTOMATION BENEFITS This is a cost-effective stand-alone (hands-off) instrument that frees personnel for other duties, performs repetitious replication, dispensing, and pipetting tasks and accurately completes large and complex pooling schemes.

DNA Preparation

LBNL is currently constructing a custom machine to prepare DNA for sequencing. We will be implementing a modified boil preparation protocol. The instrument will consist of a gantry robot equipped with a pneumatic gripper, five 8-channel dispensers, and two Beckman Biomek 8-channel, 200-ml pipettors. Centrifugation will be performed in an Eppendorf 5416 centrifuge located in the work envelope and modified for remote robot control. The system is designed to prepare 184 samples in two to three hours of unattended operation.

AUTOMATION BENEFITS The instrument will achieve consistent sample preparation relative to labor-intensive manual preps. It will run unattended, allowing personnel to perform other duties. It will have moderate throughput. DNA preparation using a boil prep protocol is very cost effective.

OTHER IMPLEMENTATIONS There are no instruments, commercial or custom, that perform high-throughput DNA preparation. Some instruments have implemented centrifugation, others depend solely on filtration, and some incorporate magnetic bead technology as part of the protocol. DNA preparation is highly sensitive to the biology, and all of the manual benchtop kits and automated procedures have required significant protocol development at each laboratory in order to achieve reliable sample preparation.

Gridding

Gridding machines spot samples at very high density on filters. Gridding applications have been implemented on an Adept robot at Los Alamos National Laboratory and on an ORCA at the Lawrence Livermore National Laboratory. Genetix Ltd. has implemented gridding applications on a relatively large gantry robot, and the Sanger Centre has implemented this application on a somewhat smaller gantry robot.

AUTOMATION BENEFITS   These instruments can grid much faster than manual methods; they can grid at much higher density than is practicable by manual methods; are cost effective, and perform repetitive tasks.

CONCLUSIONS

Much of the instrumentation developed in genome centers around the world can be described as modular. Often this has been by default, since most laboratories starting out in the new field of genomics cannot afford or risk a top-down designed automation system. In some cases a bottom-up modular automation system was created by design, as was the case at the LBNL. A key element to this approach is adaptability to change within a new field of biology.

Currently, with the rise of several multimegabase per year sequencing centers, it is conceivable that new genome centers will begin to choose from existing experimental strategies and adopt the hardware and software solutions developed elsewhere. This would be the beginning of the adoption of a rational top-down strategy. That is not to say that a new center would not bring new automation ideas into the field but that there would be great pressure to enter the field quickly at a high technical level relative to existing centers. From 1995–1996, an instrument system that can be described as a top-down implementation has been installed at the Whitehead Institute as part of the genomic mouse-mapping project. The Genomatron is a large-scale, specialized, high-throughput, and highly integrated system of machinery that will perform sample preparation, thermal cycling, and sample gridding on filters. It was developed at a mature phase in the experimental mapping protocols and justified by a cost analysis of a predictable set of laboratory protocols, goals, and stable funding. Similarly, as sequencing strategies and technology mature, there will be a tendency for new sequencing centers to adopt automation strategies based on current technology. The choice to pursue either a top-down or bottom-up approach will be based on many considerations, but some of them should be a rational understanding of what automation can do and what resources are necessary to be successful. Some of these concepts have been described here.

Large-scale genomic research is a relatively new field in biology. It will require a multidisciplinary approach to accomplish the stated goals of the HGP. While current technology should be capable of sequencing the human genome, the development of more advanced revolutionary technology will be required to bring genomic research capabilities to university and industry laboratories pursuing research in mapping and sequencing of other organisms and into the clinical health field. Engineers and biologists have not been traditional partners in research. This is beginning to change in genome research. Engineering and computer science experience and resources should be equally available to biologists to develop optimal strategies. This can only happen if each field can develop at least a working knowledge of the other fields. It is to be hoped that some of the ideas and concepts outlined here will provide a window to decades of engineering experience that can be applicable to genome research.

ACKNOWLEDGMENTS

This work was supported by the Director, Office of Energy Research, Office of Health and Environmental Research, Human Genome Program, of the U.S. Department of Energy under Contract No. DE-AC03-76SF00098.

DISCLAIMER

Reference to a company or product name does not imply approval or recommendation of the product by the University of California or the U.S. Department of Energy to the exclusion of others that may be suitable.

REFERENCES

1. F. Collins, D. Galas (1993). New five-year plan for the U.S. Human Genome Project. *Science* 262:43.

3

Large-Scale Library Characterization

ELMAR MAIER, DAVID R. BANCROFT, AND HANS LEHRACH

CONTENTS

Automation Technologies for Genome Characterization, Edited by Tony J. Beugelsdijk.
ISBN 0-471-12806-6   © 1997 John Wiley & Sons, Inc.

THE NEED FOR LARGE-SCALE TECHNIQUES

Certainly the information content of genomic DNA in higher organisms is immense. For each cell or tissue type, a very small subset of this information is read out and results in a tissue-specific population of mRNA molecules. Many of these mRNAs are then translated into proteins, forming highly complex interactions within and between cells or tissues of a developing or fully formed organism. This combination of vast tracts of genomic DNA, tissue and developmental specific transcription of small subsets of this information, plus complex interactions between gene products provides a daunting scientific and logistic challenge if a global picture of the function of gene products within an organism is to be understood.

Developing such a global understanding of gene function is not purely an academic challenge. The complex network of gene interactions can be disturbed in the case of a genetic disease, for example, by a (possibly minor) defect in the function of a specific gene or gene product. To fully understand the role of gene products in the health and well-being of an organism, biological research will have to be able to identify all the genes, to learn the signals governing their specific cell-type expression and ultimately the interactions between the genes and gene products on a functional level.

The possibility to read and understand the information encoded in the DNA sequence of the genome and its genes offers a completely new approach to biological research, with enormous consequences for medicine and industry. The identification and characterization of genes and gene products will play a major role in the understanding of the causes of major human diseases, and will provide the targets for drug development for the pharmaceutical industry. This information will therefore in many ways accelerate both basic and applied research. That research is expected to affect greatly the success of the scientific enterprise, contributing, among other factors, to human health and the competitiveness of that industry.

The analysis of gene and gene product in humans will be essential for this functional understanding. Many approaches, however, will require alternative model systems to be used. For example, organisms that offer better genetics and more easily accessible embryology (e.g., zebrafish), the possibility to construct transgenic organisms carrying extra genes (mouse, zebrafish), the possibility to eliminate specific genes by homologous recombination (mouse) or just simply a more compact genome for genomic sequencing experiments (*fugu*) [1]. Although in many respects it will be necessary to analyze human genes, parallel investigations with genomes of other organisms, utilizing the technical advantages of these model systems, will greatly enhance our understanding of the human genome.

In analyzing the functions of a gene, it will be necessary to consider information from many different experiments carried out on many genes in parallel. It is therefore essential to develop systems that combine the results of these experiments.

For certain genes, critical insights as to possible functions may come from different paths of investigation, such as by the identification of sequence homologies on the DNA or protein level; specific expression patterns of the mRNA in the developing embryo or in tissue slices; different forms of a protein by 2-D gel electrophoresis;

potential human, mouse, or zebrafish mutations in the gene showing a particular phenotype; or protein products from biochemical tests constructed *in vitro* or *in vivo*.

In this chapter we will describe our approach to large-scale library characterization on the molecular level. All the steps for providing libraries, such as high-density hybridization filters and clones suitable for large-scale characterization, can now be carried out using automated techniques. We will explain the hybridization strategies that can be employed for library characterization, the automated systems we have implemented in our hybridization-based approach, and our view of the future for large-scale characterization of genomic libraries; in particular, we will show how automation will be essential for any large-scale functional analysis of gene products.

HYBRIDIZATION AS AN EFFICIENT WAY TO ANALYZE LARGE CLONE LIBRARIES

In theory and in practice, hybridization is a rapid, simple, and reliable technique for screening DNA samples that relies on a basic property of nucleic acids: the formation of a specific duplex between complimentary sequences. Any DNA fragment, from hexamer-oligonucleotides up to YAC clones of more than one megabase, can be used as a probe or a target. Therefore many different levels of DNA manipulation can be related to one another directly. Hybridization screening seems to be an obvious strategy for comprehensive analysis of genomes and, perhaps more important, for the comparison of different genomes of interest.

Hybridization, by which many genetic samples can be arrayed on a single solid support, can be a highly efficient method for screening large genetic libraries. For example, the generation of expressed sequence catalogs of several cDNA libraries with more than 10^5 clones of different tissues or different stages of development will only be possible in the near future by hybridization fingerprinting with short oligonucleotides. [2] The efficiency of such a parallel screening approach is limited by (1) the density at which genetic material can be arrayed onto the hybridization support and (2) the analysis methods to locate and address positive signals from the results of hybridizations. Over the past years we have developed automated procedures that enhance the success of both factors.

In our approach we transfer genomic libraries onto nylon membranes at a density only possible with accurate robotic positioning systems. Approaching densities of over 10^5 clones per 22×22 cm filter means that in one hybridization experiment over 10^5 different genetic screens can be conducted.

Most of the recent and very impressive achievements in large-scale genome analysis have been based on the introduction of automation within the laboratory. As much as possible in this field, robots have been enlisted in repetitive tasks or those demanding a high degree of accuracy, and that has vastly increased the speed of data production and analysis (for a review see [3]). The success of any automated large-scale genome project requires the formation of close, interdisciplinary groups of biologists, chemists, physicists, engineers, and computer scientists. We have combined specialists from these different fields to form a group that is responsible for automation and technology development at the Max-Planck-Institute für Molekulare Genetik.

Hybridization Strategies

The flexibility of hybridization as a basic property of nucleic acids allows a broad spectrum of experimental strategies to be applied. Hybridization probes can be short oligonucleotides, single cDNAs, simple physically mapped sequences or complex probes such as whole YAC clones or mixtures of IRS-PCR products. Target DNA can also be any of these types and immobilized in unique positions on solid supports. Given that target DNA can be efficiently gridded at high density, it then becomes possible to characterize efficiently the genomic libraries of complex organisms by a variety of techniques, which we turn to next.

cDNA SELECTION AND POSITIONAL CLONING   High-density gridded libraries have proved to be an efficient means of identifying genomic regions of interest provided that a short fragment of interest can be used as the probe. Given a number of selected cDNAs, clones containing similar sequences can be rapidly identified by hybridization of cDNAs to genomic libraries. In the case of positional cloning, genetic markers that co-segregate with the phenotype (e.g., RFLPs or microsatellites) can be hybridized to genomic libraries. Depending on insert size and the density of markers, the gene of interest can be within or only a short genomic walk from the positive clone. This strategy has proved successful in many gene identification projects including the Huntington's disease mapping project [4], the obese gene [5], and others [6].

PHYSICAL MAPPING   Assembled contigs from a genomic library form an essential framework onto which genetic and transcriptional maps can be integrated. As such, many genomic studies focus on the construction of physical maps from random genomic clones. For complex organisms, a genomic library can consist of tens of thousands of individual clones. Although using YAC-based vectors can mean that overall library sizes are far smaller, the documented instability of this cloning system can mean that a substantial proportion of clones are chimeric or suffer microdeletion. [7] It is therefore necessary to integrate physical maps by using combinations of genomic libraries based on different cloning systems. Since robotic techniques are able to produce high-density hybridization grids, it is possible to include within these integrated maps genomic libraries with greater numbers of clones based on PACS, P1s, BACs, or even high-resolution libraries based on cosmid-clones. This approach has been used in the fission yeast mapping project [8, 9], the establishment of a chromosome 21 cosmid pocket map [10], and the X-chromosome mapping project [11].

GENETIC MAPPING   A hybridization approach has also proved successful within genetic mapping projects. The European Collaborative Interspecific Backcross (*M. musculus* × *Mus spretus* [12]) has been shown to be a useful resource of polymorphic inter-repeat sequence PCR (IRS-PCR) products [13]. These polymorphisms can be easily scored by hybridization and form an important framework for the mouse genetic mapping project [14].

EXPRESSION PATTERNS   Given a complex network of gene interactions, coupled with highly unequal expression levels of genes that exists within a single cell, it is understandable why tissue-specific cDNA libraries can be over 10^6 clones in size. If efficient robotic systems are available that can pick clones as individual colonies and array these large libraries at high density as gridded resources, it becomes possible to study the expression pattern of individual tissues. Repeated hybridization of these grids with different short oligonucleotides produces a characteristic "fingerprint" of individual clone inserts based on the underlying sequence of the cDNA. This oligo fingerprint can be used to normalize complex cDNA libraries and generate expressed sequence catalogues of individual tissues [2]. To fully understand the gene-function within a complex organism with different tissues, each with different developmental stages and each stage providing a different cDNA library, it is apparent why automated systems become a prerequisite for large-scale studies of gene expression.

TRANSCRIPT MAPPING   Mapping of individual or small numbers of gene transcripts back to genomic clones forms the basis of positional cloning projects. Undertaken on a large scale, the positioning of gene transcripts on to a physically and genetically mapped genomic library would produce a transcriptional map of the human genome. Approaches toward such a goal have been made with YAC [15] and chromosome-specific libraries [16]. However, the challenge to many genome centers is to produce such a transcriptional map over the entire genome and at a resolution finer than provided solely by YAC-based maps.

AN AUTOMATED HYBRIDIZATION APPROACH TO LARGE-SCALE LIBRARY CHARACTERIZATION

We believe that a large-scale hybridization approach [17] provides a necessary compliment to PCR- or sequence-based strategies within genome analysis. To improve the efficiency of this approach, we have developed several robotic devices. The first implementation, about seven years ago, was a small robot designed to transfer *E. coli* bacteria from 96-well microtitre plates onto 1–3 nylon membranes in dense arrays up to 9216 clones per 222 × 222 mm. A second generation "spotting" robot with a capacity of up to six filters followed soon after necessitated by the establishment of our reference library database system (RLDB) in 1989 (see below). The strong demand for high-density reference library filters [18] and the extension of our hybridization-based approach toward analyzing large cDNA libraries required us to generate a third-generation spotting robot with a capacity of up to 12 filters [2, 9, 19]. Further robotic developments were necessary in order to enhance the efficiency of our strategies for the characterization of genomic libraries. The current status of automation in our laboratory is based on automated clone picking, a large-scale thermocycling robot, automated clone spotting at high density, non-isotopic hybridization, and automated image analysis.

Automated Clone Picking

The first step toward the analysis of large libraries is the arraying of the clones into microtitre plates for long-term storage, analysis, and subsequent individual retrieval. Over the last few years several clone-picking systems have been designed [20, 21] that rely on the use a single- or a 4-pin picking head. Our clone-picking robot differs in several crucial aspects from these previous developments. We have integrated the clone-picking feature in our flatbed robot system of the spotting robot, which will be described later. The system avoids the complicated and fragile mechanics involved in plate stacking. We have also developed a picking manifold with 96 spring-loaded pins (see Fig. 4.5, Chapter 4), where each pin can be individually extended into a colony using a pneumatic actuator positioned by a small version of a linear drive XY table. Since plate inoculation and pin sterilization are the two steps most limiting to the picking rate, our system is several times faster than previous devices. The system is capable of picking approximately 3000 clones per hour into 384 microtiter plates. Up to two large colony trays (225 $\times$ 225 mm) and twenty-four 384-well microtiter plates can be positioned on the robot bed.

We have recently switched to servo-controlled linear drives, which provide better control during sensitive movements, a higher speed (up to 2 m/s), and repeatable positional accuracy better than 5 μm. The entire axis beam act as the motor's magnet, the drive coils being the carriage. Positional information is provided by an optical encoder running the entire length of the axis.

To select randomly arranged, sized, and shaped colonies on the basis of somewhat subjective criteria is a nontrivial task. To then integrate computer vision with robot motion and accurately pick the selected clones depends on calibration of vision combined with motion, and that was achieved by a collaborative effort with Acuity Imaging Inc. We developed a new picking application within the ImageAnalyst 8.23 and Vision-Guided Motion (VGM) software packages (Acuity Inc., USA) for high-precision clone picking. Using this user-friendly application package, we have achieved hit rates of close to 100% even with very small colonies (0.5 mm in diameter and less). To increase the success and rate of picking, calibration is undertaken automatically, and variation in background intensity is corrected individually for each of the 48 frames of an agar plate. A video camera scans a frame to identify colonies, the head of the picking device is then positioned and a pin is extended to pick the individual clone. System calibration typically takes 10 minutes after power-up, and once completed, we can pick agar plates of clone libraries quickly, reliably, and without any user intervention.

Large-Scale Thermocycling Robot

Thermal cycling, followed by detection of the amplified DNA products, forms the core of many genome analysis applications (e.g., genotyping, EST mapping, oligo fingerprinting, and sequencing). Several systems have been assembled in other laboratories, but all large-scale DNA amplification systems are based on water baths for temperature control. The Whitehead Institute (Cambridge, MA) and engineers at Intelligent Automation Systems, Inc. (Cambridge, MA) have, with considerable investment, developed a complete automatic assembly line for high-throughput screening of human genomic YAC libraries based on DNA amplification and dot-blot hybridization.

We are using a much simpler version of a large-scale thermocycling robot for DNA amplifications of, for example, cDNA libraries. Sequence-specific hybridization of short oligonucleotides (7–9 mers) to DNA targets requires purified insert DNA to avoid high-background signals due to the presence of cloning vector and host cell DNA. The most feasible way to prepare DNA from whole cDNA libraries is to perform a DNA amplification reaction on each clone separately. We have been able to adapt our reaction conditions in such a way that amplification can be performed directly in 384-well microtiter plates with a specially developed robotic thermocycler (Fig. 3.1). Using three heated 225-liter water baths, we are able to cycle up to 135 plates (51,840 reactions)

Fig. 3.1 The robot system used for large-scale DNA amplification of up to 51,840 PCR reactions in parallel. The basket can be filled with up to 135 quadruple microtiter density (384-well) plates. The polypropylene plates are heat sealed with a thin plastic sheet. Cycling is done between three water baths (225 liters each) set at the required temperatures. Amplification efficiencies compare to commercial benchtop thermocycling machines.

at a time. The basket of plates is moved from bath to bath using a pneumatic X–Z sliding configuration (Festo, Germany).

Visual Basic software controls the whole system, including robot motion, temperature probes, and water-level sensors. Settings can be changed easily in the user menu allowing different types of DNA amplifications from simple cDNA amplifications to those producing complex products (e.g., Alu-Alu or IRS amplification) from targets such as individual YAC clones or total genomic DNA.

The polypropylene plates used for thermocycling are heat sealed with a two-sided plastic film and a heat sealer. The plastic film can easily be removed after the amplification step with an especially developed heat sealer. Using this very simple system, it is now possible to amplify DNA products from several cDNA libraries in a short time. After the amplification reaction, the DNA product is sufficiently pure and concentrated to be spotted directly onto nylon membranes, in preparation for hybridization. We have found the system to be as reliable as commercial benchtop thermocycling machines. Since there has been an increased demand for thermocycling in many different applications in our laboratory (template preparation, cycle sequencing, etc.), we are currently developing another, but smaller and more mobile, version of the water-bath thermocycler to amplify up to ten 384-well plates at a time.

Automated Clone Spotting at High Density

Since the introduction of the "high-density array" concept for genome analysis by hybridization [17], several clone-spotting devices have been built in various genome centers. We have further built several spotting robots that are able to transfer clones stored in microtiter plates onto nylon membranes in high-density arrays. Recently we have integrated the picking and spotting machine into one robot. By exchanging the picking head with the spotting head, the picking robot can be easily transferred into a spotting robot, and vice versa. Already machines of our design are being used in several genetic screening or drug discovery centers that require rapid, reliable, and high-density screening of small quantities of biological or biologically active material.

We have improved these spotting robots even further (Fig. 3.2) by switching to high-quality servo-controlled linear drives. Several advantages are provided by servo control systems including better positional accuracy (5 μm), higher speeds (up to 2 ms^{-1}), and far greater control of position or speed during movement. This last factor enables the robot to handle, both gently and reliably, precious library plates and then spot clones at high-speed using combined movements of three axes. Various spotting heads can be accommodated by our system. Using spring-loaded devices with 96- or 384-pins of various tip diameters, a variety of biological material can be spotted at different densities. With these machines we routinely spot 15 filters with 57,600 clones in a duplicate pattern in about three hours.

We have recently upgraded our microtiter plate-stacking system to hold 56 plates so that the robot can run without any user intervention. A grabber attached to the spotting head of the robot takes the individual plates out of the microtiter plate rack and places them onto a plate holder where the lid is automatically removed and the bar code is read. The bar-code reader supplies unique plate identifiers to a database of

Fig. 3.2 Our fourth-generation spotting robot used for generating high-density PCR matrices as well as high-density *in situ* colony filters of cDNA-, cosmid-, P1-, and YAC-library reference filters. The head moves with a speed of up to 2 m/s in three dimensions, spotting 57,600 colonies from 384-well microtiter plates onto each of 15 nylon filters in three hours. The picture shows a front view of the machine with a 24 microtiter plate stacker in the foreground. A plate stack which holds 56 microtiter plates has been realized already and is now being used on the machines. The lid of the microtiter plates are automatically removed by a pneumatic system, and the bar codes are read into the computer. This information is essential for decoding *XY* coordinates of hybridization results within the reference library database system. By simply changing the head of the robot, this machine can be converted into a high-throughput picking robot (see Chapter 4 by Sarath Krishnaswamy and John E. Agapakis).

DNA source libraries and clones, making it easy to locate and retrieve colonies of interest. After the required number of spotting cycles, the lid is replaced, and the grabber lifts the plate with lid back into the rack system and moves on to the next plate.

For nonradioactive hybridizations in our laboratory using automated image analysis, we routinely spot 57,600 clones on a 222 × 222 mm nylon filter (100 times standard microtiter-plate density). Reference library filters that are distributed to outside users are mostly supplied at a lower density (36,864 clones), and clones are spotted in a duplicate format (Fig. 3.3) to facilitate manual scoring procedures for the individual users. With the system described here, we have now achieved experimental spotting densities of up to 147,456 clones per 222 × 222 mm membranes, equivalent to 256 times the density of a standard microtiter plate.

Non-isotopic Hybridization

We have now developed our hybridization procedures toward fluorescence-detection techniques on nylon filters [22]. Directly labeled fluorescent probes so far cannot be

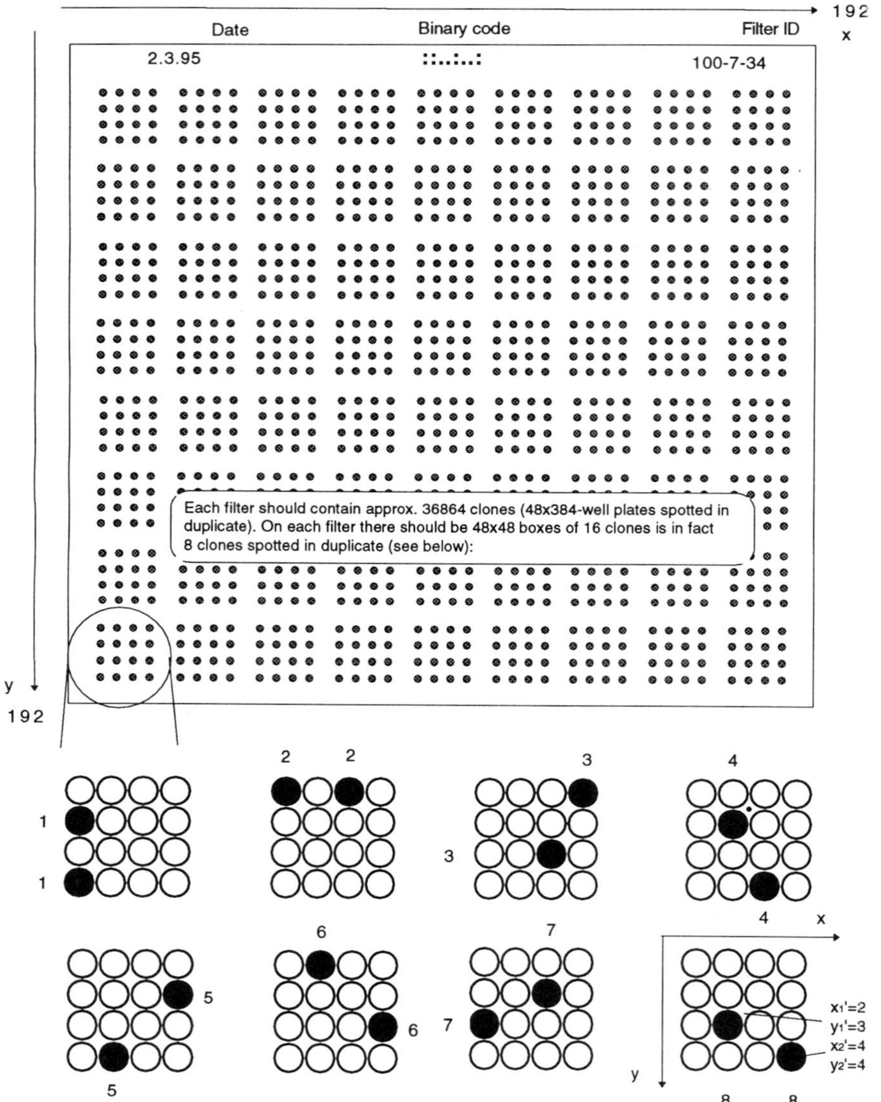

Fig. 3.3 The spotting pattern of high-density filter arrays used within the reference library database system (RLDB). As an internal control each microtiter plate has been spotted twice on a filter. This reduces the false positives rate since each clone is duplicated within a 4 × 4 block of the high-density grid. Clones can therefore be identified easily by their orientation within a block.

used as universal hybridization tools for arrayed DNA filters (for an example see Fig. 3.4). This is due to the low signal-to-noise ratio for hybridization on contemporary nylon membranes. Several enzyme-amplified detection systems are available, for example, the enzymatic conversion of BCIP/NBT or chemiluminescent compounds by alkaline phosphotase (AP). However, these systems prove inefficient if many hybridizations per filter have to be processed automatically in a short time and at high resolution. In contrast to the process of chemiluminescence, a fluorescent substrate such as Attophos™ (JBL Scientific, San Luis Obispo) is highly fluorescent after the liberation of the phosphate group, where the number of photons emitted is strongly related to the intensity of excitation light. The signal intensity per labeled probe molecule is much higher than standard fluorescent dyes such as fluorescein or rhodamine because every active centre of the alkaline phosphotase can process about 10^4–10^5 substrate molecules per minute.

The advantages of Attophos™, which is the benzothiazole derivative of 2′-(2-benzothiazolyl)-6′-hydroxybenzothiazole phosphate (BBTP), in automated DNA multiplex sequencing have been demonstrated recently [23]. We have adapted this system for hybridization techniques on gridded high-density DNA and *in-situ* colony filters

Fig. 3.4 Example of an arrayed filter a very high-density matrix with up to 112,896 spots (294 quadruple-density microtiter plates) on a 222 × 222 mm nylon membrane. Detection and analysis of hybridizations on these matrices are done based on our fluorescence-based hybridization and our automated image analysis systems.

using digoxigenin-labeled probes and their detection via the enzyme-linked fluorescence of BBTP (excitation: 420 nm (max.); emission: >560 nm). Probes of any length ranging from short oligonucleotides (8–10 mers) used for sequence fingerprinting in cDNA analysis to complex PCR products of YAC clones used in long-range mapping and pools of cDNAs for transcript mapping have been hybridized successfully.

We now analyze fluorescent hybridizations using our special-purpose detection system. For BBTP detection the nylon membranes containing the high-density clone grids are placed in a light-proof box and are illuminated with two 8 W black-light UV tubes (about 365 nm). The UV tubes are fixed at an angle of 45 degrees to the detection system to minimize scattering of the excitation light. A high-resolution CCD camera (Photometrics PXL1400) fitted with an interference filter (589 nm, bandwidth about 80 nm; Herolab, Germany) is linked to a controller (Photometrics PXL) and an Apple Macintosh PowerPC 8100. This combination of Kodak CCD chip (1317 × 1035 pixels) and fast digitizing controller system enables us to rapidly scan a hybridized filter at high resolution. To go from hybridized filter to digitized file takes a matter of seconds, and in speed and resolution our system compares very favorably to commercial fluorescent scanning systems based on photomultiplier detection. Since the spatial resolution of this detection system can be much higher than phosphor storage screens, we anticipate much higher spotting densities using these nonradioactive hybridization and fluorescence light detection techniques than by radioactive approaches. We are also investigating the use of other detection systems such as time-delayed integration (TDI) cameras (Dalsa Inc., Canada) for high-speed and high-resolution scanning.

Automated Image Analysis

An important prerequisite for large-scale characterization of genetic libraries is a reliable automated image analysis system able to localize positive hybridization signals and subsequently quantify the results. We have tested several commercial systems but none has fulfilled our requirements. Therefore we have developed software to analyze hybridization patterns of positive signals on high-density hybridization filters spotted by a robot.

Image analysis of high-density grids proceeds though a few basic steps: (1) The digitized image is read and preprocessed to provide standardized initial conditions such as rotation angle, background removal, and noise suppression. (2) The grid outline is found by localization of the basic rectangular grid shape, (3) localization of the overall block structure of the filter is determined by guide spots and characteristic gaps between the blocks, (4) the individual nodes in the hybridization grid including the negative of missing spots are localized within each block, (5) spot-nodes are quantified by evaluation of the hybridization intensity in the spot-nodes and the surroundings of the spot, and (6) spot-nodes indexes are converted to plate-ID and well positions.

The underlying algorithms of the software package have been developed over a course of three to four years and they comprise a set of C-programs managed by a shell script implemented in a UNIX environment. The core of these procedures is the grid-fitting routine, a sophisticated statistical method that allows distortions from the ideal aligned grid due to warped filter membranes, bent robot pins, missing spots, and high-intensity background regions (= false spots). The grid localization is performed on both a global and local scale, in contrasts to other methods that normally only utilize local information. This method performs both spot-finding and grid-finding simultaneously, which makes it very robust. Conventional methods search spots first and then fit the grid—an approach that often proves nonoptimal because it requires that spots be well defined in shape and intensity and that the image not contain false spots due to background noise.

In this method spot-node positions are localized by identifying a minimal energy distribution based on a Markov random field template, and that requires a trade-off between the regularity of the grid and the trust in the image data. A certain amount of deviation from the ideal grid geometry is allowed, and the grid model is fitted to the image by searching the maximum a posteriori (map) estimate through a simulated annealing scheme. This is an iterative sampling scheme from the underlying density, proposed by Geman and Geman [24] and since used frequently in image-restoration problems. The scheme is performed as a number of sweeps over the image, where each grid node is visited.

Once the positions of individual nodes have been determined, the quantification of signal intensity is relatively straightforward. The major problem is, of course, to determine the factors that decide whether an individual spot is positive or negative with respect to hybridization and works on a range of different quality images. Human judgment is so far the only basis by which automated decision making can be compared, but this cannot be presumed to be always reliable. Therefore we store the pixel values of all nodes of a hybridization image and perform a global analysis at the end of data analysis. With this approach we can repeatedly analyze the results of large-scale library characterization projects using various positive-negative thresholds and form an estimate of the confidence limits for our overall results.

The underlying C-programs of our image analysis system have been integrated with a user-friendly interface (Fig. 3.5). This allows users to easily monitor the success of an analysis, make small changes to node localizations, or re-run individual stages of the process without needing to enter complex shell commands. We are committed to further improving the reliability of the spot location algorithms and making the whole software package more user friendly and portable to different computer platforms.

SHARED BIOLOGICAL RESOURCES

A second, and unique, approach that we have undertaken toward the characterization of large genetic libraries involves distributing common genetic resources. The reference library database and distribution system (RLDB; [18]) is a model for the idea of using shared biological material. It allows participating scientists to combine the results of their individual experiments to a common pool, to connect the separate information in one database, and to analyze the data much more efficiently. Here cosmid, YAC, P1, and cDNA libraries are distributed on high-density filters to the scientific community (see also Table 3.1) and experimental results are stored in a common object-based database accessible through the internet (WWW: http://rldb.rz-berlin.mpg.de/). For

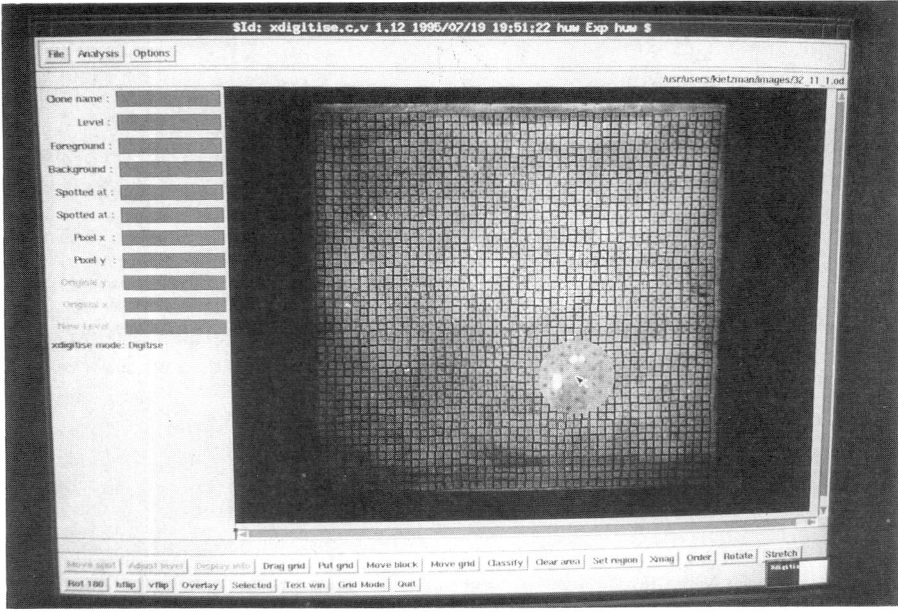

(a)

Fig. 3.5a   Interface of our semiautomated image analysis system. This version of the software allows semiautomatic scoring of the hybridization signals being used for experiments (e.g., in transcript mapping) with very few positives.

this approach it is essential that experimental results are generated using well-defined experimental materials (e.g., ordered arrays of clones or probes), since each clone is unique and can therefore only be described if the same clones or clone collections are used in the different experiments. Our lab started to generate and distribute Reference Libraries about five years ago; libraries are distributed to participating laboratories as high-density spotted filters, and collections of clones are permanently stored in microtiter plates, allowing the unique identification of each clone by its address in a microtiterplate. The number of requests for library filters and clones from the Reference Library System has continued to increase significantly over the last few years, and the success of this service has been a major driving force in the development of the automated procedures described above. Until the end of 1994 we have distributed more then 4400 reference library filters and 30,000 clone stabs (see also Table 3.2).

High-density filters are distributed and hybridized with specific probes at the recipient laboratory. The *XY* coordinates of the positive signals in the spotted grid are determined on an autoradiograph of the filter and then communicated back to the RLDB, together with a description of the probe and the id-numbers of the filters used in the hybridizations. This information is entered into the database, the microtiter plate positions of the potential positive clones are calculated, and the requested clones are picked into

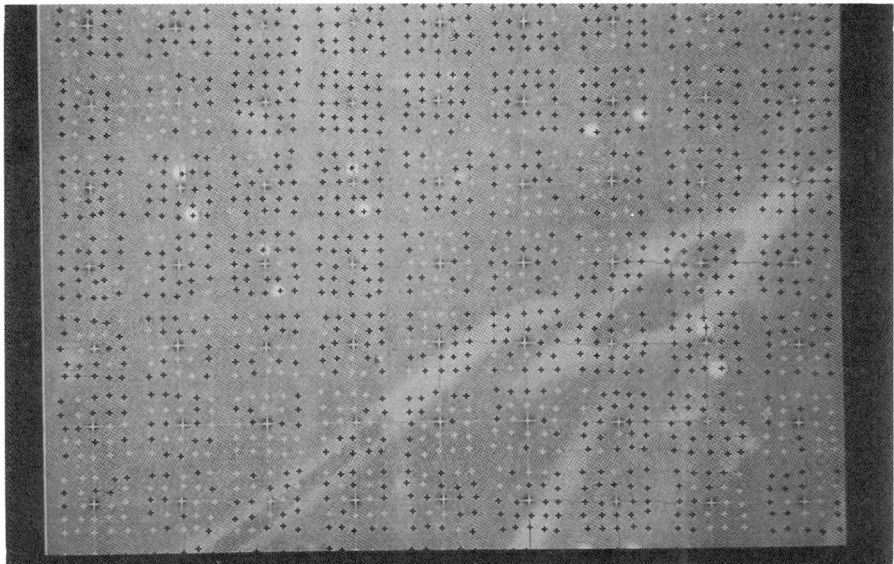

(b)

Fig. 3.5b In another version the automated grid- and spot-finding algorithms are implemented so that TIFF files can be read in and analyzed automatically without any user intervention. Image processing results in an ASCII file with clone names and corresponding pixel values. The figure shows an automatically analyzed image of a hybridization on a 74 × 111 mm filter containing 9600 PCR products (100× standard microtiter plate density). The clones are spotted in 5 × 5 boxes of a 384-well microtiter plate footprint in a duplicating pattern as described in Fig. 3.4. The positive hybridization signals are well localized with the automated image analysis software package.

agar stabs. These clones are then forwarded from RLDB to the scientist together with any nonconfidential information about previously identified probes hybridizing to the same clone. One individual colony of each clone that is subsequently confirmed as positive should be returned to the RLDB system, together with other results of this secondary screen. This allows RLDB to update the information in the database and to include the clone into a growing collection of characterized clones. To avoid duplication of work, the publicly available list of probes and clones (updated weekly, but not including information declared by investigators as confidential) can be checked for clones previously identified by other users of the system.

The concept of shared biological resources coupled with efficient automation efforts will serve as a key focus of the German Genome Project. Worldwide scientific, medical, and political interest in large-scale characterization of genetic libraries has led to the establishment of national (in the United States, France, the United Kingdom, Japan, Canada, Italy, etc.) and international genome projects, and the foundation of a number of large genome centers (e.g., in the United States, France, and the United Kingdom).

TABLE 3.1   Available Clone Libraries Through the Reference Library Database System (See Also WWW:http://rldb.rz-berlin.mpg.de/).

Cosmid Libraries

Library No. 60,61	S. pombe cosmid libraries (ordered)
Library No. 50	Drosophila cosmid library
Library No. 112	Human Chromosome 1 cosmid library
Library No. 503	Human Chromosome 4 cosmid library LA04N from LANL
Library No. 109	Human Chromosome 6 cosmid library from LANL
Library No. 113	Human Chromosome 7 cosmid library
Library No. 115	Human Chromosome 9 cosmid library LL09NC01 from LLNL
Library No. 114	Human Chromosome 11 cosmid library SRL11 from LANL
Library No. 107	Human Chromosome 11 cosmid library
Library No. 108	Human Chromosome 13 cosmid library
Library No. 105	Human Chromosome 17 cosmid library
Library No. 111	Human Chromosome 18 cosmid library
Library No. 102	Human Chromosome 21 cosmid library in ED8767
Library No. 103	Human Chromosome 21 cosmid library in DH5
Library No. 106	Human Chromosome 22 cosmid library
Library No. 100,104	Human Chromosome X cosmid library in ED8767
Library No. 500	Human Chromosome X cosmid library X8C10
Library No. 110	Human Chromosome X cosmid library LL0XNC01 from LLNL
Library No. 501	Human Chromosome Xq28 cosmid library

P1 Libraries

Library No. 705	S. pombe P1 library (ordered)
Library No. 704	Human ICRF PAC library
Library No. 700	Human ICRF Total Genomic P1 Library
Library No. 701,702	Human Du Pont Merck Pharmaceutical Company Foreskin Fibroblast P1
Library No. 703	Mouse ICRF Total Genomic C57/Black6 P1 Library

YAC Libraries

Library No. 915	S. pombe YAC library (ordered)
Library No. 907	Pig YAC library
Library No. 911	Pig YAC library LMUB
Library No. 900,901,905	Human ICRF YAC library
Library No. 904	Human CEPH YAC library
Library No. 912	Human YAC library LS4X
Library No. 906	Human ICI YAC library
Library No. 902,903	Mouse ICRF C3H + B57 YAC library
Library No. 910	Mouse Whitehead YAC Library
Library No. 909	Mouse St.Marys female C57BL/10 YAC library

TABLE 3.1 *(Continued)*

cDNA Libraries

Library No. 520	Drosophila cDNA library
Library No. 507	Human foetal brain cDNA
Library No. 515	Human HPO cDNA
Library No. 509	Human Xq28 cDNA from foetal brain and liver and adult sceletal muscle)
Library No. 516	Human chromosome 21 cDNAs from LLNL and ICRF cDNA libraries
Library No. 510	Mouse adult brain cDNA
Library No. 517	Exon Library

This international effort, carried out with public funding but also to an increasing extent with private funding, has led to major progress in the establishment of preliminary genetic and physical maps of the human genome [25, 26]. In view of the enormous implications of this basic research, a similar large-scale effort will now be started in Germany to participate in this international endeavour. As the focus for this effort a systematic functional analysis of genes, the functional units of the human genome has been proposed.

A systematic analysis of the function of genes will rely on the combination of many different sources of information and will require the implementation of new techniques and concepts in data generation based on the use of common, well-characterized biological resources. Furthermore the immense amount of information generated from analysis of tens of thousands of genes will require the development of new approaches in data storage, data analysis, and data sharing. This use of well-characterized clones and clone libraries in different experiments provides a systematic mechanism to collect and redistribute all raw data generated on a specific clone throughout the entire project. This basis describes concept of a "distributed genome center," combining the effort of groups distributed throughout the world.

In our view, a highly integrated, highly automated, and global analysis of genes based on an explicit mechanisms for coordination by the use of common resources and common databases has been less stressed in many of the other national genome projects. In contrast to other genome projects relying to a larger extent on the exclusive use of sequence information as the mechanism of coordination and data exchange, the explicit mechanisms for sharing resources and clone libraries is a central and unique feature of the German Genome Project.

THE FUTURE OF LARGE-SCALE LIBRARY CHARACTERIZATION

By definition, genetics is the study of differences, differences within an individual genome, between genomes of different species, or between genomes of different individuals from a single population. If such differences are to be studied in depth, then

TABLE 3.2 Statistics of the Library Distribution as High-Density Filters Through RLDB (See Also WWW:http://rldb.rz-berlin.mpg.de/).

Number of Filters Sent Out by RLDB

Cosmid

EXPNO	LIBNO	LIBNAME	HOST	1990	1991	1992	1993	1994	Total
3	60	S. Pombe cosmid (old)	ED8 767	20	13			2	35
4	50	Drosophila cosmid	E. coli 1						
1	100	Human Chromosome X	ED8 767	16	4				20
2	102	Human Chromosome 21	ED8 767	29	8	25	14	14	90
6	103	Human Chromosome 21	DH5 alph	6	20				26
7	104	Human Chromosome X	DH5 alph	74	170	40	44	99	427
9	105	Human Chromosome 17	DH5 alph		48	38	57	44	187
10	106	Human Chromosome 22	DH5 alph		26	8	8	19	61
11	107	Human Chromosome 11	DH5 alph		6	38	32	34	110
13	108	Human Chromosome 13	DH5 alph			19	10	24	53
14	109	Human Chromosome 6	DH5 alph			17	16	110	143
16	111	Human Chromosome 18	DH5 alph				6	19	25
17	112	Human Chromosome 1	DH5 alph				14	185	199
18	113	Human Chromosome 7	DH5 alph					72	72
				145	295	185	201	622	1448

cDNA

EXPNO	LIBNO	LIBNAME	HOST	1990	1991	1992	1993	1994	Total
20	507	Human fetal brain cDNA	XL1 blue				100	299	399
21	508	Human fetal thymus cDNA	DH10B				1	56	57
27	512	Human fetal liver cDNA	DH10B					163	163
32	515	515 Human fetal lung cDNA	DH10B					48	48
25	510	Mouse adult brain cDNA	XL1 blue				25	38	63
							126	604	730

P1

EXPNO	LIBNO	LIBNAME	HOST	1990	1991	1992	1993	1994	Total
100	700	Human P1	NS3145				264	375	639
103	703	Mouse P1	NS3145					454	454
							264	829	1093

YAC

EXPNO	LIBNO	LIBNAME	HOST	1990	1991	1992	1993	1994	Total
50	900	4X + 4Y Human YAC PCR	S. cerevi				13	1	14
201	900	4X + 4Y + HD1 Human	S. cerevi		2	200	214	100	516
202	904	CEPH Human YAC	S. cerevi			12	3	93	108
210	902	C3H + C57 Mouse YAC	S. cerevi		28	58	188	84	358
213	910	Whitehead Mouse YAC	S. cerevi				22	41	63
220	907	Pig YAC	S. cerevi						
					30	270	492	379	1171

				1991	1992	1993	1994	1995	Total
			Grand total	145	325	455	1083	2424	4442

investigations will necessarily involve large sample sizes and hence will benefit immensely from an automated approach. So far we have automated most of the obvious bottlenecks in providing resources for large-scale library characterization by hybridization such as clone picking, DNA amplifications, and clone spotting. As an immediate goal we are developing automated systems to large-scale hybridization, using novel approaches that enable us to dispense with conventional bottles or bags. Also we are in the process of increasing the reliability and automatic plate handling capacities of our existing systems. The hardware and software of our machines have already become more user friendly so that they can be used in any molecular biology laboratory with minimal engineering support. To achieve this, we have recently switched to servo controlled high-resolution linear magnetic motors (Linear Drives Ltd.). We have also developed spotting software written in LabVIEW (National Instruments), a powerful and well-supported graphical programming language. The LabVIEW software package allows much more flexibility of all processes run on the robot. Such a platform provides an framework on which novel biological screens can be integrated onto existing robotic systems with minimal reprogramming of source code.

The implementation of large plate stacking units on the robots has obviously many interesting advantages such as long, overnight runs to produce very high-density grids. Our immediate application is the production of very high-density amplified DNA grids (with up to 147,456 clones per 222 $\times$ 222 mm membrane) where the amplified products from up to 294 quadruple-density microtiter plates need to be spotted up to ten times per spot. Larger stacking systems will also increase the flexibility of the system toward the selection of sublibraries (rearraying process) and the replication of clone libraries.

We are further developing other special-purpose and very high-throughput systems for micropipetting at the microtiter plate level. Projects to automate high-throughput hybridization approaches and the subsequent detection process of fluorescent hybridization signals are also well under way. Together with our automatic image analysis programs for high-density clone grids, we are now assembling a production line toward large-scale library characterization in the lab. This is absolutely crucial in the extension of our fingerprint analysis approach of cDNA libraries up to 10^6 clones.

We are developing alternative technologies and simplifying existing molecular biology protocols to make them suitable for large-scale analysis at the microtiter plate level. Another important factor in this respect is obviously miniaturization of processes such as DNA amplification on silica wafers or oligo chips for hybridization of DNA probes for diagnostic applications [26]. Although we are mainly working on format 2 of SBH (sequencing by hybridization) [28–30] where the clones are immobilized and the oligos are hybridized, we are also involved in developing oligo matrices [31] for diagnostic purposes (format 1). We are aware that complete SBH cannot be readily accomplished with today's techniques, but we think that by combining this method with less restrictive algorithms for sequence comparisons of different hybridization experiments, SBH has tremendous potential for diagnostic testing.

Many of these developments have been carried out continually in conjunction with the data production facilities of the genome analysis projects in our lab. This ensures that relevance is maintained and that new techniques are rapidly transferred to ongoing projects. The approaches we have described are not only applicable for the rapid analysis of human and nonhuman gene function, they can also be used in other biological or medical applications where there is a need for high-density screening of small quantities of biological or biologically active material. For example, these automated high-throughput systems can be used in the pharmaceutical industry to increase screening efficiencies and to reduce the costs and labor time involved in finding new candidate drugs.

What comes next in the analysis or complex genomes? There has been considerable discussion about sequencing entire genomes using existing gel-based technologies [32]. Although some sequence approaches do not depend enormously on high-resolution physical mapping in order to produce a "sequence ready map," all sequence approaches do require efficient handling of vast presequenced genetic resources such as phagemid clones or PCR products. Here obviously the establishment of an efficient production line for sequence preparation will be necessary. This would involve new shotgun cloning procedures on a large-scale, clone-picking, template preparation by PCR, spotting of PCR products on filters with subsequent hybridization of various probes (e.g., *E. coli, S. cerevisiae* in the case of subcloning YACs [33]; oligos for clone ordering, etc.), followed by sequencing of selected subsets from these shotgun libraries.

Perhaps the bigger challenge in the years to come will be the structural and functional analysis of gene products at the genome scale. Development of efficient methods for overexpression and purification of proteins can be considered relatively straightforward compared with the subsequent steps such as protein crystallization and functional analysis. Yet, by our experiences with automation of genetic library characterization, we will have a considerable advantage in any global investigation of gene function.

ACKNOWLEDGMENTS

We are very grateful to Sebastian Meier-Ewert, Igor Ivanov, Jon Curtis, Huw Griffith, Martin Horn, Markus Kietzmann, and Sigrid Rumbaum in the Abteilung Lehrach at the Max-Planck-Institut für Molekulare Genetik in Berlin who contributed their ideas to this review. We want to thank Steve Nally from Linear Drives Ltd. (UK) for his collaboration in developing the LabVIEW spotting application. We also acknowledge the help of Sarath Krishnaswamy (USA) and Chris French (UK) from Acuity Imaging in writing the picking software application. The Max-Planck-Gesellschaft and the Bundersministerium für Bildung, Wissenschaft, Forschung und Technologie provided funding for our recent work.

REFERENCES

1. S. Brenner, G. Elgar, R. Sandford, A. MacRae, B. Venkatesh, S. Aparicio (1993). Characterization of the pufferfish (*Fugu*) genome as a compact model vertebrate genome. *Nature* 366:265–268.

2. S. Meier-Ewert, E. Maier, A. Ahmadi, J. Curtis, H. Lehrach (1993). An automated approach to generating expressed sequence catalogues. *Nature* 361:375–376.

3. J. Peccoud (1995). Automating molecular biology: A question of communication, *Bio/ Tech.* 13:741–745.

4. The Huntington's Disease Collaborative Research Group (1993). A novel gene containing a trinucleotide repeat that is expanded and unstable on Huntington's disease chromosomes. *Cell* 72:971–986.

5. Y. Y. Zhang, R. Proenca, M. Maffei, M. Barone, L. Leopold, J.M. Friedman (1995). Positional cloning of the mouse obese gene and its human homolog. *Nature* 374:479.

6. S. Parimoo, S.R. Patanjali, R. Kolluri, H.X. Xu, S.M. Weissman (1995). cDNA selection and other approaches in positional cloning. *Anal. Biochem.* 228:1–17.

7. R. Nagaraja, J. Kere, S. Macmillan, M.W.J. Maisisi, D. Johnson, B.J. Molini, G.R. Halley, K. Wein, M. Trusgnich, B. Eble, B. Railey, B.H. Brownstein, D. Schlessinger (1994). Characterization of 4 human YAC libraries for clone size, chimerism and X-chromosome sequence representation. *Nucleic Acids Res.* 22:3406–3411.

8. J.D. Hoheisel, E. Maier, R. Mott, L. McCarthy, A.V. Grigoriev, L.C. Schalkwyk, D. Nizetic, F. Francis, H. Lehrach (1993). High-resolution cosmid- and P1-maps spanning the 14-Mbp genome of the fission yeast Schizosaccharomyces pombe. *Cell* 73:109–120.

9. E. Maier, J.D. Hoheisel, H. Lehrach (1994). Ordered clone libraries as high-resolution mapping tools. In *Proceedings of the 6th European Congress on Biotechnology*, L. Alberghina, L. Frontali, P. Sensi, eds., 191–194.

10. D. Nizetic, L. Gellen, R.M.J. Hamvas, R. Mott, A. Grigoriev, R. Vatcheva, G. Zehetner, M.-L. Yaspo, A. Dutriaux, C. Lopes, J.-M. Delabar, C. van Broeckhoven, M.-C. Potier, H. Lehrach (1994). An integrated YAC-overlap and 'cosmid-pocket' map of the human chromosome 21. *Hum. Mol. Genet.* 3:759–770.

11. H. Roest Crollius, M.T. Ross, A. Grigoriev, J.C. Knights, E. Holloway, J. Misfud, K. Li, M. Playford, S.J. Gregory, S.J. Humphray, A.J. Coffey, C.G. See, S. Marsh, R. Vatcheva, J. Kumlien, T. Labella, V. Lam, K.H. Rak, K. Todd, R. Mott, D. Graeser, G. Rappold, Z. Zehetner, A. Poustka, D.R. Bentley, A.P. Monaco, and H. Lehrach, (1996). An Integrated YAC Map of the human X chromosome. *Genome Research* 6(10), 943–955.

12. M. Breen, L. Deakin, B. MacDonald, S. Miller, R. Sibson, E. Tarttelin, P. Avner, F. Bourgade, J.L. Guenet, X. Montagutelli, et al. (1994). Towards high-resolution maps of the mouse and human genomes—A facility for ordering markers to 0.1 cM resolution. Mol. Genet. 3:621–627.

13. R.D. Cox, N.G. Copeland, N.A. Jenkins, H. Lehrach (1991). Interspersed repetitive element polymerase chain reaction product mapping using a mouse interspecific backcross. *Geno-mics* 10:375–384.

14. L. McCarthy, K. Hunter, L. Schalkwyk, L. Riba, S. Anson, R. Mott, W. Newell, Ch. Bruley, I. Bar, E. Ramu, D. Housman, R. Cox, H. Lehrach (1995). Efficient high-resolution genetic mapping of mouse interspersed repetitive sequence PCR products, toward integrated genetic and physical mapping of the mouse genome. *Proc. Natl. Acad. Sci. USA* 92:5302–5306.

15. D.T. Moir, R. Lundstrom, P. Richterich, X. Wang, M. Atkinson, K. Falls, J. Mao, D.R. Smith, G.F. Vovis (1994). Mapping cDNAs by hybridization to gridded arrays of DNA from YAC clones. In *Identification of Transcribed Sequences*, U. Hochgeschwender, K. Gardiner, eds. Plenum Press, New York, 289–297.

16. M.L. Yaspo, L. Gellen, R. Mott, B. Korn, D. Nizetic, A.M. Pouska, H. Lehrach (1995). Model for a transcript map of human-chromosome-21—Isolation of new coding sequences from exon trapping and enriched cDNA libraries. *Hum. Mol. Genet.,* 4:1291–1304.

17. H. Lehrach, R. Drmanac, J.D. Hoheisel, Z. Larin, G.G. Lennon, A.P. Monaco, D. Nizetic, G. Zehetner, A. Poustka (1990). Hybridization fingerprinting in genome mapping and sequencing. In *Genome Analysis: Genetic and Physical Mapping,* K.E. Davies, S. Tilghman, eds. Cold Spring Harbor Laboratory Press, Cold Spring Harbor, NY, 39–81.

18. G. Zehetner, H. Lehrach (1994). The Reference Library System—Sharing biological material and experimental data. *Nature* 367:489–491.

19. E. Maier, S. Meier-Ewert, A. Ahmadi, J. Curtis, H. Lehrach (1994). Application of robotic technology to automated fingerprint analysis by oligonucleotide hybridization. *J. Biotech.* 35:191–203.

20. D.C. Uber, J.M. Jaklevic, E.H. Theil, A. Lishanskaya, M.R. McNeely (1991). Application of robotics and image processing to automated colony picking an arraying. *BioTechniq.* 11:642–647.

21. P. Jones, A. Watson, M. Davies, S. Stubbings (1992). Integration of image analysis and robotics into a fully automated colony picking and plate handling system. *Nuc. Acid. Res.* 20:4599–4606.

22. E. Maier, H.R. Crollius, H. Lehrach (1994b). Hybridization techniques on gridded high-density DNA and in-situ colony filters based on fluorescence detection. *Nuc. Acid. Res.* 22:3423–3424.

23. J.L. Cherry, H. Young, L.J. DiSera, F.M. Ferguson, A.W. Kimball, D.M. Dunn, R.F. Gesteland, R.B. Weiss (1994). Enzyme-linked fluorescent detection for automated multiplex DNA-sequencing. *Genomics* 20:68–74.

24. S. Geman, D. Geman (1984). Stochastic relaxation, Gibbs distributions and the Bayesian restoration of images. *IEEE Trans. Pattern Anal. Mach. Intell.* 6:721–741.

25. D. Cohen, I. Chumakov, J. Weissenbach (1993). The first generation physical map of the human genome. *Nature* 366:698–701.

26. G. Gyapay, J. Morissette, A. Vignal, C. Dib, C. Fizames, P. Millasseau, S. Marc, G. Bernardi, M. Lathrop, J. Weissenbach (1994). The 1993-94 Genethon human genetic linkage map. *Nature Genet.* 7:246–339.

27. S.P. Fodor, R.P. Rava, X.C. Huang, A.C. Pease, C.P. Holmes, C.L. Adams (1993). Multiplexed biochemical assays with biological chips. *Nature* 364:555–556.

28. R. Drmanac, I. Labat, I. Brukner, R. Crkvenjakov (1989). Sequencing of megabase plus DNA by hybridization: theory of the method. *Genomics* 4:114–128.

29. R. Drmanac, Z. Strezoska, I. Labat, S. Drmanac, R. Crkvenjakov (1990). Reliable hybridization of oligonucleotides as short as six nucleotides. *DNA Cell Biol.* 9:527–534.

30. R. Drmanac, G.G. Lennon, S. Drmanac, I. Labat, R. Crkvenjakov, H. Lehrach (1990). Partial sequencing by oligo-hybridization: Concept and applications in genome analysis. In *The First International Conference on Electrophoresis, Supercomputing, and the Human Genome,* C. Cantor, H.A. Lim, eds. World Scientific, Singapore, New Jersey, London, Hong Kong 60–75.

31. K.R. Khrapko, Yu.P. Lysov, A.A. Khorlin, I.B. Ivanov, G.M. Yershov, S.K. Vasilenko, V.L. Florentiev, A.D. Mirzabekov (1991). A method for DNA sequencing with oligonucleotide matrix. *DNA Seq.* 1:375–388.

32. R.D. Fleischmann, M.D. Adams, O. White, R.A. Clayton, E.F. Kirkness, A.R. Kerlavage, C.J. Bult, J. Tomb, B.A. Dougherty, J.M. Merrick, et al. (1995). Whole-genome random sequencing and assembly of *Haemophilus influenza* Rd. *Science* 269:496–512.

33. M. Vaudin, A. Roopra, L. Hillier, R. Brinkman, J. Sulston, R.K. Wilson, E.H. Waterston (1995). The construction and analysis of M13 libraries prepared from YAC DNA. *Nuc. Acid. Res.* 23:670–674.

4

Machine Vision and Vision-Guided Motion for Genome Automation

SARATH KRISHNASWAMY AND JOHN E. AGAPAKIS

CONTENTS

INTRODUCTION

The use of laboratory automation has brought about significant throughput and productivity increases in the field of genetics and to the human genome project. Laboratory

Automation Technologies for Genome Characterization, Edited by Tony J. Beugelsdijk.
ISBN 0-471-12806-6   © 1997 John Wiley & Sons, Inc.

automation systems today are based on principles similar to those in manufacturing: faster, smaller, more accurate, cheaper ways to make the required product—in this case valuable data. As we see these systems proliferating throughout industry and major research institutions, we can expect the use of automated image analysis for process control and machine guidance to follow, as it did in the manufacturing realm. Machine vision is already a mature technology in manufacturing for precisely the same types of applications that are now being explored by laboratory automation, and its application for genome automation work is both broad and unique.

In this chapter we will review the basics of machine vision and discuss the use of a vision system in conjunction with a motion control device in order to enhance a previously "blind" process. We will then go deeper into an application example—vision-guided colony picking—and discuss how such an application is set up and calibrated. Finally, we will discuss some future applications of vision-guided robotics for genome automation that we believe are forthcoming as applications continue to demand higher throughput, greater repeatability, and significantly higher degrees of miniaturization.

MACHINE VISION BASICS—A QUICK SUMMARY

Machine vision refers to the extraction of useful symbolic descriptions from video images. Machine vision systems are used to analyze images, produce symbolic descriptions from the scene, and use these descriptions to control or affect other equipment that is performing a task on the objects in the scene. This task may be as simple as rejecting bad parts, or it may involve more complex processes such as assembly or packaging. While machine vision systems have found widespread use in industry, they are now coming into use for laboratory automation applications [1, 2, 3, 4]. These applications can be divided into categories as follows:

- Part presence/absence detection
- Part location/orientation detection
- Part identification
- Automated inspection
- Dimensional gauging
- Process monitoring and control
- Vision-guided robotics

Four different steps are involved in automated image analysis:

- *Image acquisition.* Light reflected from the scene is converted into a digital representation upon which the computer will perform operations.
- *Image processing.* An input image is modified, enhanced, restored, or filtered to create an output image that is higher in quality or in information content.

- *Image analysis.* Symbolic descriptions (e.g., area counts, edges, feature measurements) are extracted from the image by means of various algorithms.
- *Instrument interfacing.* The symbolic descriptions above are used to initiate or modify a process.

While a detailed discussion of each of these procedures is beyond the scope of this chapter, there are several books on the subject [5, 6, 7, 8] including a recent chapter by the authors describing some of the image processing and analysis tools available in current machine vision systems [3]. Image analysis algorithms commonly employed in laboratory automation applications include the following:

- Binary image analysis, including
    - area counting
    - connectivity analysis
- Grayscale analysis
- Edge detection
- Template matching by normalized correlation
- Detection of boundaries of known shapes by Hough Transform methods

The sample application we describe in this chapter makes use of connectivity and grayscale analysis algorithms. Connectivity is a binary image analysis algorithm. It requires a threshold value to separate pixels as belonging to the foreground or to the background. Once this value has been selected, the algorithm finds regions of connected pixels (either above or below the threshold) that form closed boundaries. These regions are called *blobs.* If several blobs are present in the image, a hierarchical list (the *blob tree*) is formed which identifies each blob and interrelationships among blobs (e.g., a blob within a hole within another blob). A variety of topological, geometric, and statistical features of the blobs can be measured and used to classify or inspect objects in the image. See Fig. 4.1.

Grayscale algorithms operate on the gray values of the pixels in the image. The simplest grayscale analysis functions are used to compute statistical properties of pixels in a given region of interest (ROI). These can include the average gray value, standard deviation, minimum or maximum gray value, and the like.

VISION-GUIDED MOTION

One of the most prevalent uses of machine vision is to guide motion devices (e.g., robots, *XY* positioners, conveyors) based on part orientation or positioning data extracted by the vision system. For example, machine vision might be used to identify parts as they come down an assembly line and guide a robotic gripper to the best possible orientation needed to pick up the part. Another example would be vision-guided robotic welding

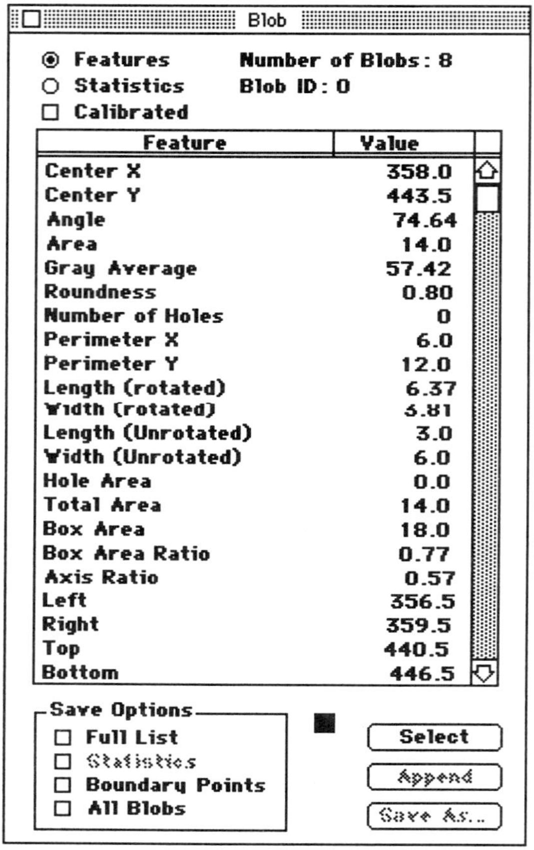

Blob

◉ **Features**   **Number of Blobs : 8**
○ **Statistics**   **Blob ID : 0**
☐ **Calibrated**

Feature	Value
Center X	358.0
Center Y	443.5
Angle	74.64
Area	14.0
Gray Average	57.42
Roundness	0.80
Number of Holes	0
Perimeter X	6.0
Perimeter Y	12.0
Length (rotated)	6.37
Width (rotated)	3.81
Length (Unrotated)	3.0
Width (Unrotated)	6.0
Hole Area	0.0
Total Area	14.0
Box Area	18.0
Box Area Ratio	0.77
Axis Ratio	0.57
Left	356.5
Right	359.5
Top	440.5
Bottom	446.5

Save Options
☐ **Full List**
☐ Statistics
☐ **Boundary Points**
☐ **All Blobs**

[**Select**]
[Append]
[Save As...]

Fig. 4.1 Image Analyst® blob tool shows the many features returned by connectivity.

processes in which the machine vision system is used to ensure that the welding torch is positioned properly with respect to the seam.

There are three frequent examples of tasks in which vision and motion are used together [9]:

- *Step-and-repeat* tasks in which a machine vision sequence is repeated at each of a series of part locations. This type of task is useful when inspecting a tray or grid of samples, such as in a microtiter plate.

- *Extended field-of-view* tasks in which measurements greater than the camera's field of view can be made by taking multiple images at different locations and accounting for motion of the camera or part when calculating measurements. This procedure eliminates having to back the camera away from the part in order to fit it into the field of view, which would mean lowering resolution.

- *Vision-guided offsets,* where the results of the machine vision sequence are used to modify motion coordinates in order to account for uncertainty or change in the work space.

Vision-guided motion is a means by which the vision and motion systems can share information to accomplish a task. For this procedure to be effective, their coordinate systems must be calibrated to one another. This calibration is accomplished by determining a set of coordinate transformations in which positional representations in a given reference frame (e.g., the camera's field of view) are transformed into another reference frame (e.g., the world frame). In this section we will give a very brief overview of how these transformations are used to represent positions in world space. More detailed discussions on transform mathematics can be found in the literature [10, 11].

Each subsystem has its own way of representing position. The camera identifies and reports positions of salient features in pixel space, that is, in the two-dimensional plane of the image. On the other hand, motor positions are given in reference to some zero location (either the site at which the robot is calibrated or some fixed feature of the system, e.g., a hard stop). The user may define positions in a completely different manner by defining coordinate axes with some natural origin point and logically defined axis directions. Typically the camera and both the object location and orientation are known in some other global coordinate system. To determine the image location in an arbitrary coordinate system, we need notation and tools that allow us to relate positions and orientations in different coordinate systems as well as describe the change in location and orientation of an object when it is translated or rotated. Calibration of the subsystems will therefore consist of determining these relationships experimentally.

The coordinates of a point are typically represented in the form of a column vector in homogeneous coordinates. Representations of this point in different reference frames will involve different column vectors. A *transformation* matrix is a matrix which, when applied to the vector representation of a point in one reference frame, produces a new representation in some other reference frame.

The sequence of operations that map an arbitrary point to its image can be characterized as a sequence of such transformations. For example, for the arrangement of Fig. 4.2 this sequence includes a translation matrix $\mathbf{G}$ representing the offset between the global coordinate system and the point where the pan and tilt axes of the camera mounting intersect, a rotation matrix $\mathbf{R}$ capturing both pan and tilt, another translation $\mathbf{C}$ between the previous axis intersection point and a coordinate system at the center of the imaging sensor, and finally a perspective transformation $\mathbf{P}$. These transformation matrices can be determined from the exact geometry of the optical system. However, this process is tedious and error prone. It is hard to measure accurately the physical configuration of the camera or the precise location of the CCD array inside the camera. A simpler approach does not require calculation of the individual matrices but rather lumps all the necessary transformations in a single transformation $\mathbf{A}$. This is referred to as the *camera matrix,* which is the product of all of the above transformations, as shown in the following equation:

$$s\,[u\ v\ w\ 1]^{T} = \mathbf{P}\,\mathbf{C}\,\mathbf{R}\,\mathbf{G}\,[x\ y\ z\ 1]^{T} = \mathbf{A}\,[x\ y\ z\ 1]^{T}$$

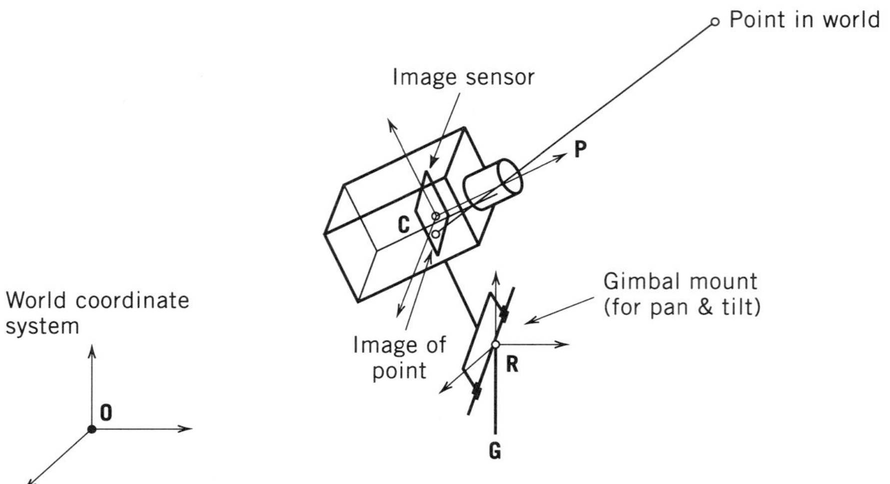

Fig. 4.2 Camera model configuration.

This equation transforms a point in the image plane (which we represent as (u, v, w), multiplied by a scaling factor s) to the corresponding point in the world frame. Using the equation, we can therefore determine the image coordinates (u, v)* of any world point (x, y, z) by simple matrix multiplication, as long as we know the camera matrix.

We can construct this matrix by looking at a few pairs of corresponding points whose positions in both world space and camera space are known. For every pair we can derive three scalar equations from this single matrix equation. Solving for the scaling factor s, we can then get two scalar equations for each point. If we have a sufficient number of object-image point pairs of known coordinates, we can solve the resulting overdetermined set of linear equations and determine the camera matrix.

This procedure is in fact how machine vision systems are traditionally calibrated. An eight-dot calibration target is presented to the camera. This target consists of a matrix of dots whose dimensions and relative positions are very precisely known. The vision system determines the image plane coordinates of each of the dot centers and is given the corresponding world positions, as shown in Fig. 4.3. The system can then use the pairs of points to calculate the terms of the camera matrix. Eight dots are used in order to produce an overdetermined set of equations for solving the matrix using some form of mathematical fit. The system then reports a *residual* term indicating the goodness of the fit to the user. In this way the transformation between the image plane and the world can be calculated directly without having to determine the camera orientation or similar parameters.

This camera matrix is not actually sufficient for the precise location of an object given its image. The coordinates of the image point only give us a line in space along

---

* Note that we are dealing with a 2-D image plane, so we drop the third coordinate w.

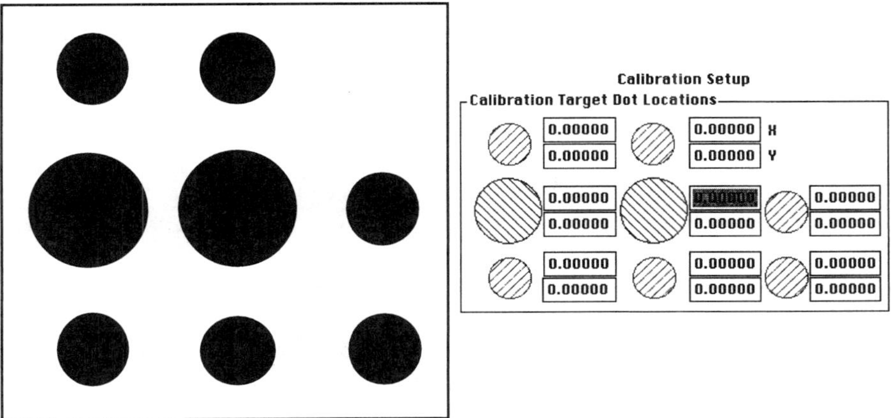

Fig. 4.3 *An image of an eight-dot calibration target, and a user interface for entering the world locations of the dots. The image plane locations are determined automatically by the vision system.*

which the object point lies, since we cannot use only the 2-D image plane data to find 3-D coordinates in the real world. An additional constraint is needed to determine where exactly along this line the object lies. This constraint can be provided by a second image—stereo vision—or by constraining the illumination to be along a known geometric structure—controlled illumination [6].

Similarly we can represent points in any other coordinate system by means of an appropriate transformation matrix. For convenience we can create new coordinate systems that better represent data being reported by the system. For example, we might choose to place coordinate axes at the zero point of a motor in the system. The position of the motor is most conveniently represented in this reference frame. We could also consider a coordinate system based on the part to be inspected—say, at the top left corner of a microwell plate. This way we can refer to the positions of wells in the plate without needing to refer back to a world coordinate system.

The matrix relating the vision system to the motion system is, like the camera matrix, a relationship between the camera and the world frame. However, in this instance the world coordinates of the calibration target are provided by the motion system rather than input by the user. A single-dot target is moved by the motion device in front of the camera (or, alternately, the camera is moved over the calibration dot). At each of several (usually eight) locations, the positions of the motion axes in the world frame and the image of the dot in the camera frame are recorded. Assuming that the motion axes have been calibrated in world coordinates, this will produce a set of corresponding pairs of coordinates between the camera and the world, just as if we had entered the dot positions of an eight-dot calibration target. Solving this set of correspondences yields a camera matrix that is calibrated to the motion axes. In the simplest case the motion axes are aligned with the world coordinate system, and hence the camera will also be calibrated to the world frame; otherwise, the system must be programmed to include the transform between the motion axes and the world frame when calculating the world frame position of objects in the image.

Once this calibration has been made, it is straightforward to include machine vision measurements in motion commands. For example, a vision offset to a motion command can be made by including a point in the image whose world frame coordinates are known. This "anchor" point serves as a bridge between the camera and world frames. Motion commands to other objects in the image can be issued by calculating the distance between the anchor and the goal object in image space and then transforming this distance to world space via the camera matrix. The transformed distance is added to the world space coordinates of the anchor to yield the world space coordinates of the goal, which may then be transmitted to the robot (Fig. 4.4).

Again, with a single camera, only two-dimensional positional information is recoverable from the image of an object. Hence for applications involving 3-D motion, the user must either use multiple cameras or act on other information about the object or the task to provide the extra dimension. For example, vision can be used to locate colonies on an agar plate in the plane; however, the user will need extra information about the height of the colonies in order to move a robot to pick them up, as is shown in the example of the next section. The user should also note the residual term provided

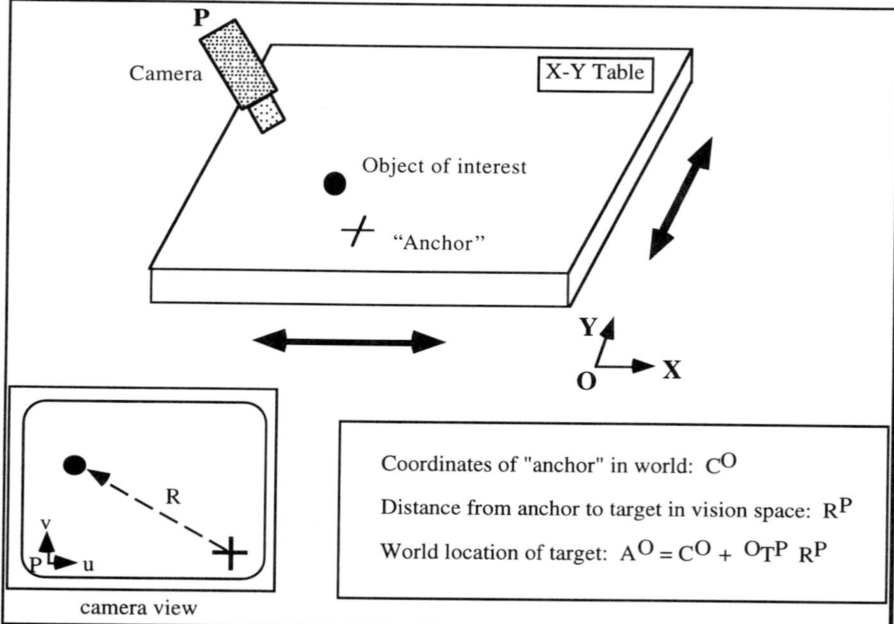

Fig. 4.4 Locations of objects found by the vision system can be transformed to world space locations for use by the robotic system.

by the calibration, which is indicative of the maximum potential error (in pixels) reported by the vision system based on the least-squares (or other) fit of the world frame and image frame coordinates of the calibration dot. A little arithmetic can be used to calculate the maximum error in world units based on the residual and knowledge of the field of view of the camera. This error should be smaller than the repeatability of the drives and must be smaller than any critical dimensions of the objects to be imaged in order for data from the vision system to be relied upon.

Once the system has been properly calibrated, it is possible to coordinate motion and vision system information. Measurements taken by the machine vision system can be passed on to the motor in world coordinates as goal positions in order to execute vision-guided moves. Conversely, motion of the robot can be taken into account when the vision system is making measurements, as for an extended field-of-view application.

AN EXAMPLE: VISION-GUIDED COLONY PICKING

Overview

Acuity Imaging has recently completed collaborative work with the Max Planck Institute for Molecular Genetics in Berlin, Germany, and Linear Drives Ltd. of the UK for automation of a colony picking task [12]. In this task individual clones are transferred from large colony trays (each approximately 22-cm square) into wells of 384-well microtiter plates as the first step in their hybridization approach for large-scale library characterization. (For more details with regard to the science behind the automation, see Chapter 3 by Meier, Bancroft, and Lehrach.)

Fig. 4.5 Detail of picking head and tool mount on Planck colony picker.

Colony picking is a clear example of how machine vision systems ensure accuracy in a laboratory task. Colony plates contain an unknown number of colonies in random locations. Ideally all of these colonies would be transferred to microwell plates for storage in order to eliminate waste of time, material, and effort in preparing plates. On the other hand, some discrimination must be employed so that unsuitable colonies are not transferred: Colonies that have grown into each other are contaminated and must not be picked. Hence some advanced sensing and processing is required to locate colonies, filter out colonies with certain characteristics, quantify the remaining colonies, and transfer them to whichever microwell plates are currently available.

While the machine vision sequences involved in colony picking are straightforward, calibration of the multiple subsystems to one another is not. In designing the software for this application, it is necessary to make sure that the system incorporates repeatable calibration procedures. Further, these procedures have to be executable by the end user without requiring complex programming skills. They have to be fast in order to minimize losses in throughput and to encourage users to calibrate the system often. As we will discuss in the next section, a point-and-click application was developed to meet these needs that allows for system calibration in under 15 minutes and can be done daily.

In the following subsections we will describe the techniques used for calibration, image analysis, motion control, and programming that went into developing this application.

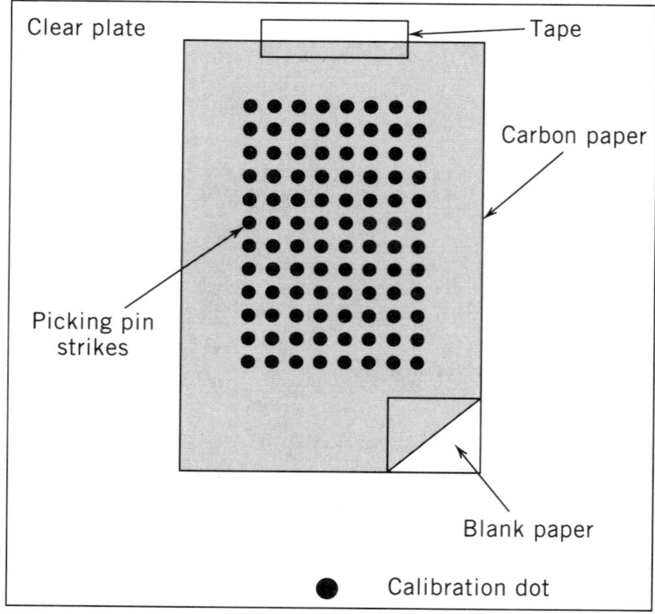

Fig. 4.6 *Schematic of calibration fixture layout.*

System Calibration

The colony picker is composed of a number of subsystems, each of which must be accurately calibrated in order for the system to be able to pick with the 0.5-mm accuracy required of it. The system includes five axes of motion: A large-scale, servo-controlled *XYZ* set of axes in gantry formation, which position a picking head, and a smaller serially controlled *xy* system for moving an air-actuated pushing pin above the picking head. The picking head itself consists of 96 spring-loaded pins arranged in a 12×8 array and spaced similarly to a 96-well microtiter plate (Fig. 4.5). The head is screwed into a fixture just below the *xy* serial axes such that the pushing pin can be moved above any of the 96 picking pins. Air actuation is used to move the pushing pin up and down, and thereby to extend or raise a given picking pin.

The locations of the picking pins themselves vary considerably. The picking head is secured in its fixture by thumbscrews, which must be tightly fastened to ensure reasonable repeatability of placement. In addition individual pins can become bent during use or in accidents (e.g., dropping the picking head or crashing into the work space somewhere). Replacing the pins takes time and materials; a slightly bent pin can still be used for picking as long as the bend is accounted for.* Enough variation can also exist between different picking heads that some means had to be provided for positively determining the location of each pin end point.

The vision setup is similarly complex. The camera is bolted to the robot *Z* axis, slightly forward of the picking head. A metal hood prevents casual adjustment of the camera aperture or focus. However, the camera is frequently unbolted and bolted on again when required so that misalignment between the camera axis and the robot *Z* axis is possible. Further the distance between the camera and the gel surface is variable because of variations in the amount of gel poured, gel shrinkage, and different labware. The field of view is approximately 28×37 mm (i.e., 48 frames in a 6×8 pattern are required to image the entire plate). Initial experiments showed that the depth of field was shallow enough that a variation of even 1 mm in the height of the agar would distort the image of the colonies sufficiently to throw picking off by millimeter or more. Hence a repeatable, simple camera calibration was to be essential for accurate picking at the scale and speed required. Further system specifications required that a technologist be able to calibrate the entire system quickly and without significant training, prior programming, or mathematical knowledge.

Calibration of the colony picker involves three separate calibration procedures:

- Motion axes
- Vision to motion
- Picking pins to vision

---

* That is, as long as the pin does not rotate. The pins being cylindrical in cross section, pin rotation is possible, which would move the tip of the pin from its predetermined location. However, for slight bends, very little pin rotation seems to take place during system operation. Noncylindrical pins would also help matters significantly.

Calibration of the *motion* axes involves repeatably setting a zero point at which the axes are based. To accomplish this task, the system goes through a startup procedure whereby all of the gross motion axes are moved against hard stops at the far limits of their travel (this being the corner of the system furthest from the user). The system then sets a zero reference point at the location of these hard stops, which is used as the world frame origin. All of the system positions (the wash station, the microwell plate locations, the air dryer, etc.) are taught with respect to this zero point (although they are accessible for modification); therefore it is critical that a repeatable zero, such as the hard stop, be used for calibration. Calibration of the motion axes takes about one minute and is performed every time the system is started up or the colony-picking program is restarted.

Once the motion axes are calibrated, the next step is to calibrate the camera to motion in order to relate the image space to the world. A special calibration target was designed by the Planck staff for this purpose (Fig. 4.6). The target consists of a 22 cm square piece of clear Plexiglas that fits into the same fixture used to hold the culture plates. The calibration target is kept clean and clear aside from a small black dot centered in the lower portion of the square, as shown. A sheet of plain white paper is then taped to the upper portion of the target, with a sheet of carbon paper laid out on top of the white paper. The target is slid into position above the light table, which is turned on for calibration in order to highlight the calibration dot.

The camera is calibrated by moving to eight distinct locations over the black calibration dot. As discussed in the previous section, at each location the dot centroid is located using a connectivity algorithm. A calibration matrix is then established with the average residual reported to the user, who has the option of accepting the calibration or repeating the procedure until the residuals are within a manageable tolerance. We found that a residual of below 0.4 pixels would give reasonable picking accuracy.* Vision calibration takes approximately one minute and can be performed whenever the user desires. Typically we recommend recalibration if the camera is adjusted or moved, or if the performance of the system degrades.

The initial height of the camera above the calibration dot is preset so that the dot is in clear focus; this height is critical because, as mentioned, slight changes in the distance between the camera and the object being imaged produce significant distortions from focusing errors. Once vision calibration is completed, the user has the option of confirming the height between the camera and the calibration target. This task is accomplished by extending a single picking pin. The user then lowers the Z axis of the robot until the pin is just touching the calibration target, at which point the system records the distance the Z axis has traveled. In this manner a repeatable measure is taken of the distance between the camera and the object in the image. Camera height calibration can take about five minutes and can be performed whenever the system camera is recalibrated. All other system camera heights are then referenced to this

---

* This means that measurements reported by the vision system may be in error by at most 0.4 pixels, which equates to approximately 0.02 mm at the current resolution. Since the drives are repeatable to 0.005–0.01 mm, we would prefer a residual below 0.2 pixels; howver, since the minimum colony diameter to be picked was 0.5 mm, we found the maximum error to be tolerable in these circumstances.

camera height; then, if the user wishes to pick from an agar plate with agar of a varying height, the appropriate modifications in camera height can be made on the fly.

Once the vision system has been calibrated to the motion axes (and hence to world coordinates), it makes an excellent tool for calibration of the picking pin locations. The robot moves above the calibration target so that the 96-pin picking head is located above the carbon paper. Each pin is then popped down briefly so that it strikes the target, which leaves a dark mark on the white sheet beneath the carbon paper. When the carbon paper is removed, an 8×12 grid of pin marks is left on the white paper. These marks are clearly visible to the camera when the light table is on and can be used to positively locate the pins with respect to the camera. After the user removes the carbon paper, the robot moves to 96 positions such that the pin marks are each close to the center of the field of view of the camera. Connectivity is used to find the centroid of the marks and store the relative distance between the mark centroid and the camera position when the marks were made. In this manner a lookup table is created that contains the 96 true pin locations and that can be used to modify picking commands to account for slight variations in the pin positions. This part of the calibration procedure can be performed whenever desired and takes about five minutes.

After all calibration procedures are executed, the system has established connections among the camera field of view, the picking pin tips, and the world reference frame. The entire calibration procedure, start to finish, might take the average user about 15 minutes, and it is performed via point-and-click buttons in a graphical user interface (GUI). While simple enough for a new user to operate, calibration is also quick enough that system calibration at the beginning of each day is feasible. In addition, should any picking problems develop during use, it is a straightforward task to halt system execution and recalibrate a system component by means of the same GUI.

System Operations

After system calibration has been completed, the user can go straight to picking colonies from plates. First, the work space must be loaded up with empty microtiter plates into which the picking head will deposit colonies. At present the system can accommodate up to 24 plates in the 96-well or 384-well configuration. The user configures the initial microplate setup on the system through a set of check boxes, after which the system will keep track of which sectors and which plates have been filled. The user must also decide whether to pick from a single agar plate or two plates side by side (the maximum capacity of the system).

In loading the agar plates, the user has several choices to make. First, since the height of the agar may vary from plate to plate (with volume of agar poured, shrinkage, plate size, etc.), the user can train the system to the true height of the agar. This training is performed in the same manner as the camera height calibration: A picking pin is extended, and the user lowers the Z axis until the tip of the pin just touches the agar surface. The system then stores this height and makes corrections to both the camera height when imaging colonies, and to the height of the picking head when picking colonies. This way the user can ensure proper picking depth without either missing colonies completely or driving too deep into the agar. The system can also be trained to accommodate sloping agar (e.g., when a plate is stored on a tilted table while the agar sets) by training the agar height at three fixed locations. The plane defined by these three heights is calculated, and the height of the picking head adjusted at each picking location at no significant cost in additional picking time.

The user can then go on to pick the entire plate, or can "sample" any of the 48 sectors of the plate, through an interactive dialogue. In sampling, the robot is moved to the sector chosen, and the image analysis algorithms are run. The user then gets a chance to see the result of the image analysis and determine if the system is reasonably discriminating among the colonies. If this is not the case, the user can then adjust any of the image analysis parameters and reexamine the vision results until a suitable set of sorting parameters is determined. Individual users can also store and retrieve their own preferred imaging parameters to meet specific needs. After confirming the image analysis, the user can go on to pick a single sector of the plate or any set of sectors desired.

Machine Vision Details

Three image analysis sequences are used in the colony picking application:

- *A vision calibration* sequence consisting of a single picture in which a connectivity algorithm is used to locate a calibration dot
- *A pin calibration* sequence in which connectivity is used to locate the centroids of pin markings in order to create a lookup table with precise pin location values
- *A colony-finding* sequence containing two pictures:
    A picture in which a spatially fixed fiducial mark is located and used as a reference point

    A picture in which a connectivity algorithm is used to locate colonies (blobs) and sort them according to user-defined parametric limits. A grayscale algorithm is also used for automatic threshold setting to account for variations in agar darkness or lighting.

While the two calibration sequences were discussed in the previous section, we will discuss the actual colony-finding sequence in detail here.

The first job of the colony-finding sequence is to establish a world space location to which to reference the positions of the colonies. While this could be a stored location in memory, for the sake of accuracy we locate this "anchor point" before starting imaging on a given agar plate. A fiducial reference mark in the robot work space (located between the two agar plates for convenience) is located by the vision system using a connectivity algorithm, providing a translation between image space and world space. When colonies are found their positions can be referenced to this known location. The subsequent robot motion can then be modified by the offset between the colony and the fiducial reference, enabling the robot to move to a previously undefined location.

When imaging colonies, the system actually performs two rounds of processing on the image. The first round runs a grayscale algorithm in which the average gray level

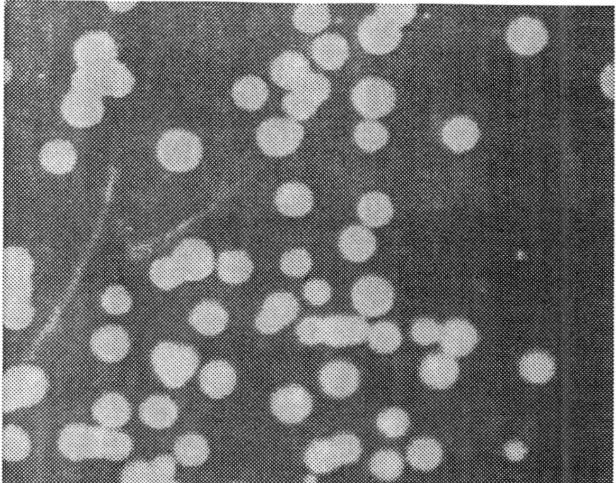

Fig. 4.7a   Image of colonies on an agar plate.

of the image is calculated. This average gray level is then used to adjust the threshold level of the connectivity algorithm used to locate blobs (Fig. 4.7*a*). Recall that a binary algorithm such as connectivity finds blobs in the image by determining their boundaries, making this determination based on a threshold value that separates dark pixels from light ones. When the lighting of the image varies significantly, the algorithm can produce varied results for a given threshold. In the case of the picking robot the brightness of the image varies significantly—as much as 40 gray levels—across the agar plate because of unevenness of the light sources and reflections from the clear walls of

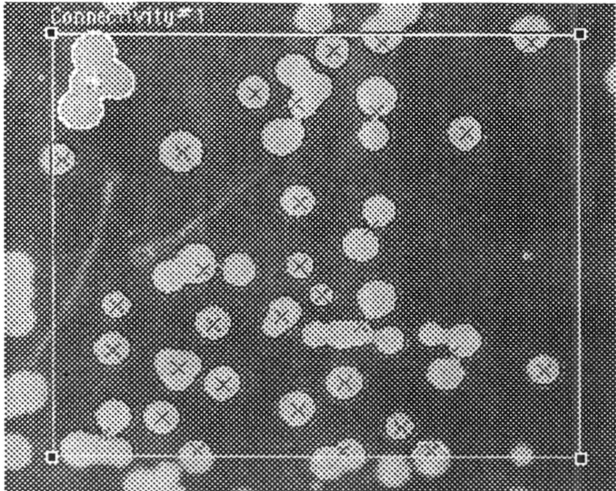

Fig. 4.7b   Same image with connectivity algorithm identifying blobs.

the picker. Thus a fixed threshold cannot be used to determine the colony boundaries. By using the autothresholding algorithm and placing baffles around the light source, we were able to light the agar plate reasonably evenly and account for any unevenness in software.

Once the new threshold for the image has been computed, a connectivity algorithm determines the location of the various blobs in the image (Fig. 4.7b). To differentiate between colonies eligible for picking and other entities (clumps of colonies, the edge of the plate, reflections, etc.), a blob-sorting algorithm was developed. The algorithm sorts the blob list, using a set of parameters with user-definable bounds. These parameters include the following:

- Blob roundness (valued between zero and one, with one being a perfect circle)
- Blob size (area in pixels)
- Blob axis ratio (ratio of the largest and smallest axes of the blob)
- Blob gray average

The blob gray average measurement can also be used to discriminate colonies in blue/white picking situations, although generally the blue color needs to be reasonably developed before this discrimination can be relied upon, given the variation in background illumination. We are currently experimenting with other methods such as using color filters or dynamically updating the grayscale sorting algorithm to account for shifts in illumination.

Results

The system is currently picking quite accurately: colonies 0.6 mm in diameter can be repeatably picked at full speed. System throughput is approximately 3000 colonies per hour, and the system can be operated unattended for several hours before requiring refill of either microwell plates or colony plates. The system's burst picking speed (i.e., not accounting for the time required to wash and dry the picking head) is approximately 76 colonies per minute with the drives at about 75% of top speed. Wash-and-dry time has a significant effect on system throughput and is being examined further by the Planck Institute staff.

All system operation and programming is performed through a point-and-click GUI developed by Acuity Imaging to specifications set by the Planck staff. This interface sits on top of Acuity Imaging's Image Analyst® machine vision software and VGM® utilities for vision-guided motion control.

We are also in the process of completing additions to the vision-guided motion code that will allow serial communications to other third-party motion control devices. This change will allow us to take advantage of more complex motion options such as profiled velocity and acceleration commands, while also allowing the end user more flexibility in choice of motion vendors.

FUTURE APPLICATIONS

As applications in genetics become smaller in scale yet larger in throughput, vision guidance is becoming more and more essential to automating tasks. One area related to the colony-picking task is automated reading of spotted filter sheets. Image analysis algorithms have already been developed that will look at the grids laid down on these sheets and provide a report on grids or duplicates that show as positive. With the advent of faster, more accurate motion control systems used to spot the filters automatically, it is possible to design a system that can lay down very large grids on a very small surface. The Planck picker/spotter is capable of creating a 25 × 25 grid in the same area that it now produces 5 × 5 patterns, allowing for significantly more data to be incorporated on a single filter sheet, provided that the robot is fast enough to create the sheets in reasonable time. This step can be followed up with an automated scanning phase in which an image of the sheet is sent to a machine vision system that can identify hits and forward these on to a database.

We have also seen multiple applications involving inspections in microwell plates and anticipate that vision guidance will be needed as the scale of these experiments continues to drop. Applications are now being explored in which nanoliters of fluid must be manipulated inside similarly small containers. Aside from highly specialized drives and stages, advanced sensing is required in order for these processes to involve any feedback. As in the electronics manufacturing industry, machine vision-guided systems will play a vital role in enabling automation of tasks on the nanoscale.

ACKNOWLEDGMENTS

Funding for some of the work described in this chapter was provided in part by a Small Business Innovation Research grant from the National Center for Research Resources, an institute of the National Institutes of Health, Bethesda, Maryland.

The authors wish to thank Charlie Winterford of Acuity Imaging Ltd. and Hans Lerach, David Bancroft, Jon Curtis, Martin Horn, Elmar Maier, and the rest of Dr. Lerach's group at the Max Planck Institute for their assistance in designing and developing the colony-picking application. Particular thanks go to Chris French of Acuity Imaging Ltd. for considerable effort in application development and engineering to make the system—and the project—work.

REFERENCES

1. K.B. Ward, M.A. Perozzo, W.M. Zuk (1988). Automated preparation of protein crystals using laboratory robotics and automated visual inspection. *J. Cryst. Growth* 90:325–339.

2. W.J. Martin, R.M. Walmsley (1990). Vision assisted robotics and tape technology in the life-science laboratory: applications to genome analysis, *Bio/Tech.* 8:1258–1262.

3. S. Krishnaswamy, J.E. Agapakis (1996). Machine vision in laboratory automation: Fundamentals and applications. In *Handbook of Clinical Automation, Robotics, and Optimization,* G. Kost, ed. John Wiley & Sons, Inc., New York.

4. J.E. Agapakis, W.S. Cole, S. Krishnaswamy, P.A. Lamoreaux (1994). A vision-integrated robotic system for laboratory applications. Poster presentation. *Third Annual Conference on Automation, Robotics, and Artifical Intelligence Applied to Analytical Chemistry and Laboratory Medicine.* San Diego, CA, January 25.

5. A.K. Jain (1989). *Fundamentals of Digital Image Processing.* Prentice Hall, Englewood Cliffs, NJ.

6. D. Ballard, C. Brown (1982). *Computer Vision.* Prentice Hall, Englewood Cliffs, NJ.

7. S. Inoue (1986). *Video Microscopy.* Plenum Press, New York.

8. R.C. Gonzalez, R.E. Woods (1992). *Digital Image Processing.* Addison-Wesley, Reading, MA.

9. S. Krishnaswamy, J.E. Agapakis (1994). Vision-guided motion applications in laboratory automation. Poster presentation. *International Symposium on Laboratory Automation and Robotics,* 1994, Boston, MA.

10. J.J. Craig (1989). *Introduction to Robotics: Mechanics and Control.* Addison-Wesley, Reading, MA.

11. B.K.P. Horn (1985). *Robot Vision.* MIT Press, Cambridge, MA.

12. C. French, S. Krishnaswamy, H. Lehrach, A. Ahmadi, J. Curtis (1995). A vision-guided robotic colony picking/spotting system for genetic research. Poster presentation. *Fourth International Conference on Automation, Robotics, and Artificial Intelligence Applied to Analytical Chemistry and Laboratory Medicine,* Montreux, Switzerland, February 7–10.

II

CONTROL SYSTEMS

A Script-Directed Controller of Modular Automation (SDCMA) for Genome Laboratories

BOBI K. DEN HARTOG AND PATRICIA A. MEDVICK

CONTENTS

Automation Technologies for Genome Characterization, Edited by Tony J. Beugelsdijk.
ISBN 0-471-12806-6   © 1997 John Wiley & Sons, Inc.

INTRODUCTION

"Change" is in the air in today's biological and chemistry laboratories. To accommodate constant change, we have designed a flexible, user-programmable controller that allows plug-and-play exchange of equipment. This chapter discusses the problem and motivation for building the SDCMA controller, the components of the controller, and an example of the execution of a user-written script by SDCMA.

PROBLEM DESCRIPTION

Since 1992 Los Alamos National Laboratory's (LANL) Gridding Robot has been making hybridization membranes [1, 2] from cloned fragments of DNA for LANL's Center for Human Genome Studies (CHGS) project for mapping chromosomes of the human genome. As CHGS automation needs became more sophisticated and mapping neared completion, it became evident that redesigning control software for each new application was prohibitively time-consuming and expensive. We decided to use the opportunity of building a new gridding system to construct a more flexible controller that permitted faster throughput. The new controller also had to be applicable to construction of an automated sequencing system. We called the new controller a Script-Directed Controller of Modular Automation (SDCMA).

Nearly all molecular genetics research relies on the availability of cloned fragments of DNA. To clone pieces of DNA, enzymes are used to cut the original DNA into fragments. These fragments are marked and put into host cells such as bacteria or yeast. The host cell replicates the inserted DNA as if it were its own. Thus the marked fragment from the original DNA is amplified by the host. When all the DNA in a genome has been cut into fragments and cloned, the result is a genome library. Libraries can also be formed from one chromosome of an organism.

Libraries are stored in covered 96-well microtiter plates. Each well contains one amplified fragment and is identified by row number and column letter. Each plate has a barcode label of its library name and plate within the library.

Through the efforts of the National Laboratory Gene Library Project, many libraries of human DNA-cloned fragments have been created for developing physical maps of the human genome. To assist the mapping, LANL created a biologist-friendly automated system to transfer clone contents from the microtiter dishes onto nylon membranes and thus produce gridded membranes. The location of the deposit is vital, since it ties the plate and well number to the eventual results.

These gridded membranes, containing copies of the DNA clone fragments, look like matrices of dots on glossy paper.

To reduce the sample size necessary for each dot, the gridded membranes are incubated overnight. The underlying agar supplies nutrients to the host and the "dots" enlarge.

The DNA on the gridded membranes is denatured, that is, the two strands are separated. A solution containing radioactively labeled single-strand DNA with known sequence is washed across the membrane. Exact complementary strands renature. The excess solution is removed. By putting X-ray film on top of the membrane, an autoradiograph is produced. This picture, which looks like a photographic negative, highlights the matches, which can be traced to their original microtiter plates and specific wells. By exhaustive repetition the contents of most wells in all the plates can be matched with DNA probes.

We have automated the gridding and autoradiograph interpretation steps in this process. Because of the accuracy and precision of automation, higher-density grids are possible than by manual gridding. Higher-density grids result in faster throughput in many steps of the mapping process and less radioactive waste from the radioactive labeling steps.

Our new robotics system for automating gridding consists of a robot with a gripper for holding 96-pin tools, a wash station for sterilizing the tools, a plate stacker for presenting microtiter plates and removing their lids; a barcode reader for reading the plate labels; and a database for storing clone library information, plate-gridding locations and system information. Control software at many levels ties these pieces to the user interface.

When gridding, the robot picks up a clean tool from the wash station. The plate stacker pulls a plate containing DNA clone fragments out of the ungridded plate stack. The barcode reader reads the barcode and the database is queried about the barcode. If the plate is in the library and has not yet been gridded, the plate stacker removes the plate's lid. The robot dips the clean 96-pin tool into the 96-well microtiter plate, moves to the location on the membrane specified by the database and touches the tool to the membrane. The plate is gridded in as many locations as the library information in the database describes. Then the plate's lid is returned, and the plate put onto the used plate stack. The robot places the dirty tool into the wash station's dirty-tool port and picks up a clean tool from the wash station's clean-tool port. The wash station sonicates the dirty tool in an alcohol bath to clean it, passes the tool through the drying station, and leaves it on the wash station's clean-tool port.

The prototype gridding system proved the concept and led to the plan for this higher-throughput, more flexible version.

Requirements for Speed and Physical and Process Flexibility

1. The control software (controller) must be inexpensive to change for different hardware configurations.

2. The controller must be easy to change among processes. The laboratory user should be able to create and change the processes.

3. The operating speed must be increased due to the magnitude of the task of mapping and sequencing the human genome; however, accuracy cannot be sacrificed.

The resulting controller design was a scaled-down version of the architecture proposed by the Department of Energy's (DOE) Contaminant Analysis Automation (CAA) project [3]. CAA's goal is to provide a reusable infrastructure for an environmental chemistry laboratory. The reusable infrastructure includes a task sequence controller, human-computer interface, database, and communications and control protocols. By making the specialized equipment necessary for a specific chemical procedure follow the CAA standard communications and control protocols [4], new equipment modules can be plugged into the CAA controller, user interface, database, and communications without rewriting the framework, permitting work to continue with a minimum of disruption. In both the environmental chemistry and biological research laboratories, a flexible control system was desirable, since the program could not be locked into a permanent hardware configuration and could not be reengineered with each new piece of hardware or new process. By standardizing the look and feel of a piece of laboratory equipment, the control system could use standard protocols and expect standardized behavior from any piece of equipment.

CAA is a large project, involving five national laboratories, several universities, and industrial partners over a period of five years. The CHGS controller and system could not be of such a time, cost, or conceptual magnitude. The twin features of script-based control and standardized hardware modules using standardized interfaces were retained as the keys to the CHGS's flexible, fast system. Hardware standardization is based on CAA's Standard Laboratory Module™ (SLM™) concept. A Standard Laboratory Module, in addition to following standardized communication and control protocols, advertises its operations in a "Capability Dataset." This set of operations includes standard commands such as "enable" or "status" as well as equipment-specific commands such as "wash" for our wash station or "opengripper" and "move" for the robot. These operations, or capabilities, then form the basis for the programming language for the system. Stringing together a sequence of actions—"move" the dirty tool to the wash station, and "opengripper" to release it, then "wash"—creates a program called a *script*. These scripts are stored and reused.

In the CHGS system we found we needed some application-specific data servers. To avoid confusion with the definition of an SLM, our system has, instead, "Services." We extended the SLM model to include Services that are software programs. For our gridding task, a report Service and a membrane Location Server Service have capability datasets, and their operations can be used in the scripts in the same manner as the operations of the hardware modules.

To start a task, the user needs to specify which script to run and what set of Services are in the current configuration. For convenience, these are stored as "tasks"—each containing a script name and a list of Service names.

Our Strategies to Meet the Requirements of Flexibility and Speed

To obtain the flexibility and speed desired, a common strategy was developed for the system's hardware and software:

1. Flexibility in configuration and process was achieved through standardized modularity and dynamic descriptions. By using standard protocols for Services, a new Service can be introduced to the system without changing the control and communication commands. Services are either traditional physical hardware or Virtual Services. Virtual Services are data servers that use the same language interface as the hardware. By describing each Service in a Capability Dataset and by storing scripts in the database, the control software can switch between laboratory configurations and processes.

2. Speed was obtained by the removal of serial-only execution. In the previous gridding system a PC serial controller allowed only single-file execution of instructions: An instruction had to be dispatched and completed before the next step could be issued. This requirement ignored whether the operations needed to be done as a series or could run in parallel. For instance, the gridding task ideally would be putting a used plate away and getting a new one ready while cleaning the used tool, but in the prototype system, due to single-thread, synchronous execution, each operation was done in sequence and held the full attention of the single-thread controller throughout its performance.

By operating in parallel, the process gains speed on steps that can run simultaneously. The user specifies the step dependencies explicitly in the script, as in most parallel computing languages. Due to the plug-and-play paradigm, physical bottlenecks can often be eliminated by adding another Service module to the laboratory. The control software looks for the first idle Service that matches a script step's operation. If no idle Service is found, the step is placed on a matching but busy Service's work list. The system uses both true parallelism, since various hardware Services can be working at the same time, and simulated parallelism due to the control software's round-robin time allocation to different, parallel-appearing software processes.

The Controller Is a Key Part of an Integrated Approach

The common strategy was used in each part of the system's development.

1. Sophisticated custom equipment such as wash stations and plate stackers were built to operate from integrated control and have adequate sensor and alarm notification capabilities for integration.

2. Software bridges were developed for each piece of equipment to provide the standard interface protocol. All Services understand the standard set of Service commands (Fig. 5.1).

3. Service, task, and script descriptions are added to the database rather than coded as a part of the control system.

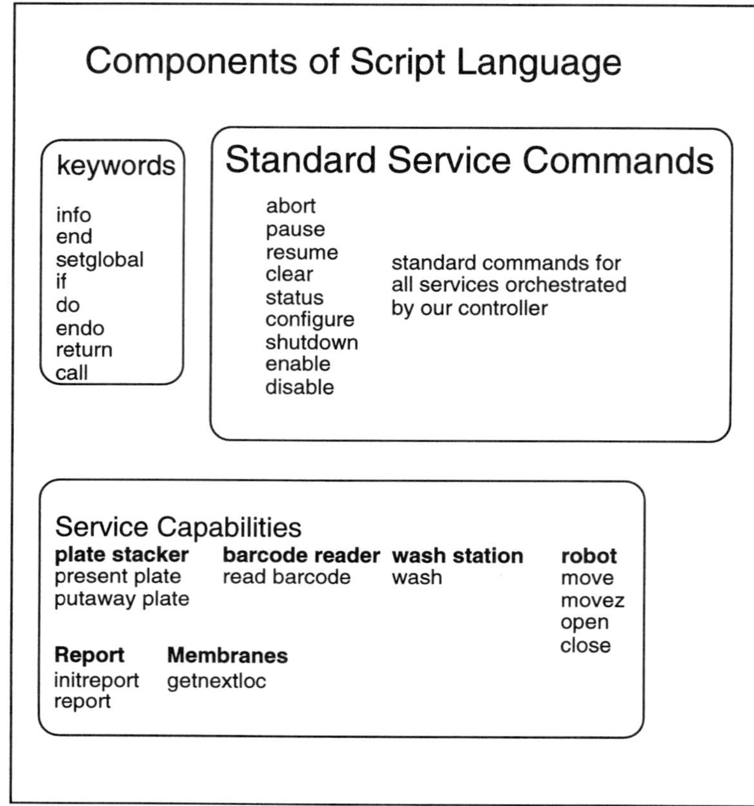

Fig. 5.1 *Three sets of commands that compose the script language.*

4. Control software integrates and controls the laboratory system, based on the programmed scripts, descriptions of the Services, and the task stored in the database.

METHODS

Capability-Based Scripting for Process and Physical Flexibility

A user writes a script that describes the process the system should perform. The script is composed of steps. Each step contains an operation and the information needed to do it, and information about which steps to perform next. An operation can be (1) a capability of some Service in the laboratory configuration for this script, (2) a standard capability that all Services support, or (3) a keyword the control software provides (Fig. 5.1). Some operations use input and output arguments. The control software provides for named variables to be declared and used throughout the script. Some operations set simple yes or no return codes.

The steps to be performed next are listed in a yes-set, a no-set or an error-set in the step (Fig. 5.2). Each step can have all three sets. If an error is detected, the control software will provide default error handling if there is no error-set indicated on the step. If an error-set is listed, the user-defined error handling is used. The user-defined error handling is just another subscript of steps. In normal operation the yes-set or no-set is used, depending on the value of the return code set by the step's operation.

Components of the Controller

CONTROL INTERFACE   The Control Interface (CI) is the controller's connection to the user interface and can also be connected to a higher level controller (Fig. 5.3). It is a daemon program that waits for user commands and executes them. The CI displays status, error messages, and system requests. It creates the Task Manager (TM). The CI creates the work list for sending CI commands to the TM, and the message list for carrying information from the TM to the CI (Fig. 5.4). The CI recognizes the commands control, status, start, pause, resume, clear, abort, and disconnect. Operating states of *nocontrol, inactive,* and *active* indicate whether or not this interface port is a control port or a silent monitor, and whether or not a task is in execution. Error states of *error* or *ok* indicate whether an error has been detected. Pause states of *paused* or *ok* indicate whether the system has been paused. The Operating, Error, and Pause states determine whether or not a command is valid. For example, abort is valid whenever a task has started, but start would be invalid if a task were already in execution.

TASK MANAGER (TM)   The Task Manager, TM, is a general manager initiated from CI's "start *taskname*" command. The *taskname* indicates a task in the database. The

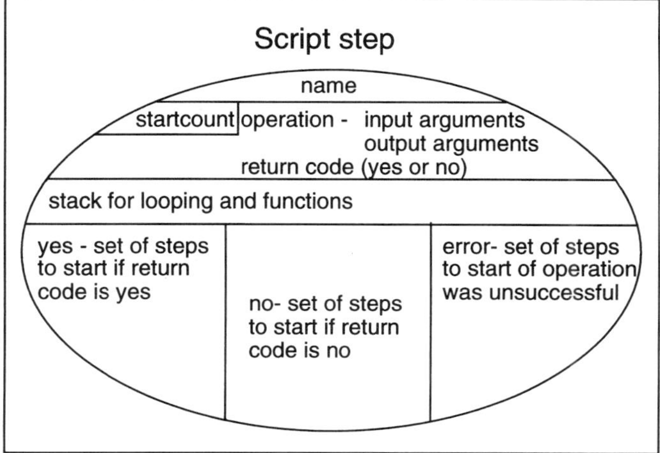

Fig. 5.2 Contents of a script step.

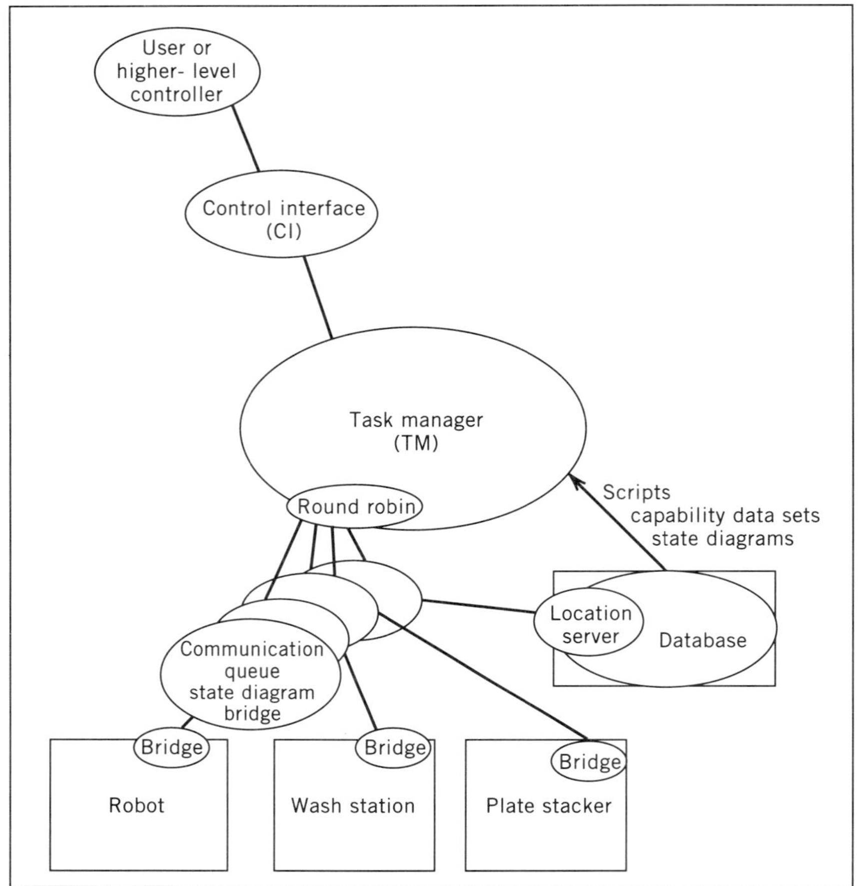

Fig. 5.3   *Run-time architecture of a task, showing only some of the Services used.*

task describes which Services to use and which script to execute. Once the Services and first script step of a task are read from the database, the script execution begins. The TM reads Service parameters and capabilities from the database. For each Service the TM builds a work list to send commands to the Service, as well as a message list to return information from the Service (Fig. 5.4). The TM provides a central ready-to-run queue for steps whose conditions are met for execution. The TM interprets commands from the CI, operations from the script steps, and messages from the Services. Through a keyword server, the TM provides keywords **info, end, setglobal, if do, enddo, return,** and **call** (Fig. 5.1) as operations available to script writers. Another part of the TM, the argumentManager, manages named variables in the script. Operating states (Fig. 5.5) of *configuring, configured, executing, shuttingdown,* and *shutdown* indicate whether all Services have completed their configuration or

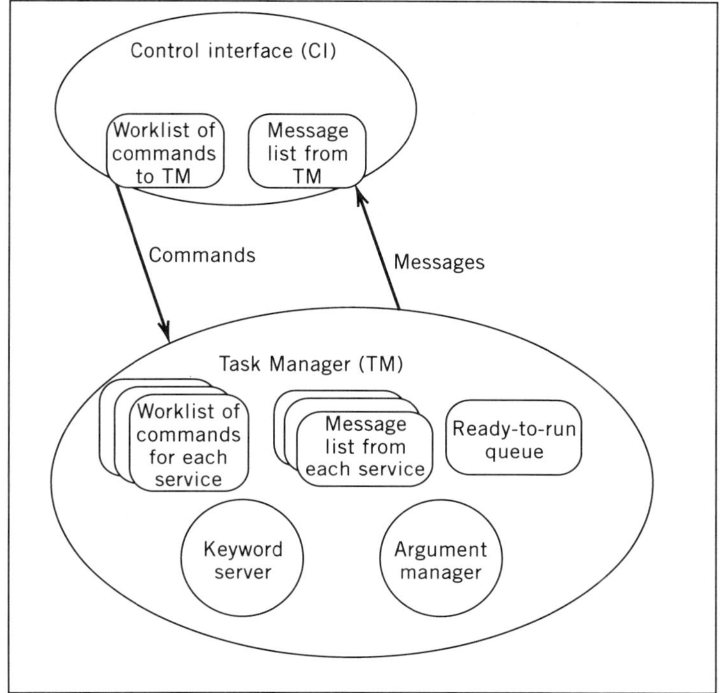

Fig. 5.4 *Information handling, passing, and storage.*

shutdown procedures. The TM stays in the *executing* state until the last step of the script is read, indicated by the keyword **end**, and begins shutting down the Services. Error and Pause state diagrams for the TM (Fig.5. 5) differ from the CIs by having additional transition states (e.g., *pausing*) during which the TM issues the appropriate command to the Services (e.g., *pause*) and collects all confirming responses before moving to the confirmed state (e.g., *paused*). Commands are verified by appropriateness to current Operating, Error, and Pause states. For example, **clear** would be inappropriate if no error had been detected.

SERVICES   Service descriptions are read from the database and include the Virtual Service indicator, the synchronous indicator, and the list of capabilities and arguments that the Service supports. Virtual Services differ from their physical counterparts in that they exist only in software. Our membraneLocationServer and reportServer Services are Virtual Services. Each Virtual Service has a set of operations called its Capability Dataset, and a user utilizes these operations in script steps in the same manner as operations from physical Services. By including the Virtual Service switch in the description, the control software can make minimal differences in the way it treats physical and virtual Services, thus keeping the overall controller structure cleanly han-

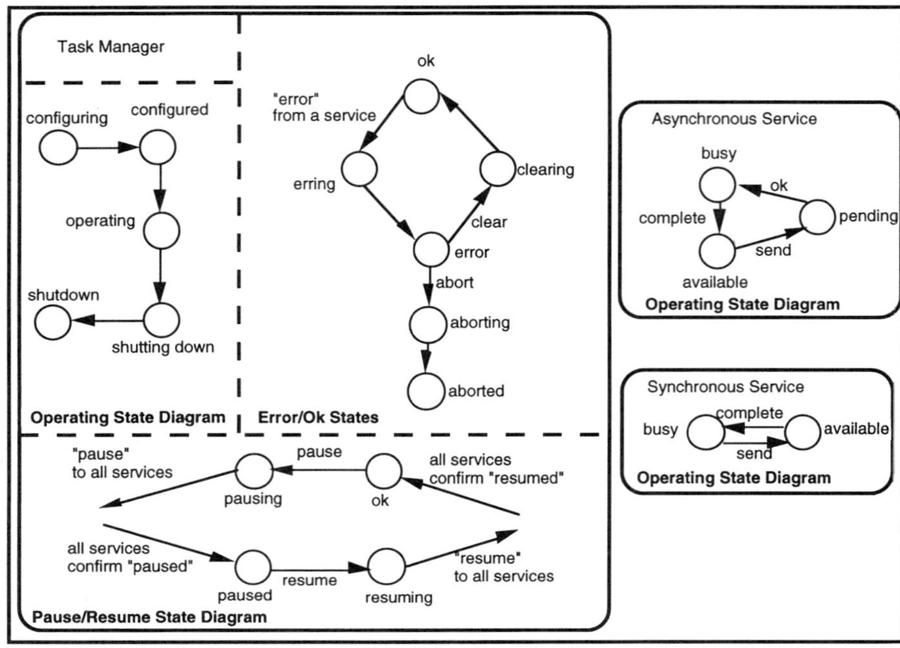

Fig. 5.5 State diagrams for defining Task Manager and Service behavior.

dling any Service. All Services support the standard Service operations of **abort, pause, resume, clear, status, configure, shutdown, enable,** and **disable** (though a Service may do nothing with some of these instructions, it must return proper acknowledgment and completion responses). Each Service's Capability Dataset includes the specific operations of the Service. State diagrams differ for synchronous and asynchronous services (Fig. 5.5) by the omission of the *pending* state between the *idle* and *busy* states for synchronous services. The controller uses the synchronous indicator to speed execution by grabbing a response immediately from a synchronous Service's execution, thus bypassing the Service's message queue and not waiting for the next time in the round-robin cycle to read the response. Error and Pause states are identical to those in the CI. Like the CI and TM, Services verify commands for appropriateness to the Service's Operating, Error, and Pause states.

Tasks Enable Last-Minute Binding of Processes and Laboratory Configurations

Because of the flexibility to switch configurations and processes, the user indicates these when starting a run. The user tells the controller which Services are currently available to be used. The user indicates which process to execute. For convenience, a user ties a configuration and a script together in a "task." To make a new task, the user selects the first step of a stored script and a subset of the predefined Services. This can be visualized as making choices from a menu of existing scripts and Services, with the additional convenience of storing and recalling the choices.

Script Interpretation and Execution by the Controller

Once the CI has received the "start *taskname*" command from the control port and built the TM from the task description in the database, the TM configures the Services and becomes the driver for all activity. Because we are simulating parallel processes on a single-process machine, control is passed round-robin for short time slices among these functions:

1. The CI reads commands from the users. If any, it puts them onto the TM's work list. The only commands allowed during the execution of the task are **abort, pause, resume, status,** and **clear. Disconnect** is not technically allowed but cannot be stopped. If a **disconnect** is received, a cleanup routine inside the CI sends **abort** to the TM and stops an executing task.

2. The TM reads its work list. If an instruction is present, it is executed.

3. The TM reads from the queue of steps that are ready to run, matching them to keywords or capabilities and putting them on the work list of the appropriate Service. To keep as many steps queued as the script parallelism allows, multiple steps are read, matched, and dispatched.

4. The TM gives each Service a slice of time to execute commands from its work list, to change its Operating or Error status, and, for physical services, to read and to process responses from the bridge code.

5. The TM reads the responses from each Service's message list. If the response is for a successful completion, the TM marks the step just completed by that Service as "done." If there were problems in the completion of a step, the TM begins error handling.

6. The CI reads, processes, and displays messages to the user from the TM's message list.

When the script finishes (signaled by a script step whose operation is **end**), the cycle continues, but the TM issues shutdown instructions rather than operations from script steps.

The marking of a step as "done" is worth describing. Step classes have subroutines for "start," "done," and "error" attached to them. The "start" subroutine in turn has two subroutines. The "start" routine always calls "ok-to-start," which decrements a step's start-count (Fig. 5.2). When a step's start-count reaches zero, all required immediate predecessor steps have finished, and the step can be released to the TM for matching and dispatching to a Service. This release is done by the second subroutine of "start," "add-to-ready-to-run-queue," by adding the step to the TM's queue of ready-to-run steps, and the step's start-count is reset to its initial value.

The step's "done" subroutine issues "start" to all steps named in the yes-set or no-set, depending on the return code set by the step's operation.

The step's "error" subroutine issues "start" to all steps in the error-set, if any; otherwise, default error handling is provided by the TM. If a known fatal error occurs, such as the attempted execution of a step whose operation does not match any keywords or any capabilities, the TM informs the user through the CI and aborts the process. For most errors, though, the default error handling pauses the system, informs the user of the problem, and asks whether or not to continue. If told to abort, the TM aborts the task. If told to continue the task, the control software tries to reexecute the step whose operation failed. For example, the level of the wash bath can become too low during the course of a gridding run because of evaporation, and the **wash** command will fail. The system is paused for safety since the user will have to enter the system envelope to refill the bath. The TM does not issue any more commands to the Services until the control is explicitly returned to the automated system by the user.

Step operations can use input and output arguments. The TM maintains the task's Global Argument List through the argumentManager (Fig. 5.4). When an operation has an input variable, the argumentManager searches the Global Argument List for a variable by the same name, and copies the master value for that variable from the Global Argument List to the step's input argument. When an operation has an output variable, the argumentManager searches the Global Argument List for a variable by the same name. If there is one, the argumentManager copies the step's output argument value to the Global Argument List, to become the new master value for that variable. If there is no variable by that name in the Global Argument List, one is created, with the value indicated by that step's output argument value. Variable values can also be set by the keyword **setglobal,** which establishes a new variable and its value or changes the value of an existing variable in the Global Argument List.

Keywords such as **call, end, do,** and **enddo** use a stack to indicate the step to **return** to, or the top of the **do** loop for the **enddo** instead of using the yes-set, no-set, or error-set. Parallel branches in the script can be doing stack manipulations at the same time if each path has one of these four operations. Because of this, each path maintains its own stack, passing it from step to step (Fig. 5.2). Called subroutines give organization and reusability to a script. The script writer gives a step's **call** operation the name of the first step in the subroutine script. The subroutine script ends in a step with a **return** operation. By using the temporary memory in a step's stack instead of permanently coding the next destination in the main script, frequently used routines can be written once and called over and over without modification.

A SAMPLE SCRIPT AND ITS EXECUTION

In Fig. 5.3 the task "grid" calls for many Services, and three are shown in detail—the robot, the wash station, and the membraneLocationServer. The robot is a synchronous, off-the-shelf machine with a standardized Service interface. The wash station is an asynchronous, custom-built machine, also with a standardized Service interface. The membraneLocationServer is a synchronous, custom-built data server. We will follow the sample script for the task grid (Fig. 5.6).

After the TM has read, initialized, and received confirmation that all the Services listed in the task are configured, the TM's Operating state changes to *executing*, and the first step of the script is read and started (Fig. 5.7).

The TM reads step S1 from the ready-to-run step queue. The TM matches step S1's operation, **initreport,** to the reportServer and issues the **initreport** instruction to that Service.

Successful completion of **initreport** sets step S1's return code to yes, and the "done" procedure for step S1 issues a "start" for each step in step S1's yes-set of steps. Steps S2, S3, S4, S5, S6, S7, and S8 are started. The TM matches their **setglobal** operations to the keyword server built into the TM and issues these instructions to the keyword server. When steps S2-S8 are done, the Global Argument List contains the declared and initialized variables. As each of steps S2-S8's **setglobal** operations finish successfully, the "done" procedure for each step invokes the "start" procedures for each step on the yes-set. Each yes-set in Steps S2-S8 points to step S9.

But step S9 should not run 7 times. Step S9 has a start count of 7. When S9 receives "starts" from all seven steps S2-S8, only then is step S9 placed on the ready-to-run step queue. The TM takes step S9 off the ready-to-run step queue and matches the step's operation, **call.**

When the subroutine **washtools** is completed, its **return** step pops the top name (here, step S9) off its stack and issues the "done" for S9.

Step S10's operation, **presentplate,** can have two successful responses: no (no more plates in the input stack), or yes (a plate was available and its lid removed and the open plate presented to the robot envelope). When the "done" procedure runs for step S10, the step's return code is set to no or yes.

If step S10 resulted in the no-set being started, step S11's operation **info** is a keyword that displays a message on the user interface and can take a yes/no response from the user. Here the script writer asks the user whether or not plates have been added for the script to continue. If not, the gridding run ends by calling steps S12 and S13. When both of these steps complete, the **end** keyword in step S14 signals the control software that the script is finished. The control software can begin the standard shutdown steps required to finish the task and prepare to accept another task.

```
Task name:  grid
Services required:      robot
                        washstation
                        platestacker
                        barcode reader
                        report
                        membraneLocationServer
First step of script:   S1
```

Fig. 5.6 *Task description for gridding.*

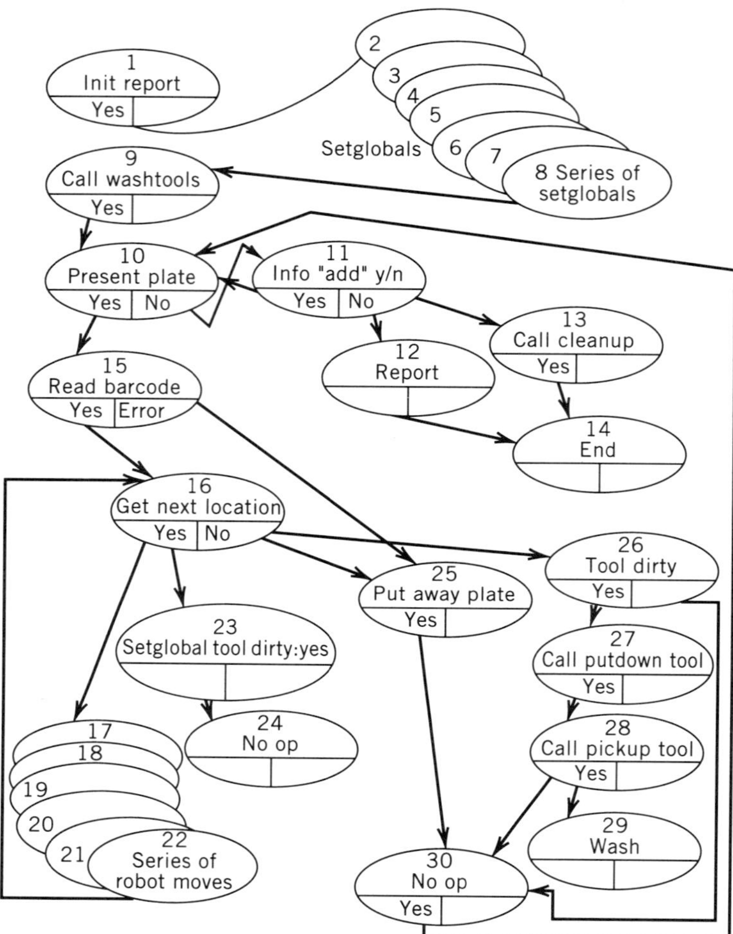

Fig. 5.7 *Example of a gridding script.*

When there are plates available to grid, however, the user's response of yes to the **info** query sets step S15 is started. The **readbarcode** operation rejects unreadable barcodes as errors. Not wishing to use the default error handling provided by the TM, the script writer gives explicit instructions in the script for error handling if step S15 encounters an error. The TM's default error handling requires the user to clear errors through the user interface. In this case the writer prefers that the system put away a plate and tries to read another one if it cannot read the barcode. Step S15's error-set starts a script path that does this.

Steps S15 and S16 have variables in them. Step S15's **readbarcode** operation returns a string that is parsed into the output variable named *BC*. The argumentManager updates the master value of the variable named *BC* on the Global Argument List.

In step S16, the **getnextlocation** operation takes a barcode as input, and returns a gridding location and a return code. The TM, upon seeing an input variable, calls the argumentManager to supply the master value for the variable named *BC*. The *BC* value is sent to the membraneLocationServer and the gridding location *GL* returned. The argumentManager updates *GL* on the Global Argument List.

Speed is increased when the physical services run in parallel. When a plate has been gridded completely, the tool must be exchanged for a clean one and cleaned, the plate's lid replaced, the plate stored in the used plate stack, and a new plate's barcode read. In Fig. 5.7 the step to cover and put away a used plate (step S25), and the step to put down the dirty tool (step S27), are parallel paths in the script, since the script writer knows the steps can run at the same time. Steps S27 and S28, however, would be senseless in parallel, since one puts down a dirty tool and the other picks up a clean tool. Parallelism in the script comes from the script writer, who understands the true dependencies of the process.

RESULTS

This controller was implemented with the new robotics system in January 1995 and yielded a sixfold increase in gridding throughput over the original LANL gridding robot. It was written in C++, using the object-oriented database ONTOS(R) and was developed under Purify©. Further work planned for the controller includes construction of an editor for scripts and device configurations.

CONCLUSIONS

Our new modular automation controller provided some key advantages:

- Adding parallelism to the gridding system controller pushed the runtime bottlenecks to the real process—the gridding or sequencing task, instead of imposing false single-file execution.
- Intelligent Services with Good Citizen SLM behavior [2] provide known communal behavior through the Standard Command Set while keeping individual capabilities. This permits a uniform control structure that does not need to be changed on addition of more Service modules.
- Storing Service descriptions and scripts in a database enables the control structure to switch between configurations and processes by simply reading a new description from the database.
- Task-level organization provides a simple way for a user to specify the Services needed for the configuration and to point to the first script step describing the desired process.
- Supplying keywords in the scripting language gives the user the convenience and power of a high-level language.

- Interpreting Service operations in the scripting language allows unlimited flexibility by depending only on the capabilities of the Services chosen for the current configuration.

By pushing the complexity and flexibility into the control software, the necessary flexibility and speed were obtained to meet the demands of today's biological and chemical research.

ACKNOWLEDGMENTS

Support was provided by the Human Genome Program of the Health and Environmental Research Branch of the Department of Energy. The complete hardware and software delivery included the efforts of Jerome Chen, Len Stovall, and many others in Tony Beugelsdijk's automation section.

Standard Laboratory Module and SLM are trademarks of SciBus Analytical, Inc. ONTOS is a registered trademark of ONTOS, Inc. Purify is a registered trademark of Pure Software, Inc.

REFERENCES

1. P.A. Medvick, R.M. Hollen, R.S. Roberts, D. Trimmer, and T.J. Beugelsdijk (1992). Automated DNA hybridization array construction and database design for robotic control and for source determination of hybridization responses. *Int'. J. Genome Res.* 1:17–23.

2. P.A. Medvick, R.M. Hollen, R.S. Roberts (1991). Development of an automated workcell for DNA hybridization array construction. *J. Lab. Robotics Automation* 3:169–173.

3. J.M. Griesmeyer, T.D. Urenda, R.M. Pacetti, J.J. Ferguson (1994). A standard control system for modular automation of chemistry. *Lab. Robotics Automation* 6:79–84.

4. T.H. Erkkila, R.M. Hollen, T.J. Beugelsdijk (1994). The standard laboratory module: An integrated approach to standardization in the automated laboratory. *Lab. Robotics Automation* 6:57–64.

5. P.A. Medvick, J.T. Chen, B.K. den Hartog, F.S. Smutniak Submodule approach to building automated systems for mapping and sequencing genomes. *Lab. Robotics Automation,* Vol 7, No. 2, 81–4.

6

GUILE: A Laboratory Automation Software Framework

SCOTT P. HUNICKE-SMITH

CONTENTS

INTRODUCTION

The level of automation at the Stanford DNA Sequence and Technology Center (SDSTC) is typical of automated sequencing labs of similar size; some new and some

Automation Technologies for Genome Characterization, Edited by Tony J. Beugelsdijk.
ISBN 0-471-12806-6　© 1997 John Wiley & Sons, Inc.

custom instruments are available for specialized tasks, each requiring a particular software and hardware interface. Even as these devices' performance and specifications are improved, they remain intended for stand-alone use and their interfaces vary greatly in quality and flexibility. Further it is exceptionally rare to find a complete set of instruments (e.g., sample prep, analysis, postprocessing) which can share a common computer platform (e.g., Macintosh, MS-DOS, MS Windows, or UNIX) and then might have some hope of interoperability. GUILE (Graphical User Interface to Laboratory Equipment) was created within this environment to provide a single, uniform, extensible, and efficient system to take full advantage of dedicated instrumentation. Figure 6.1 illustrates the diverse collection of devices at SDSTC and how GUILE is used among them.

One's perspective defines exactly what GUILE is. An end user will view GUILE as an attractive user interface providing a consistent method of access to defined protocols as well as data associated with and information about these protocols. The protocol developer would see GUILE as an interactive framework to establish new protocols with the ability to draw on library functions for code re-use. The instrument designer or integrator sees GUILE as a flexible socket into which a new instrument can be inserted using provided interface routines and then immediately used. Thus GUILE is not merely a program; it is a collection of tools, a directory structure, and a set of user interfaces that offers many branches for the addition of custom "leaves." The net affect is a tool that can be applied to diverse instruments, developers, users, and purposes.

The core of GUILE is the well-accepted scripting language Tcl [1] and its companion graphical toolkit Tk [2]. This is important because while the Stanford DNA Center does not have the resources to support GUILE directly, the Tcl/Tk community has well over 100,000 users, active Internet discussion, and many companies providing training and support [3]. There are over 390 extension packages freely available for Tcl and Tk that can be linked with GUILE to provide additional functions ranging from advanced mathematics to image processing. There are also 449 freely available applications and many commercial applications written using Tcl and Tk, including GUILE (available via anonymous ftp from **genome-ftp.stanford.edu**). While GUILE is currently only implemented on a Sun workstation running the UNIX operating system, Tcl and Tk are both being ported to MS-DOS/Windows and the Macintosh operating system, and thus GUILE can be ported fairly easily.

This work serves to describe GUILE at a fairly high level while providing concrete examples of its use for evaluation; since GUILE is freely available, it may be further investigated simply by acquiring a copy and reading the detailed documentation. To this end, the chapter is divided into four sections.

STRUCTURE

Regardless of how GUILE is to be used in a laboratory environment, it is important to understand the three "interaction levels" (I levels) available between a user and GUILE. These conceptual levels are (1) source code, (2) script, and (3) graphical user interface.

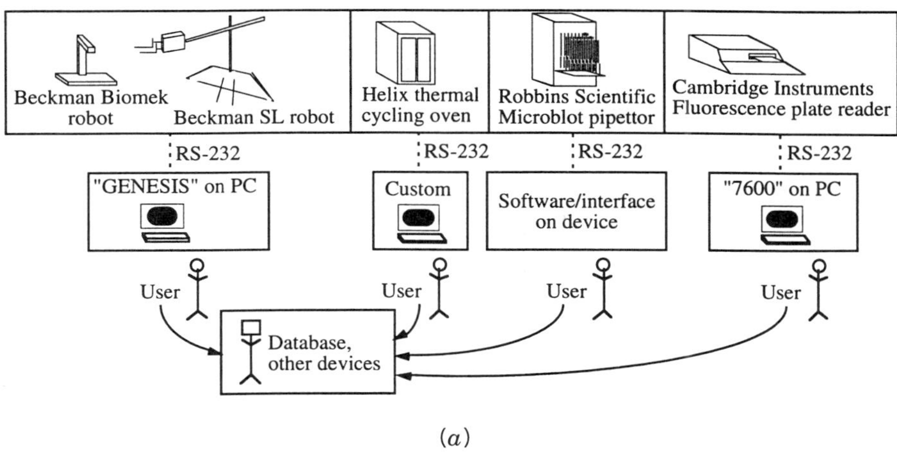

(a)

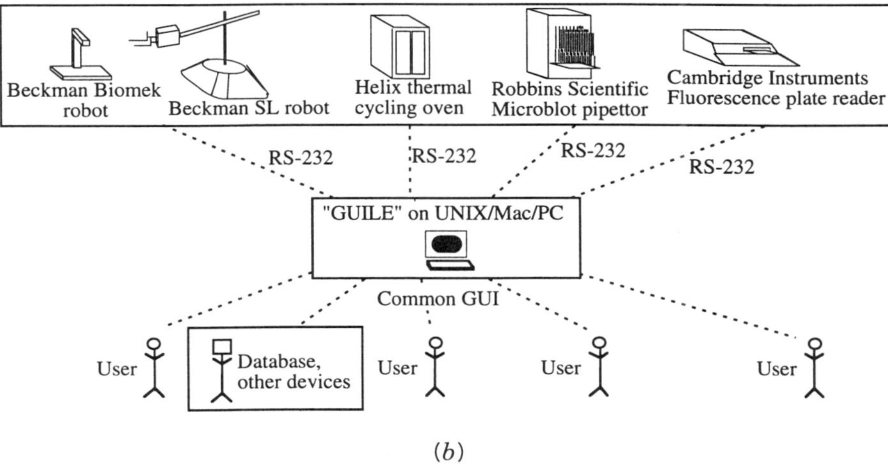

(b)

Fig. 6.1 (a) A typical diverse hardware and software environment; (b) the environment easily unified and standardized by GUILE.

This structure serves several purposes. First, changes made in one level do not affect the performance of other levels, so additions and changes are made more quickly and robustly. Second, modularity is encouraged within each level to further simplify changes and reduce the possibility of problems associated with changes. Finally, many different specialized individuals may work together in creating modules or using the system so that one "super user" is not required for development.

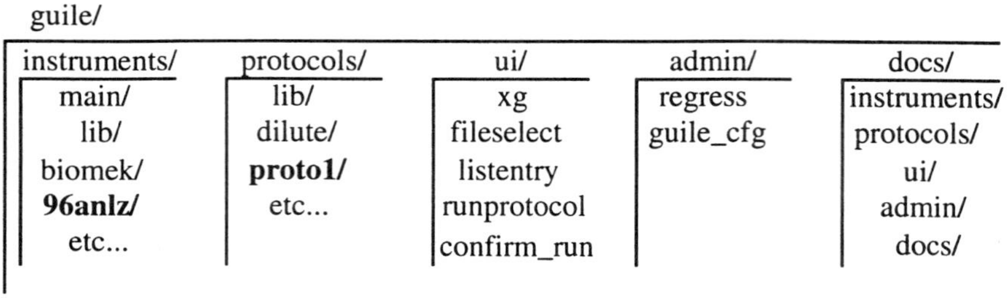

guile/

instruments/	protocols/	ui/	admin/	docs/
main/	lib/	xg	regress	instruments/
lib/	dilute/	fileselect	guile_cfg	protocols/
biomek/	**proto1/**	listentry		ui/
96anlz/	etc...	runprotocol		admin/
etc...		confirm_run		docs/

Fig. 6.2 GUILE file structure; boldfaced items are examples and not included in the distribution.

These three I levels are realized in the directory structure of GUILE (see Fig. 6.2). For source code changes, the **instruments** directory contains all source code and provides means for addition, since instruments tend to be the primary motive for new source code. Scripted programs, typically laboratory protocols, are maintained in the **protocols** directory. User interface scripts, intended for the production user, are kept in the **ui** directory. In addition to these, the **admin** and **docs** directories provide locations for system administration functions and full documentation on all system components. Each of these directories will now be discussed in some detail; this should further convey why this structure exists, what is provided within it, and how it can be best used. Specific examples will be shown in the next section.

Instruments

There are many stand-alone instruments and OEM components available for specific tasks that are designed for computer control. An incomplete but typical list is provided in Table 6.1.

TABLE 6.1   Typical Instruments Suitable for Use with GUILE

Instrument	Trade Name	Manufacturer
1/8-channel pipetting/pin transfer robot[a]	Biomek 1000, 2000	Beckman Inst.
Thermocycler	PE 9600	Perkin Elmer
96-channel pipetting device[a]	Hydra	Robbins Scientific
Single- and multichannel syringe pumps	Various	Cavro, Hamilton
96-well spectraphotometer	SpectraMax 250	Molecular Devices
96-well fluorometer[a]	Microplate Fluorometer	Cambridge Instruments
Pipetting/thermocycling robot	Catalyst 800	Applied Biosystems
Fluorescent sequencer	ABI 373	Applied Biosystems
Multi-port switching valves	XL Smart Valve	Cavro

[a] Devices currently integrated with GUILE.

The instruments listed in the table are used in many genome centers, in methods ranging from individual sample processing to high-throughput, multiple-sample automation. The challenge is to interconnect these devices to form a workable, efficient system. To see how GUILE can connect these devices via I level 1 (source code), Table 6.2 lists some common features and how GUILE provides for them.

The devices selected for illustration are RS-232 instruments, and GUILE currently only provides interface functions for RS-232. However, devices with other communication interfaces can still be added in a modular fashion, though they would require additional low-level code specific to the new interface.

The first step in adding a new instrument in a modular fashion is to create a new subdirectory in the **instruments** directory. Within that directory, one writes low-level code in C or a similar language to establish the instrument's command set within GUILE. This amounts to creating short code fragments that parse a command line, interpret the arguments suitably for each instrument command, and send the information to the instrument. GUILE provides low-level C language RS-232 communication routines (currently UNIX specific) so that the system integrator can begin by immediately selecting how the machine's command set (e.g., "move" or "home" commands for a robot) will be mapped to the command space in GUILE. The most straightforward implementation is to map each instrument command to a GUILE command. As will be seen later, for performance reasons one might choose to combine several instrument commands into one GUILE command, and vice versa. This is convenient because often the functions available in the instrument are not exactly what will be most useful in the lab; they may form a superset or subset of I level 2 (script level) commands.

The code now in that directory should be compiled into a binary library. If a makefile exists in the new instrument's subdirectory, simply typing "make" from the **instruments** directory will cause compilation of the new instrument's source code (note that while many systems provide a "make" function, this implementation is UNIX

TABLE 6.2   GUILE Features Corresponding to Common Instrument Features

Common Device Feature	Easy Integration with GUILE
Commands available remotely via RS-232 interface	Serial communication function library
Command sets are small (usually less than 12 commands)	Commands may be made available, combined, or masked at any of the three levels of interaction
Data are generated or used in their operation	File and user I/O and string parsing/assembly functions are built-in
Instruments may be used in one or multiple operations	Independent scripts may be written for particular tasks (protocols) using same instruments
Instruments may depend on one another	GUILE allows multiple concurrent sessions with interprocess communication or single-session, multi-device implementation

specific). If the library then exists in the instrument's subdirectory, it will be linked into the main GUILE executable (located in the **instruments/main** directory). Since each device's source code is contained within one directory, the code can be easily shared among labs or provided by instrument manufacturers. Each instrument should only reference files in the **instruments/lib** or **instruments/main** directories to maintain modularity.

Once a device's commands have been registered into GUILE's framework by the compile and link steps just discussed, those commands can be evoked by entering the command to the GUILE command line. The arguments to the command are passed to the code written for the device and that code handles device control. GUILE provides routines that allow logging of machine errors to a **log** file both in the instrument's source code directory and in the current protocol directory. If the programmer makes use of the provided RS-232 interface routines and establishes a **connect** command for the instrument, the code will automatically be tested as part of the administration regression test facility **regress.**

For common use in the laboratory, commands would be packaged into a script, or protocol, via I level 2 for the robotic "actors" to execute (see the Protocols section). Once the command set is put into GUILE, low-level changes made to it will be independent of any protocol using those commands, and vice versa. For example, one could enhance a device's performance at the C source code level with no changes to a lab protocol, or write a new lab protocol without ever needing to recompile source code. This serves to keep source code reliable, decreases protocol development time by forcing modularity, and reduces the need for a programmer to do protocol development.

To link multiple robots, a multi-tasking operating system such as UNIX allows several GUILE sessions to run simultaneously and interact with one another via Tcl's internal **send** command at I level 2. One session will serve if the instruments are to be run sequentially and not simultaneously. The key point here is that design of more complicated control between devices is well separated from the core device driver, simplifying implementation and reducing the possibility of problems while running.

Protocols

With devices in place, the next step is to develop protocols. Typically protocols are written as scripts in the Tcl language. This language has all the functionality one would expect from a programming language, including file input and output, conditional and looping constructs, and scalar and array variables. In addition all the commands established for the various integrated instruments are available.

The difficult aspect of creating protocols remains: translating user requirements, which are often vague, into a final scripted program. This is where I level 2, the script level of interaction, can prove invaluable. If devices have been integrated at the source code level to be self-protecting, there is little that a user can do at the script level to damage people or equipment. Aside from the increased user and equipment safety this provides, protocol development can be done interactively by entering commands and verifying results. Successful command series can then be saved as **procedures.** Procedures act exactly as built-in commands but are saved and loaded in modifiable text files. This provides the key to efficient protocol development: creation of a library of commonly used functions with still higher-level procedures acting as individual protocols. Any of these files can be made read-only to allow limited access for protocol revision.

Once library and protocol procedures are developed, GUILE takes over the rudimentary tasks of providing user access to them as well as structure for documentation, on-line help, and data logging. By placing library procedures in the **protocols/lib** directory, GUILE will automatically ensure that they are available to all protocols. Completed protocols should be given their own subdirectory in the **protocols** directory and the protocol and protocol directory must share the same root name. This ensures modularity of protocols within I level 2; a common library with a single protocol file is all that is necessary to exchange protocols, and there should be no interdependence among protocols.

With this structure, GUILE will create three new files in the protocol's subdirectory when started: a **setup** file, a **gen_info** file, and a **log** file. The **setup** and **gen_info** files created are template documents that the user must fill in to describe the setup required for and general information about the protocol. These files are themselves hypertext scripts (still I level 2) that may have buttons linked to further information as needed. The **log** file will contain any logging information generated explicitly during a protocol with the **logfile** command or implicit information written by low-level machine interface code should an error occur. Machine errors are also logged in the instrument's source code directory.

The extensive list of I level 2 GUILE commands simplifies many of the critical details in protocol development. For example, it is quite easy to call external UNIX programs requiring input from the protocol and to use their output within a protocol. These could be data-processing programs, special-purpose data input functions, or, perhaps most important, sample tracking databases. The full power of automation can now be used because the system does not simply replace menial human tasks; it enhances the usefulness of the data being acquired during a protocol or allows for protocols requiring sophisticated individual sample information.

This last point of open, intelligent protocols is perhaps the most important single aspect of GUILE. Even if an equipment vendor or designer were to develop the perfect automation system for the genome project today, it may be superseded shortly thereafter. This is due to continuously changing sequencing and mapping strategies and protocols; in the drive to reduce cost while increasing speed and maintaining accuracy, strategies will become more sophisticated, and data exchange throughout any process will be more critical. GUILE can accommodate both the rapid changes and the increased data exchange necessary as the genome project progresses.

As mentioned earlier, GUILE also provides the Tk graphical toolkit via I level 2. With the ability to quickly create graphical widgets in I level 2 such as list boxes, buttons of various sorts, and text entry and display elements, protocols can be overlaid with point and click interfaces. This presents a much more appealing interface to the end user or developer (at I level 3), limits damage that can be done by incorrect entries, and lowers the barrier to a new system for most users. Some generally useful, protocol nonspecific GUILE user interfaces written with this tool kit will be presented in the next section.

User Interface (ui)

The **ui** or user interface directory contains one main application and many generic tools. These user interfaces, or I level 3, provided by GUILE tend to be minimal and generic. This is to promote customization necessary for acceptance of GUILE in a lab environment. The danger in this minimalist implementation is in loss of modularity; since it is so simple to create interfaces, one can easily create extensive, nonmodular interfaces unique to particular tasks instead of investing in short, generic, modular units.

There is only one primary interface to GUILE named **guile.** It simply presents a list of all available protocols in the **protocols** directory and allows the user to select one to run or show setup or general information. If the **connect** command is created for instruments involved in the protocol and the provided RS-232 commands are used, the protocol may also be run in test mode, which executes all the protocol's commands without actually communicating with the device. This is very handy for debugging. The other tools available and their functions are listed in Table 6.3.

This set of tools is intended to be sufficient to allow one to create and run protocols as quickly as possible after installing the system. By examining the files (all these tools are themselves simple textual scripts), it is easy to see how to modify them or create new interfaces that are more suited to the particular lab environment. Again, the danger is in creating new interfaces that are not modular and become "code dinosaurs".

Admin

As for the user interfaces, the administration section contains a minimal set of tools for administering GUILE's interfaces, protocols, and instruments. A regression test script **regress** is provided that will automatically run all recently modified protocols in test mode so that the code is tested without actually running any of the attached instruments. Tests of all protocols will be made if the source code in the **instruments** directory is newer than any of the protocols available. Recall, however, that this test mode is only available on instruments that have implemented the **connect** command and use the provided RS-232 communication routines.

Also provided in the **admin** directory is a global configuration script **guile_cfg.** By modifying this script, an administrator can set global variables defining the location of

TABLE 6.3 UI Tools and Functions

Tool	Function
fileselect	Presents a list of files present in a directory or from a file and allows the user to select one or more
listentry	Presents entry templates for the user to enter one or more text items
runprotocol	Given a protocol name, runs a protocol and displays information as protocol runs
confirm_run	Presents a list of information and asks user to confirm; also has buttons for general info and setup info access

GUILE's protocols, primary program, and user interfaces, as well as any locations of data files necessary for particular protocols. Multi-user systems simply need to have each GUILE user reference this configuration script to correctly set their environment to run GUILE.

Further tools that might be useful would be a revision control system for instruments and protocols, a user log specific to the actions of privileged users, or graphical extensions of the tools provided.

Docs

Since it is unlikely that one person can administer and use all parts of the GUILE system, documentation has been somewhat localized to its point of need. Each protocol subdirectory should contain its own documentation on how to set up and run that protocol as well as general information about the protocol. Since the documents are hypertext, links to separate documentation on library routines, instruments, or other general information can be incorporated. Further, there is data logging of both protocol operation and device failures if implemented by the user at I level 1, so all system use is documented. Source code for instruments should be documented within the code with a separate **overview** file to describe the general code structure.

The **docs** directory itself contains a mirror image of the GUILE directory tree. Within each directory are text files generally describing the directory's function and contents (**overview** files) as well as any detailed information on what the user/programmer should do in that directory (**details** files). Since this chapter is intended as a more descriptive overview to GUILE, these directories should be consulted for full details on any topic.

EXAMPLES

The above discussion gives but an abstract overview of several aspects of the GUILE system. Taking a simple example device to illustrate these concepts, we will now show how to implement device integration, protocol creation, and user interface addition. Figure 6.3 schematically illustrates the process. The device will be a simple 96-well microtiter plate analysis system. While similar to several commercially available devices, this device is fictitious and has a set of commands representative of almost any laboratory device.

Integrate a New Machine

Suppose that the device has the command set shown in Table 6.4. Communication with the device always begins and ends with special characters. Some commands return data and some do not, but all at least acknowledge completion of a command.

To integrate this device, one first creates the directory **guile/instruments/96anlz.** Since this device's command set is fairly small and straightforward, each command with the exception of **init** will map to a unique GUILE command. The next step is

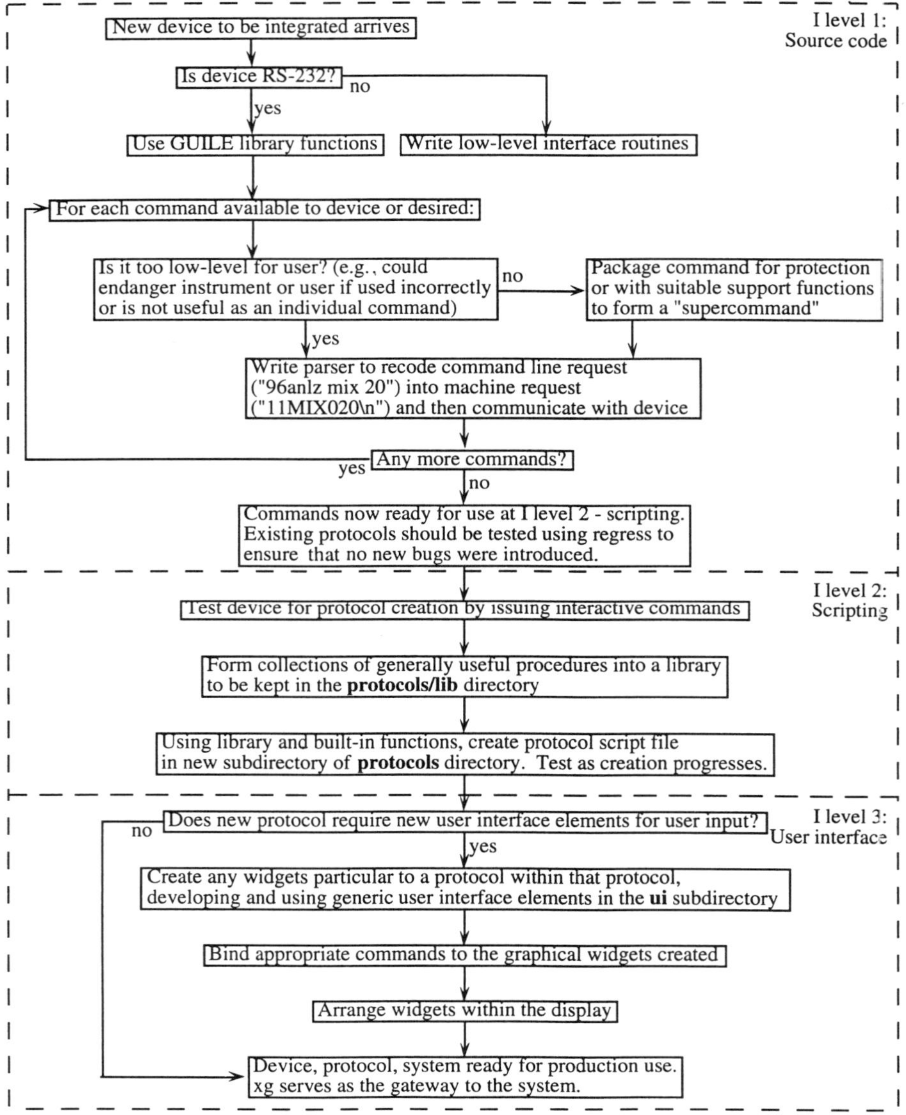

Fig. 6.3 *Schematic of development and integration for new device, protocol, and user interface.*

TABLE 6.4   Example Device Command Set

Command	Arguments	Function	Data Return
init	None	Initializes device	Yes; error codes if exists
load	None	Loads plate to analyze	Only if failure
add	Volume, microliters	Adds bulk reagent to plate before analysis	Only if failure
mix	Time, milliseconds	Mixes plate for time given	Only if failure
read	None	Analyzes all wells individually	96 integers in ASCII, tab delimited, or 0 on error
eject	None	Ejects plate from device	Only if failure

to write source code that will take command line arguments like **"96anlz mix 20,"** convert them to correct representations for the device and communicate them to the device, returning information if available. GUILE will recognize the **"96anlz"** portion of that command and send it with the other arguments as strings to the source code. For good form, one function should be written to interpret what machine command was requested (**load, mix,** etc.). This function must have a particular form as specified in GUILE documentation. Separate device functions are then written in separate files to check the user input, interpret it, communicate with the device using the provided RS-232 commands, and return any results specific to that command. Figure 6.4 shows what this code would look like for the **load, mix,** and **read** commands. Note that these commands should be written such that, if the device is not **connected,** all error checking of user input is done without communicating with the device. This requires special attention when information is expected back in response to a command.

There must be two additional commands added to interface with GUILE: the **96anlz connect** and **96anlz disconnect** commands. The **connect** command should associate a logical file with the device's RS-232 port and then initialize the device, in this case by sending the **init** command and checking its return. Since the logical file is NULL when GUILE begins, commands to a device should still be active but not communicate until after a valid **connect. Disconnect** serves the opposite function in order to close the device's logical file and take care of any other steps necessary to disengage the device. In this example, the user might want the **disconnect** command to check that the **eject** command has been executed if the **load** command was used so that a plate has not been accidentally left in the device. GUILE will also check that **disconnect** has been executed on any connected devices before allowing the user to quit.

With this coding done, GUILE must be made aware of the new **96anlz** command and the global logical file object. To do this, edit the file in **guile/instruments/main** called **add_dev_here.c,** and follow the instructions included. Briefly this entails adding three lines of C code to let GUILE know about the function and its logical file. Assuming that the code works, a **makefile** should be created in the **96anlz** directory to build the library **96anlz.a.** Typing **make** in the **guile/instruments** directory should

```c
int load_96anlz(void) {
   int i, tmp;
   char *sendstr="load", ansstr[2];
   if (96anlzFile==NULL) return(0); /* If dev. not connected... */
   for(i=0;i<strlen(sendstr);i++) { /* send the command */
      send_char(sendstr[i],&tmp);
   }
   send_char(tmp, &tmp); /* Send checksum */
   send_char(ENDXMIT, &tmp); /* Mark end of communication */
   get_data(ansstr, 10); /* Rcve response, timeout in 10 secs. */
   if(ansstr[0]==1) return(1); /* If error... */
   else return(0);
}

int mix_96anlz(int time) { /* time in integer milliseconds */
   int i, tmp;
   char *basestr="mix", tempstr[12];
   if (96anlzFile==NULL) return(0); /* If dev. not connected... */
   sprintf(tempstr, "%s%6d", basestr, time);
   for(i=0;i<strlen(tempstr);i++) { /* send the command */
      send_char(tempstr[i],&tmp);
   }
   send_char(tmp, &tmp); /* Send checksum */
   send_char(ENDXMIT, &tmp); /* Mark end of communication */
   get_data(tempstr, 10); /* Rcve response, timeout in 10 secs. */
   if(ansstr[0]==1) return(1); /* If error... */
   else return(0);
}

int read_96anlz(Tcl_Interp *interp) {
   int i, tmp;
   char *sendstr="read", ansstr[];
   if (96anlzFile==NULL) { /* If dev. not connected... */
      for(i=1;i<=96;i++) { /* then make up response of all 0's */
         Tcl_AppendResult(interp, "0\t", (char *) NULL);
      }
      return(0);
   }
   for(i=0;i<strlen(sendstr);i++) { /* send the command */
      send_char(sendstr[i],&tmp);
   }
   send_char(tmp, &tmp); /* Send checksum */
   send_char(ENDXMIT, &tmp); /* Mark end of communication */
   get_data(ansstr, 90); /* Rcve response, timeout in 90 secs. */
   if(ansstr[0]==0) return(1); /* If error... */
   else {
      Tcl_AppendResult(interp, ansstr, (char *) NULL);
      return(0);
   }
}
```

Fig. 6.4 *Source code in C for **load, mix,** and **read** commands.*

cause the new instrument's command set to become a part of GUILE. GUILE comes with several examples that can be followed to integrate new instruments.

Create a New Protocol

With the **96anlz** functions available, protocols can be created. Suppose that one user has developed a protocol he or she wishes to use repeatedly on the device consisting of a 20-μl addition of reagent to each well in the plate, a 10-second mix, and then a reading. This protocol should be coded in GUILE's scripting language and saved in the directory **guile/protocols/proto1,** for example. The full protocol is shown in Fig. 6.5. Note that since the analysis results are returned as the result of a command, they can be manipulated within GUILE. This would allow mathematical processing of the numbers as well as reformatting of the output. In this example, the raw data are multiplied by a conversion factor and saved in tab delimited format to a file.

To run the new protocol, the user simply runs **guile** from the UNIX command line and selects **proto1** from the menu. GUILE then prompts the user to edit the template **setup** and **gen_info** files created when GUILE first encounters **proto1,** entering information on the setup and general purpose of the protocol. After this first run these files, which are kept in the protocol's subdirectory, will be easily accessible through

```
proc proto1 {} {
    set cnv_fact 15.8; # Set the data conversion factor
    # First, present a  dialog box telling user to place plate.
    gprompt "Hit .any key when plate is in place" tmp
    96anlz load
    96anlz add 20
    96anlz mix 10
    set raw_values [96anlz read]
    set raw_list [split $raw_values \t]; #divide data into an array
    set proc_values ""; # initialize the proc. value list
    for {set i 0} {$i<96} {incr i} {
        set proc_values "$proc_values \
                    [expr $cnv_fact*[lindex $raw_list $i]] \t"
    }
    set fileout [open "output.dat" a+]
    puts $fileout $proc_values
    close $fileout
}
```

Fig. 6.5 *Scripted version of protocol ready for use in GUILE.*

GUILE's graphical interface. Should a user desire to keep a protocol for personal use only, GUILE has the ability to directly start running a protocol file that is not in the GUILE directory structure. Simply type **guile <protocol path and name>**, and GUILE skips the general menu system and starts execution of the given protocol. While this form is not encouraged, it may be useful for less often used or unusual protocols.

Add a New User Interface

Now suppose that instead of a set of fixed protocols, the users want a generic interface that allows them to interactively operate the device. This should allow for loading the plate, adding and mixing the reagent, and finally reading the plate and saving the processed values to a file. The plate should automatically eject after reading. Creating this interface is conceptually a three step process: First, create the graphical elements or widgets necessary; next, bind the appropriate commands to these widgets; and finally, arrange the widgets on the display. The result is a single script file that is placed in the **ui** directory and run simply by entering the file's name on the UNIX command line.

For brevity we give the final script file result for this entire interface in Fig. 6.6, and a black-and-white copy of the interface in Fig. 6.7. Note that binding commands to graphical elements are fairly straightforward, particularly in this case where one graphical element corresponds exactly to one command. Virtually all aspects of the presentation are easily modified, including the color of the widgets, font styles and sizes, and arrangement of widgets.

CONCLUSIONS

There are now several subtle points to consider in a GUILE implementation. First, it is important to understand the various trade-offs in using GUILE in terms of execution speed, development time, and end usability. The lab automation tools used in the genome project are not innately very high-throughput devices (e.g., >10 samples per second) and so do not require a very high-speed control system. At the same time, since the laboratory protocols and sequencing and mapping strategies are progressing so rapidly, development time must be kept to a minimum, and end users must begin using the system even during the development. In this niche GUILE fits nicely. As with most interpreted languages, scripted GUILE code is at least 100 times slower than similar code written and compiled directly into machine code. However, there are advantages bought with scripting:

- Software portability across computers and instrument hardware.
- Code reuse available at two distinct levels—script and source code.
- Protection of users and equipment by limiting their routes of interaction.
- Plain, easy to understand text files for protocols.
- Easy to implement graphical elements to entice users at the GUI level.

```
#################################################################
# Step 1: Create all the graphical widgets
#################################################################
# First make all the buttons
button .b_load -text "Load Plate"
button .b_add_mix -text "Add Reagent and Mix"
button .b_read_save -text "Read Plate"
button .b_quit -text "Quit"

# Now make the entry areas for users to enter values
entry .e_vol -textvar volume -relief sunken
entry .e_time -textvar time -relief sunken
entry .e_file -textvar filename -relief sunken

# and make some helpful labels to identify the entry widgets
label .l_vol -text "Volume in microliters:"
label .l_time -text "Time in microseconds:"
label .l_file -text "Save data in:"

# and finally make a frame to put all the widgets into
frame .outerframe  -height 5i -width 7i

#################################################################
# Step 2: associate commands with the buttons
#################################################################
.b_load configure -command {96anlz load}

.b_add_mix configure -command {
   if {[expr $volume<=0||$time<0]} {
      if {![winfo exists .err1]} {
         message .err1 -text "Volume and time must be > 0" -aspect 550
         pack .err1
      }
      blt_bell
   } else {
      if {[winfo exists .err1]} {
         pack unpack .err1
      }
      96anlz add $volume
      96anlz mix $time
   }
}

.b_read_save configure -command {
   set cnv_fact 15.8
   set raw_list [split [96anlz read] \t]
   set proc_values ""
   for {set i 0} {$i<96} {incr i} {
      set proc_values \
         "$proc_values [expr $cnv_fact*[lindex $raw_list $i]] \t"
   }
   set fileID [open $filename a+]
   puts $fileID $proc_values
   close $fileID
   96anlz eject
}
```

Fig. 6.6 *Script for generic user interface to 96anlz.*

```
.b_quit configure -command {destroy .}; # This should check status before
quit.

##############################################################################
# Step 3: arrange the elements on the display
##############################################################################
# Arrange things into the frame as though they were elements in a matrix:
blt_table .outerframe \
    .b_load 0,0 -columnspan 4 -fill x \
    .b_add_mix 1,0 -rowspan 2 -anchor w -fill x \
    .l_vol 1,1 -anchor e \
    .e_vol 1,2 \
    .l_time 2,1 -anchor e \
    .e_time 2,2 \
    .b_read_save 3,0 -anchor w -fill x \
    .l_file 3,1 -anchor e \
    .e_file 3,2 \
    .b_quit 4,0 -columnspan 4 -fill x

# Now display the frame and it's contents
pack .outerframe
```

Fig. 6.6   *(Continued)*

Note that should specific processes become limiting in speed, they can always be recoded at the source code level. This, however, may reduce the usability or extensibility of that process. In a nutshell, this is the innate trade-off in Tcl/Tk implementation.

Second, besides these trade-offs in performance and implementation, the natural extensibility of GUILE ensures its success. With Tcl/Tk at its core, the system is open at both the source code and script level, and this allows for implementation of one of the most forgotten elements of laboratory automation: sample tracking. Even with the modest-scale automation currently in use, sample tracking is critical to the success of any aspect of the genome project; unlike many other conventional biological research projects, the primary product here is textual data. Each sample contributes its own unique piece of data and must be tracked throughout processing. While many good database systems exist on various computer platforms, forcing users to manually enter data on an otherwise automated process does not ensure a good database. GUILE's ability to interact with the computer system can easily be used to exchange information directly with any type of database. The SDSTC relies on a Sybase database on a UNIX workstation to track sample information. Since GUILE and the database already share

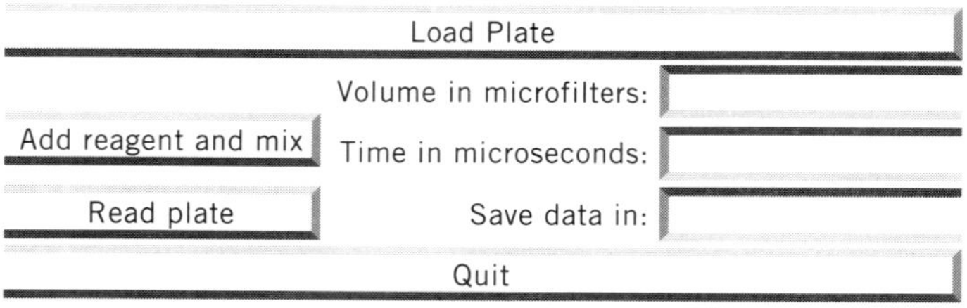

Fig. 6.7   *Black-and-white display of 96anlz user interface as scripted in Fig. 6.6.*

the same computer system, their interaction is as simple as calling an external UNIX command. A protocol can thus package information for the database simply by assembling a valid UNIX command from the abundant string manipulation functions in GUILE. Similarly it can parse the databases response quite easily without any user interaction.

GUILE has been in production mode use at SDSTC since 1994. While it has changed considerably in that time, it has remained stable enough for regular use because the code and protocols are both modular. We have experimented with having another programmer develop instrument interfaces and protocols. This programmer was experienced with the stack-based programming language Forth and had done a little work with procedural languages like C and Tcl but was still able to deliver working protocols for two new instruments in less than two months. The lab technicians have had no trouble learning the system and even making protocol changes themselves.

While GUILE comes prepackaged and ready to use in a proven structure, there is nothing particularly magical about it's arrangement or implementation. It is a bundle of tools for several different levels of interaction and so may be disassembled and scavenged at will. This might prove useful in some circumstances but could make upgrading and software exchange with other labs more difficult. If widely accepted, GUILE could serve as an efficient means for exchange of laboratory protocols and implementation of custom or commercial devices. Here again, the openness of the system could be abused such that portability or modularity are lost; only the contributing user/programmer can control for this.

Perhaps the best way to become further acquainted with GUILE is to install it and try it out. GUILE is freely available via anonymous ftp from **genome-ftp.stanford.edu** in the **pub/guile** directory. Installation instructions are included in that directory (**readme** file). Should someone choose to start with separate installations of Tcl and Tk, these are both available via anonymous ftp from **ftp.cs.berkeley.edu** in the **pub/ tcl** directory. Reference [3] provides an excellent road map to the several hundred extensions and applications written with or for Tcl/Tk. It is hoped that keeping GUILE freely available will increase its acceptance in the biological community. While freely available software is often criticized for lack of support, the size of the Tcl/Tk community can provide abundant support for many key GUILE features. With the power available and ease of use, GUILE can serve well as a fundamental automation tool in the genome project community.

REFERENCES

1. J.K. Ousterhout (1990). Tcl: An embeddable command language. *Proc. USENIX Conference,* Washington, D.C. 133–146.

2. J.K. Ousterhout (1991). An X11 toolkit based on the Tcl language. *Proc. 1991 USENIX Conference,* Dallas, TX 105–115.

3. Anonymous ftp to **ftp.aud.alcatel.com** provides the 5 part Tcl FAQ (frequently asked questions) in the **tcl/docs** directory (internet directions for readers to retrieve information indicated).

III

ADVANCED TOPICS

CHAPTER

7

Capillary Gel Electrophoresis for Large-Scale DNA Sequencing: Separation and Detection

NORMAN J. DOVICHI

CONTENTS

Introduction
Instrumentation
Separation Matrices
    Capillary Gel Electrophoresis—Crosslinked Polyacrylamide
    Capillary Electrophoresis—Noncrosslinked Polymers
    Long Read Length Sequencing by Capillary Electrophoresis
Multiple Capillary Systems
    Scanning Detectors
    One-Dimensional Detector Arrays
    Two-Dimensional Arrays—The Genosequencer
Future Prospects
Acknowledgments
References

Automation Technologies for Genome Characterization, Edited by Tony J. Beugelsdijk.
ISBN 0-471-12806-6 © 1997 John Wiley & Sons, Inc.

INTRODUCTION

DNA sequencing is a fundamental tool in the biological sciences. Two powerful sequencing techniques were reported in 1977, one a chemical degradation method developed by Maxam and Gilbert and the other an enzymatic method developed by Sanger's group [1, 2]. In both cases, a nested set of radioactively labeled DNA fragments is generated in four reactions. The reaction products are separated by size in adjacent lanes of a high-resolution polyacrylamide gel and detected by use of autoradiography. The sequence is interpreted from the pattern of alternating bands in the lanes corresponding to the terminal base of the fragment. Sequence determination by these now classic techniques is an important, albeit tedious, and labor-intensive task.

Advance in sequencing technology occurred in 1986–87 when Smith and coworkers in Hood's laboratory, workers in Ansorge's group, and Prober and coworkers at DuPont reported DNA sequencers that replaced the radioactive labels and autoradiography in Sanger's method with fluorescent labels and laser-based detection [3–5]. In Smith's four spectral channels, single-lane sequencer, four fluorescently labeled primers are associated with the terminating dideoxynucleotide through use of separate chain terminating reactions. Fluorescence, excited by two laser lines and detected in four spectral channels, is used to identify the terminal nucleotide. In Ansorge's single spectral channel, four-lane sequencer, a single fluorescent label is used with each dideoxynucleotide chain terminating reaction, and the products are run on separate lanes of a slab gel; sequence identification is similar to the classic autoradiography technique. In Prober's two spectral channel, single-lane sequencer, one of four fluorophores is associated with the terminating dideoxynucleotide through use of distinctly labeled dideoxynucleotides in a single-reaction mixture. Fluorescence, excited by a single laser line and detected in two spectral channels, is used to identify the terminal nucleotide. Smith's, Ansorge's, and Prober's fluorescence sequencers have been commercialized by Applied Biosystems, Pharmacia, and DuPont, respectively. As an example, the Applied Biosystems model 377A instrument runs 36 lanes simultaneously on a slab gel to produce sequencing rates of 60 bases/hour/lane or 2200 bases/hour/slab. Similar sequencing rates are produced by other instruments. Sequence may be determined, by use of computer algorithms, to about 500 bases, which is limited by the resolution of the gel.

Increased sequencing rate is produced by operating the gels at high electric field strength. Unfortunately, the finite resistance of the separation buffer leads to unacceptable heating of conventional 200- to 400-μm thick gels at electric field strengths much greater than 50 V cm^{-1}. Gel-filled capillaries have attracted interest because their high surface-to-volume ratio provides excellent heat transport properties, allowing use of very high electric field strength, including the use of fields over 1000 V cm^{-1} for DNA sequencing applications. Typical capillaries are 50 to 75 μm inner diameter and 25 to 100 cm long, although early work on RNA separations used cellulose fibers of 10 μm diameter and 25 mm length [6]. Finally, capillaries are quite flexible, which facilitates automation of DNA sequencing and interface with microtiter plates.

INSTRUMENTATION

The first separations of DNA sequencing fragments by capillary electrophoresis were reported in 1990 by groups at DuPont, Utah, Wisconsin, Northeastern, and Alberta [7–11]. These initial reports tended to focus on the serious detection issue associated with capillary electrophoresis. While the narrow diameter capillary minimizes Joule heating during electrophoresis, it also minimizes the amount of sample that can be loaded onto the gel. As a rough estimate, a few hundred attomoles of total DNA can be loaded onto the capillary [12]. Assuming that the DNA is distributed across 1000 different fragments, a few hundred zeptomoles of DNA is present in an average band. Finally, weak bands migrating late in the electropherogram are expected to be present at ~10% of the average concentration. Clearly high detection sensitivity is required to obtain satisfactory sequencing data by capillary electrophoresis.

Zagursky at DuPont modified a standard DuPont sequencer to run with an array of 500-μm ID capillaries; a single laser beam was scanned across the capillaries and fluorescence was detected in a two-color spectrometer [7]. The instrument operated at 50 V cm^{-1}; 9.5 hours were required to separate fragments 500 bases in length. Sequencing accuracy was less than 97% for fragments ranging from 29 to 512 bases in length. Unfortunately, the relatively large diameter tubes used for the separation appear unable to support high electric fields for high-speed separations. Much more interesting results were reported by Swerdlow and Gesteland at Utah, who presented the first high electric field separations of DNA sequencing fragments in capillary electrophoresis [8]. Capillary separations were three times faster, with better resolution (2.4 ×), and higher separation efficiency (5.4 ×) than a conventional automated slab gel DNA sequencing. This single-color instrument used an on-column fluorescence detector for analysis of fragments generated with a single-base termination; the system generated detection limits of 6000 molecules injected onto the capillary. Drossman and coworkers in Lloyd Smith's laboratory in Wisconsin also reported the separations of a single-reaction mixture in gel-filled capillaries [9]. Their system operated at 400 V cm^{-1}, which produced an order of magnitude increase in separation speed compared with conventional slab gels. Their on-column detector could study attomoles of sequencing fragments. Barry Karger's laboratory at Northeastern reported separation of DNA fragments, labeled with a single fluorophore, at 350 V cm^{-1}; base line resolution was obtained for fragments up to 330 nucleotides in length [10]. My research group at Alberta reported the use of the sheath-flow cuvette for detection of sequencing fragments separated by capillary electrophoresis; detection limits of 10^{-20} moles of fluorescent primer were reported [11]. The postcolumn sheath-flow cuvette generates much lower light scatter than on-column detectors, which results in improved sensitivity.

The first high-speed capillary electrophoresis system for DNA sequencing was reported by Lloyd Smith's group in 1990 [12]. They used the ABI fluorescent labeled primers in four sequencing reactions. The products were pooled and separated in a single capillary. A multiline argon ion laser was used to illuminate the narrow capillary simultaneously with light at 488 and 514 nm. Fluorescence was collected at right angles from the capillary. A set of beam splitters was used to direct the fluorescence to a set of four photomultiplier tubes, each equipped with a band-pass spectral filter. This system recorded fluorescence simultaneously in the four spectral channels. At an electric field of 300 V cm^{-1}, 80 minutes were required to separate fragments up to 360 bases in length. A 3%T, 3%C gel was used for the separation. No errors were observed for fragments ranging from 80 to 356 bases in length.

In 1991 my research group reported three different sequencing techniques by capillary gel electrophoresis [13]. In the first method four-color fluorescently labeled primers were used. Fluorescence was excited with two lasers, an argon ion laser operating at 488 nm, and a helium-neon laser operating at 543.5 nm. The two beams were sequentially focused onto the capillary, with the 488-nm beam exciting fluorescence from FAM and JOE and the 543.5-nm beam exciting fluorescence from TAMRA and ROX. The fluorescence was collected with a single microscope objective and discriminated with a filter wheel. A 6%T, 5%C polyacrylamide gel was used for the separation at 150 V cm^{-1}. Sequence was identified for fragments over 550 bases in length in a two-hour separation. A second detector was built for the DuPont fluorescently labeled dideoxynucleotides. A 488-nm argon ion laser excited fluorescence, and a two-channel direct reading fluorescence spectrometer discriminated fluorescence between the dyes. The poor incorporation rate of the highly modified dye labeled deoxynucleotides, along with the severe spectral overlap, lead to poor sequencing performance. A 4%T, 5%C polyacrylamide gel was used at an electric field of 465 V cm^{-1}. A third instrument was built for the four-level peak-height encoded sequencing technique. A helium-neon laser operating at 543.5 nm was the excitation source for a single-channel fluorescence spectrometer. TAMRA labeled primers were used for the sequencing reaction. The four-level sequencing protocol required excellent electrophoretic separation between adjacent peaks. Any overlap between the peaks leads degraded the estimation of peak height, leading to poor sequence accuracy.

Also in 1991 another sequencer was reported for the ABI fluorescently labeled primer system. Ray Gesteland's group at the University of Utah used a spectrograph and charged coupled device (CCD) camera to resolve fluorescence from the four dyes [14]. At an electric field of 190 V cm^{-1}, roughly 60 minutes were required to separate fragments up to 340 bases in length in the 4%T 5%C gel-filled capillary. Sequencing accuracy of 96.5% was observed for fragments ranging from 60 to 350 bases in length.

Four-color sequencing was challenging to implement in capillary electrophoresis. The small amount of sample and challenging optics limited the sensitivity of the early four-color instruments. As a result sequence read lengths were limited to ~360 bases. My group reported a two-color, peak-height encoded sequencing protocol in 1992 [15–18]. In this technique fluorescein-labeled primer is used with ddA and ddC in a 3:1 ratio for one sequencing reaction, while rhodamine-labeled primer is used with ddG and ddT in a 3:1 ratio for a second reaction. The pooled products are separated on a single capillary equipped with a two-color fluorescence detector. Large peaks associated with fluorescein are A, small fluorescein peaks are C, big rhodamine peaks are G, and small rhodamine peaks are T. Sequence information is based on both spectral and amplitude information. The technique has been used to generate over 550 bases of sequence in capillary electrophoresis [16, 18]. Fluorescent label may be incorporated with either dye-labeled primers or dye-labeled *deo*xynucleotides [15]. These modified deoxynucleotides are very inexpensive and represent a simple method for the incorporation of fluorescent label into sequencing reaction products. The accuracy of the two-color, peak-height encoded sequencer was evaluated with several clones from the malaria genome. Sequencing accuracy ranged from 97.7% to 99.6% for read lengths ranging from 500 to 600 bases in a two-hour separation; both read length and accuracy were superior to those produced by commercial sequencers [16]. Alternative two-dye sequencing protocols have been reported based on binary coding [9–20]. One based termination was performed with the first dye, the second termination was done with the second dye, the third termination reaction was performed with both dyes, and the fourth base was null. Unfortunately, the use of a null signal to encode the presence of a nucleotide produces poor sequencing accuracy [21]. The binary coding approach has not achieved sequence read length of over 350 bases.

Steve Soper has recently applied near-infrared emitting dyes to DNA sequencing by capillary electrophoresis [22]. The dyes have much better photostability than fluorescein-based dyes and the near-infrared portion of the spectrum has relatively low fluorescence and Raman scatter background, which improves detection limits. Although only one dye has been used in a peak-height encoded sequencing protocol, the combination of similar dyes should result in a powerful tool for two-color, peak-height encoded or four-color DNA sequencing.

SEPARATION MATRICES

A key characteristic of an electrophoresis system is the sequencing read length. The number of primers required for directed sequencing strategy and the number of templates generated and sequences in shotgun sequencing are inversely proportional to the read length. Long read length minimizes sample preparation. More subtly, long reads minimize the computational effort required to assemble shotgun generated data into finished sequence. Finally, long read length allows sequencing of difficult templates that contain long stretches of repeated sequence.

Capillary electrophoresis can operate at very high electric fields. There is little data in the literature on the behavior of DNA during separation at these high fields. As a result it has been necessary to perform a number of fundamental studies to characterize DNA separations in capillary electrophoresis. A few results are summarized here.

Capillary Gel Electrophoresis—Crosslinked Polyacrylamide

Early use of capillary electrophoresis focused on the use of crosslinked polyacrylamide at high electric fields to produce rapid separations of DNA sequencing fragments. Unfortunately, operation of gels at high electric fields decreases the read length. The phenomenon of biased reptation with orientation causes longer fragments to migrate with identical mobility, destroying the separation for those fragments [23–25].

We have studied the effect of electric field on the separation of DNA sequencing fragments for both crosslinked and noncrosslinked polymers in capillary electrophoresis. For longer fragments, the mobility is given by

$$\mu \approx \chi \left[\frac{1}{N} + \frac{1}{N^*} \right] = \chi \left[\frac{1}{N} + \alpha E \right]$$

where μ is the mobility of a single-stranded piece of DNA containing N bases, χ is a constant related to the free solution mobility of DNA, N^* is the fragment size for which biased reptation with orientation becomes significant, α is a constant, and E is electric field. The onset of biased reptation with orientation is inversely proportional to electric field; the number of bases determined in a separation decreases inversely with field strength.

Figure 7.1 presents a summary of our experimental results for 4%T Long-Ranger gels in capillary electrophoresis. As the electric field goes to very high values, the separation time becomes very short. As an extreme example, we have determined 200 bases of sequence in 3.5 minutes from injection at an electric field of 1200 cm^{-1} [26]. Biased reptation with orientation limits the read length at this high field to only 200 bases.

The migration time for a given fragment decreases inversely with electric field. Similarly the read length decreases inversely with electric field. As a result there is a trade-off between separation time and read length. Figure 7.1 presents a plot of the minimum migration time versus read length for this gel and temperature. In general, the minimum possible migration time increases quadratically with sequence length because lower electric fields are required to obtain longer sequence. The asterisks show our experimental results, while the smooth curve summaries our theoretical prediction.

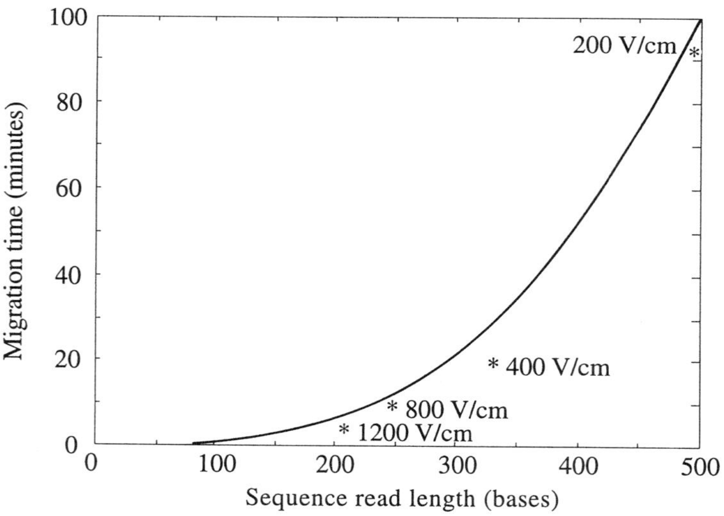

Fig. 7.1   Trade-off between separation time and sequence length for 35-cm long 4%T LongRanger gels at room temperature. The smooth curve is the result of a model, while the asterisks denote experimental results.

The total monomer concentration influences the onset of biased reptation with orientation. Figure 7.2 presents a plot of mobility versus fragment length for DNA sequencing fragments at an electric field of 800 V cm^{-1} and room temperature. This high field was employed to accentuate the effects of biased reptation. Note that the mobility of longer DNA fragments is essentially independent of fragment length for the 6%T material, while the mobility monotonically decreases for the 3%T material for fragments up to 600 bases in length. This continuous decrease in mobility with fragment length allows long sequence read length for low %T polymers [26].

Long-Ranger gels are also stable at elevated temperatures. We have evaluated the gels for separation of sequencing fragments at temperatures ranging from 23°C to 50°C and at an electric field of 300 V cm^{-1}, Fig. 7.3 [27–28]. The mobility of the sequencing fragments increases by 2.2%/°C. Separation at 50°C is roughly twice as fast as separation at room temperature. Roughly 500 bases of sequence was obtained in 50 minutes at 40°C [28]. Unfortunately, read length does not seem to improve with temperature for the crosslinked polymer at an electric field of 300 V cm^{-1}. We have obtained sequence from fragments 570 bases in length in a room temperature separation at an electric field of 200 V cm^{-1} [29].

Capillary Electrophoresis—Noncrosslinked Polymers

The use of crosslinked polymers in capillary electrophoresis leads to a serious limitation: The entire capillary must be replaced when the separation medium has degraded. This replacement can be quite tedious because of alignment constraints of the optical system

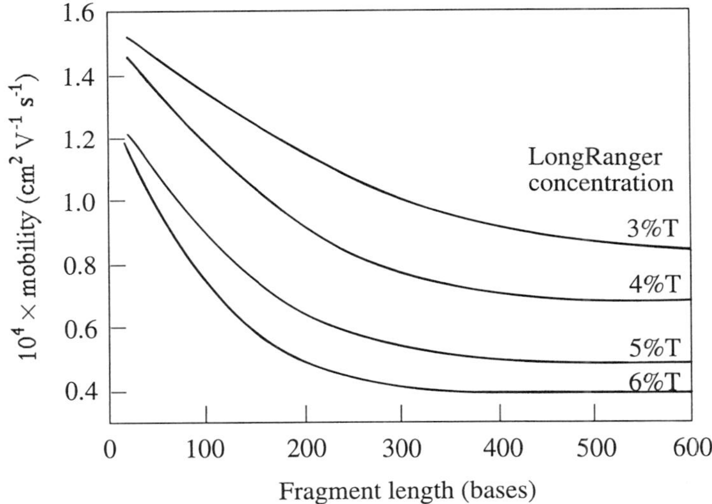

Fig. 7.2　Mobility for DNA sequencing fragments of 800 V cm^{-1} and room temperature. The smooth curves are experimental results for the indicated LongRanger concentration.

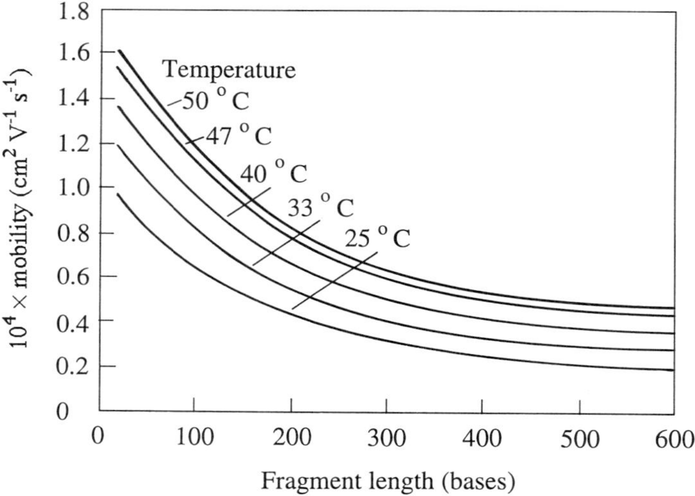

Fig. 7.3 *Mobility as a function of temperature for 5%T LongRanger gels at 300 V cm^{-1}. The smooth curve is generated by data at the indicated temperature.*

with the narrow diameter capillaries. Instead, it is attractive to use low-viscosity polymers for DNA sequencing by capillary electrophoresis. Low-viscosity polymers may be pumped from the capillary and replaced with fresh matrix without replacement of the capillary or realignment of the optical system.

It may be useful to review the terminology used in describing polyacrylamide. The weight percentage of monomer plus crosslinker in the polymer is denoted %T. The mole fraction of crosslinker to monomer is denoted at %C. Polymers with no crosslinkers are called *entangled polymers, syrupy solutions, linear polymers,* or *noncrosslinked polymers.* We prefer the latter terminology.

Bode first used noncrosslinked polyacrylamide in combination with agarose for separation of proteins and double-stranded oligonucleotides [30]. Crambach and co-workers compared the performance of 0%C and 5%C gels at various total acrylamide concentrations [31–33]. More recently Karger and coworkers compared 0, 0.5, and 5%C gels for separation of double-strand DNA in capillary electrophoresis [34]. Bocek and coworkers reported that 0%C polyacrylamide could be pumped from a capillary through use of a special high-pressure syringe [35]. However, Righetti has demonstrated that the high viscosity of 10%T noncrosslinked polyacrylamide makes refilling of capillaries extremely difficult [36]. Guttman and coworkers reported the use of noncrosslinked polyacrylamide for separation of double strand DNA; they claimed that the capillary could be reused 100 times without replacement of the separation medium [37]. We have demonstrated that a capillary can be reused 18 times without replacement of the separation medium for DNA sequencing [38]. To reuse the separation medium, care is required to minimize the loading of template and to avoid the loss of ions from the capillary tip due to differences in transference number.

Polyacrylamide is not the only noncrosslinked polymer used for DNA sequencing. Yeung has reported the use of noncrosslinked polyethylene oxide [39]. This material is commercially available and, as a result, should generate more reproducible separations compared with locally prepared polyacrylamide. Also of interest, Yeung reported that a simple acid-wash can be used to treat the capillary walls to eliminate electroosmosis. This development is quite exciting, and it will be very interesting to see the sequence read length and sequence speed that are produced by the novel material.

There have been applications of noncrosslinked polyacrylamide for separation of DNA sequencing fragments. Unfortunately, early reports on the use of noncrosslinked polyacrylamide were disappointing. In one report less than 200 bases of sequence were determined with 9%T polyacrylamide [40]. Pentoney and coworkers reported DNA sequencing in 10%T, noncrosslinked polyacrylamide at an electric field of 300 V cm^{-1}; sequence could not be determined for fragments longer than 300 bases [19]. Similar results have been reported by Mathies's group with 9%T noncrosslinked polyacrylamide gels [20]. Karger and coworkers have shown separation of fragments 350 bases in length with 6%T noncrosslinked polyacrylamide, again at an electric field of 300 V cm^{-1} [41]. These early reports on the use of noncrosslinked polyacrylamide appeared to use high concentration of polymer and high electric field. It appears that this combination results in short read lengths.

We have investigated the effect of electric field on the onset of biased reptation with orientation for DNA separations in noncrosslinked polyacrylamide. As in the crosslinked material, N^* increases linearly with electric field; low electric fields are required to obtain long sequencing read length. We have also investigated the mobility of DNA fragments as a function of %T; as with the crosslinked material, longer sequencing read length is obtained for low %T noncrosslinked polyacrylamide, Fig. 7.4 [26]. These results have been confirmed by Terabe and coworkers [42–43].

Finally, we have investigated the effect of temperature on DNA separations with noncrosslinked polyacrylamide, Fig. 7.5. Unlike the behavior of the crosslinked material, the onset of biased reptation with orientation is delayed at higher temperatures. Longer sequencing read length is obtained at elevated temperature with the noncrosslinked material.

Operation of gels at elevated temperatures increases the separation speed. The viscosity of the medium decreases by about 2% per °C, which produces a proportional increase in mobility and decrease in separation time [27–28]. Separation at 60°C should be roughly twice as fast as separation at room temperature.

More important, separation at high temperature improves accuracy. Determinate errors occur in DNA sequencing due to the secondary structure of the single-stranded fragment, which can fold back on itself and self-hybridize to generate a hair-pin structure. This structure migrates faster than relaxed DNA, resulting in overlapped electrophoretic peaks, which are called *compressions*. These compressions can be relaxed fully through separation at elevated temperatures in the presence of a denaturing reagent, most commonly 7-M urea. In capillary electrophoresis, formamide has been used as an additional denaturant in room temperature sequencing to reduce but not eliminate

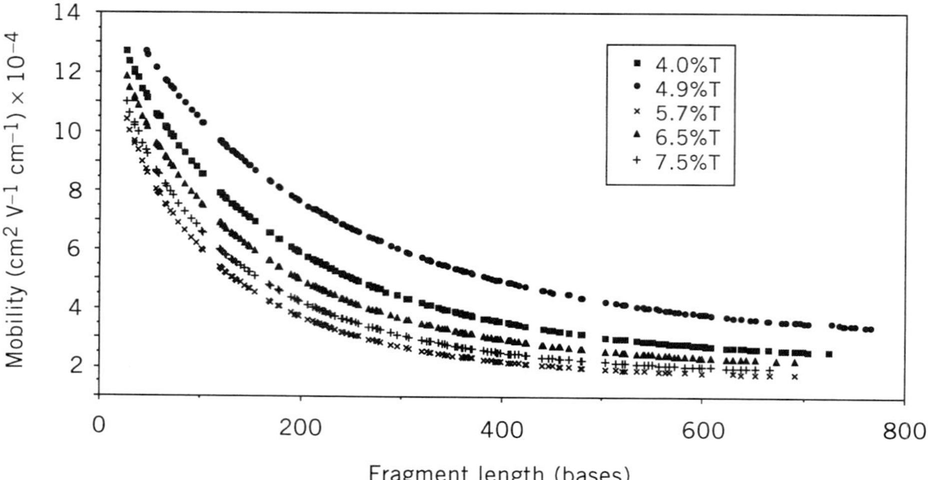

Fig. 7.4 *Effect of total monomer concentration on DNA mobility for noncrosslinked polyacrylamide at 200 V cm⁻¹. The symbols are experimental data generated at the indicated polymer concentration.*

the effects of compressions [13, 44]. Compressions apparently can only be avoided by operation of the capillary at high temperatures. We have found that separation at 60 to 70°C at the use of 7-M urea essentially eliminates compressions in our capillary electrophoresis data [29].

Long Read Length Sequencing by Capillary Electrophoresis

The results of the previous section may be summarized here. Long sequencing read length is produced by use of low electric fields, low %T polyacrylamide, and the use of high-temperature separations. The use of high temperatures and low %T polyacrylamide are also advantageous for high-speed separations; the only apparent trade-off comes in the use of high electric fields, which produce high-speed separations but decrease the read length. We have chosen a simple and arbitrary condition to deal with this trade-off. We strive to maximize the read length that is achievable in a two-hour separation; this period is sufficiently short that several runs can be performed per day, accelerating the optimization cycle. In a fully automated instrument, this separation period allows up to 10 sequencing runs per day in an automated instrument.

In 1994 we reported the separation of fragments 570 bases in length in two hours by use of 6%T noncrosslinked polyacrylamide at an electric field of 200 V cm⁻¹ and at room temperature [18]. It has become clear that use of lower %T and separation at higher temperature would improve the separation speed, while a decrease in electric field would increase the read length. In 1995 we reported the separation of fragments 640 bases in length in two hours by use of 5%T noncrosslinked polyacrylamide at an

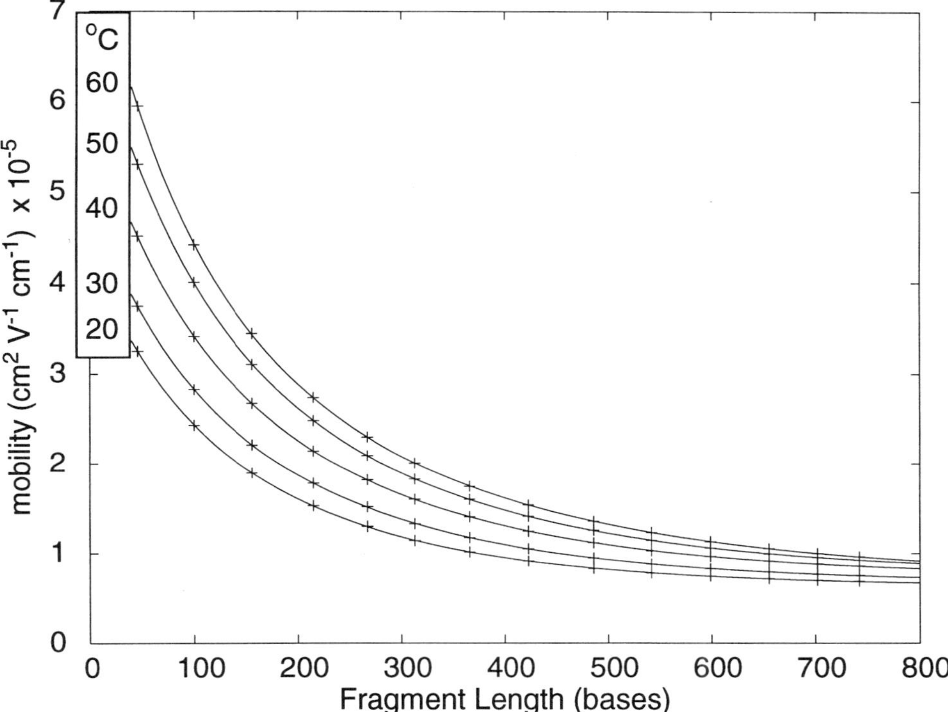

Fig. 7.5 Effect of temperature on DNA mobility in 5%T noncrosslinked polyacrylamide at an electric field of 150 V cm⁻¹. The smooth curve is the mobility generated at the indicated temperature.

electric field of 150 V cm^{-1} and at 60°C [29]. This report was the first demonstration of the use of noncrosslinked polyacrylamide at elevated temperatures. There had been a fear that the low-viscosity material would extrude from the capillary at elevated temperatures. However, by carefully coating the interior capillary wall, any residual electro-osmotic pumping does not pump the material from the capillary.

Very recently we have obtained separation of fragments over 1000 bases in length in two hours. This separation required the use of very low concentration noncrosslinked polyacrylamide, 3%T and the use of relatively high temperature, 70°C at an electric field of 100 V cm^{-1}. It appears that capillary electrophoresis can provide very long read lengths in relatively short periods. It is also interesting that the optimum electric field is dropping to relatively low values. The efficient cooling provided by capillary electrophoresis is less important at this modest electric field. Instead, the small dimensions of the capillary are important to retain the very low viscosity ($\sim$1000 cp at room temperature) separation material within the capillary. Figure 7.6 presents a portion of the separation achieved under these conditions.

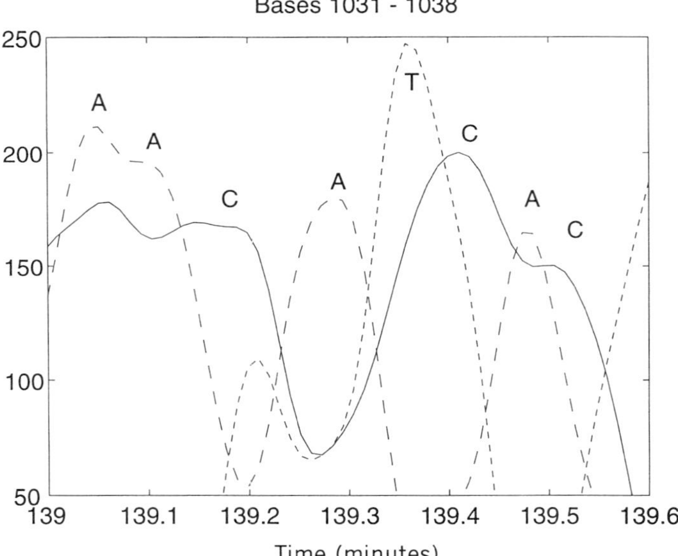

Fig. 7.6 *Separation of DNA sequencing fragments 1031 to 1038 bases in length. The separation was performed in a 40-cm long, 50-μm ID capillary at an electric field of 100 V cm^{-1}. The M13mp18 sample was prepared with the Thermosequenase kit according to the manufacturer's instructions. The sample was injected at 100 V cm^{-1} for 20 seconds. The capillary was filled with a 3%T noncrosslinked polyacrylamide solution in 6-M urea. The separation was performed at 60°C.*

MULTIPLE CAPILLARY SYSTEMS

Capillary electrophoresis provides rapid separation of long DNA sequencing fragments. However, it is necessary to operate many capillaries in parallel to compete with slab-gel electrophoresis systems. A number of capillary array instruments have been reported. These systems are not simple. The low sample loading in capillary electrophoresis produces heroic demands on detector sensitivity; reproducing this detection sensitivity in an array format is a formidable challenge. The systems fall into two classes: Systems built around a scanning optical detector and systems that employ a detector array.

Scanning Detectors

There have been two scanning detector systems for capillary array electrophoresis. The first was reported in 1990. Zagursky modified a commercial DuPont Genesis 2000 sequencer to run with 500-μm ID capillaries [7]. The instrument operated at 50 V cm^{-1}; 9.5 hours were required to separate fragments 500 bases in length. Sequencing accuracy was less than 97% for fragments ranging from 29 to 512 bases in length. The low electric field, and slow separation rate, probably result from the use of large-diameter capillaries, which are unable to support the higher electric field.

In 1992 Mathies and coworkers reported a scanning instrument to image an array of 100-μm ID capillaries [20]. In this system an array of capillaries is scanned under a confocal microscope, which is a two-color version of an instrument manufactured by Europhore [45]. The instrument produced sequence information for fragments 320 bases in length. The confocal design reduces the detection of light scatter from the capillary walls.

Both of these scanning systems suffer from a fundamental limitation in sensitivity. Because the detector interrogates each capillary sequentially, the detection duty cycle is low. This duty cycle is simply the fraction of time that each capillary is illuminated. For example, in a 100 capillary instrument, each capillary is probed, on average, for 1% of the time. To first approximation, the fluorescence intensity from each capillary will, on average, be reduced by the duty cycle; in a 100-capillary instrument the fluorescence signal will be reduced to 1% of the value generated in a single-capillary instrument.

One-Dimensional Detector Arrays

To eliminate the severe duty cycle penalty paid by the use of a scanning detector system, several systems have been built with detector arrays. In these systems one detector, or detector element in a camera, continuously monitors fluorescence from each capillary. Because each capillary is continuously monitored, there is no penalty in duty cycle.

In 1993 Yeung's group reported a multiple capillary DNA sequencer. Here a ribbon of capillaries was illuminated with a line-focused laser beam. Fluorescence was collected at right angles and imaged onto a CCD camera. The use of the CCD camera ensured that all capillaries are monitored simultaneously [46–47].

A similar detector was developed by a group at Hitachi, wherein a linear array of capillaries was placed within a sheath flow cuvette [48–49]. Fluorescence was collected at right angles and imaged onto a CCD camera. Because of relatively poor collection efficiency, relatively large inner diameter capillaries were required to load sufficient analyte for detection. The wide-bore capillaries lead to large intercapillary spacing, which decreased the detection efficiency further. As a result both the optical and the heat-transfer properties of the system were not ideal, and 245 bases of DNA sequence was generated in roughly two hours.

My group has developed several multiple capillary sequencing instruments [50]. In one case, Fig. 7.7, a laser beam illuminates analyte migrating from a linear array of capillaries in a sheath flow cuvette. Two laser beams are used to excite fluorescence from DNA sequencing fragments prepared with Applied Biosystem's dye-labeled primers or dye-labeled dideoxynucleotides. Fluorescence is collected at right angles with a microscope objective, which provides a high-efficiency image of the fluorescent analyte. In this system, fluorescence is imaged onto a set of fiber optics, with one fiber per capillary. The image size from the capillary is about 1 mm in diameter, which is larger than the

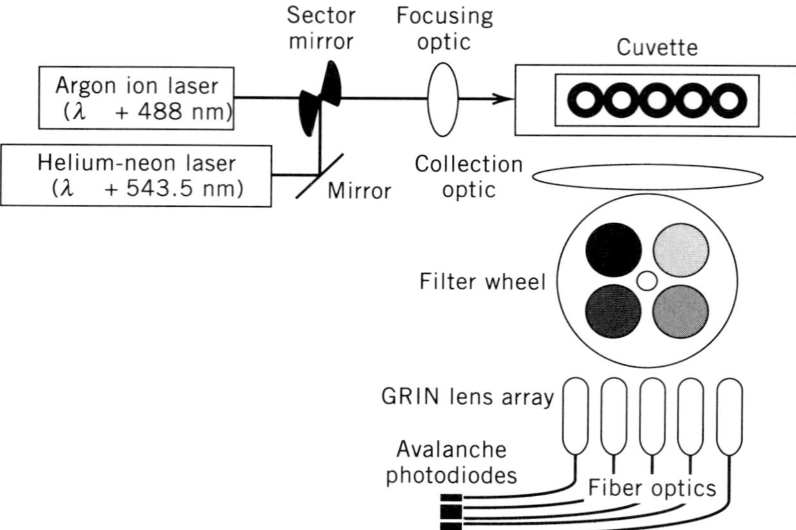

Fig. 7.7 *Linear capillary array instrument. The light from the two lasers is alternately focused into a rectangular sheath flow cuvette. Fluorescence is collected at right angles, spectrally filtered, and imaged onto an array of GRIN lenses. Fluorescence is then transmitted through fiber optics to a set of avalanche photodiodes.*

core of the optical fiber. A 1-mm diameter GRIN (GRadient INdex) lens is used to collect the fluorescence and couple it with the optical fibers. A filter wheel is used to resolve fluorescence into four spectral bands. The filter wheel is synchronized with a sector wheel so that fluorescence is monitored in the appropriate spectral band for each laser wavelength. For example, the argon ion laser (λ = 488 nm) is used to excite fluorescence from FAM and JOE. Fluorescence is detected at 540 and 560 nm during the argon ion laser excitation period. The helium-neon laser (λ = 543.5 nm) is used to excite fluorescence from TAMRA and ROX. Fluorescence is monitored at 580 nm and 610 nm during the helium-neon laser excitation. Finally, fluorescence is detected with avalanche photodiodes. These solid state devices, from EG&G Canada, have high quantum efficiency and are extremely rugged.

A detailed drawing of a cuvette designed to hold five capillaries is shown in Fig. 7.8. The capillaries fit as a linear array within a rectangular glass chamber. The depth of the flow chamber matches the outer diameter of the capillaries, while the width matches the width of the array of capillaries. Sheath fluid is pumped through the interstitial space between the capillaries. As DNA fragments migrate from the capillary tips, they are entrained within the sheath fluid, forming distinct streams within the cuvette. The laser beam passes through the cuvette, below the capillary tips. Because the beam passes through the sheath fluid, rather than the capillaries, its passage is unimpeded by any curved glass surface, which would act to deflect or defocus the

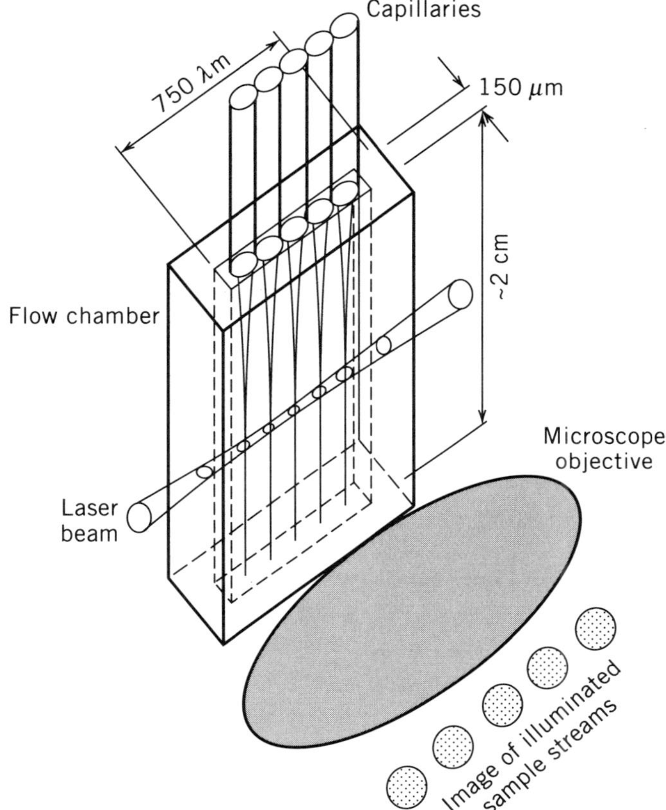

Fig. 7.8 Sheath flow cuvette for linear array of capillaries. The cuvette snugly holds the capillaries within a rectangular flow chamber. Sheath fluid passes through the interstitial space between capillaries, drawing DNA as individual streams from each capillary.

beam. Because the DNA concentration is very small and the stream diameter is very narrow, there is very little loss in the beam intensity while traversing each sample stream.

The total number of capillaries that can be used a linear array is limited by two optical properties of the system. First, width of the array should not be larger than the size of the photodetector. It is a fundamental law of optics that a demagnifying optical system can not efficiently collect fluorescence. If the capillary array is larger than the photodetector array, then the collection optics inherently are inefficient. For example, a 1-cm^2 CCD camera can be used with at most 67 capillaries of 150-μm outer diameter in a unit magnification optical system before significant optical losses are generated in the system.

Of course discrete photodetectors can be used over a larger area. A second optical property limits the number of capillaries that can be used efficiently. In general, the laser beam spot-size should be relatively uniform across the array. If the beam is much larger or smaller than the size of the same stream, then analyte molecules will be inefficiently excited. Unfortunately, diffraction prevents collimation of a laser beam over a large distance. At best, the beam will remain roughly collimated over a distance given by twice the Rayleigh length about the beam focus. The more tightly focused the beam, the shorter is the Rayleigh distance. If the beam size at the focus is matched to the sample stream radius ($\sim$25 μm), then the Rayleigh length for an argon ion laser beam is about 4 mm. The beam will remain collimated over about 50 capillaries before diffraction causes the beam to become significantly larger than the sample stream.

In any event, optics limit the efficient excitation and collection of light from an array of 50 to 75 capillaries, and only if the capillaries are closely packed and 150-μm outer diameter. Linear arrays of this size are useful instruments for medium-scale sequencing projects that require a sample throughput that is a few times larger than current commercial slab-gel instruments. A completely different optical design is required for larger-scale sequencing.

Two-Dimensional Arrays—The Genosequencer

We have developed several designs for very large-scale capillary arrays. In the only one to be presented here, a two-dimensional array of capillaries is inserted into a square flow chamber, Fig. 7.9. In this instrument the sheath flow is directed upward to efficiently sweep bubbles from the chamber. Several rows of capillaries are inserted into the flow chamber, and these rows are offset to form a staircase. Sheath fluid is pumped in the interstitial space between capillaries, drawing the DNA fragments as discreet streams, one stream from each capillary.

To ensure that each row of capillaries is illuminated, an elliptically shaped laser beam forms a line-focus slightly downstream from the capillary tips. The laser beam generates a two-dimensional set of fluorescent spots, one spot downstream from each capillary. Fluorescence is collected from the side by a high-efficiency optical condenser that generates a unit magnification image on a CCD camera. To avoid overlap of the spots, the rows of capillaries are offset, so the last row of capillaries is farthest downstream. When imaged from the side, each illuminated sample stream forms a discreet spot at the image plane. However, the first row of capillaries is imaged through a very small thickness of sheath fluid, while the last row is imaged through the entire thickness of the fluid. To equalize the optical paths, an aberration-correction prism can be inserted in the optical train so that the optical distance is equal for all rows of capillaries.

A 32 by 32 array of capillaries terminates in a 4.8 by 4.8 mm array. This array of 1024 capillaries may be imaged efficiently onto a 1-cm^2 CCD camera. Fluorescence is spectrally resolved by use of a rotating filter wheel. If necessary, the system could be expanded to a 64 by 64 array of capillaries for the simultaneous determination of sequence from 4096 clones.

We have constructed 25 and 96 capillary instruments; a 576 capillary device is under construction and a 1024 capillary system is in design. The 576 capillary instrument mates

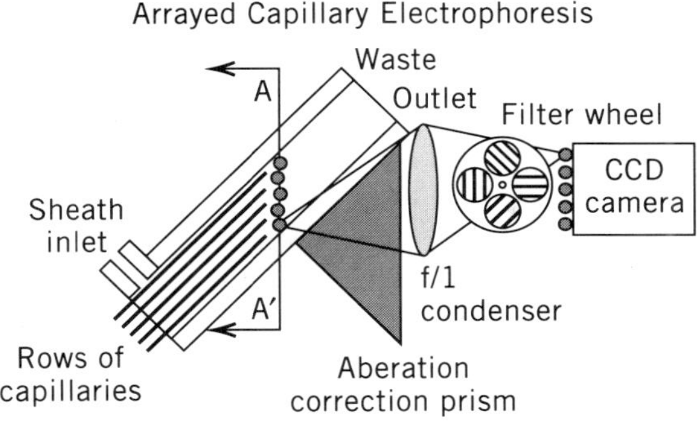

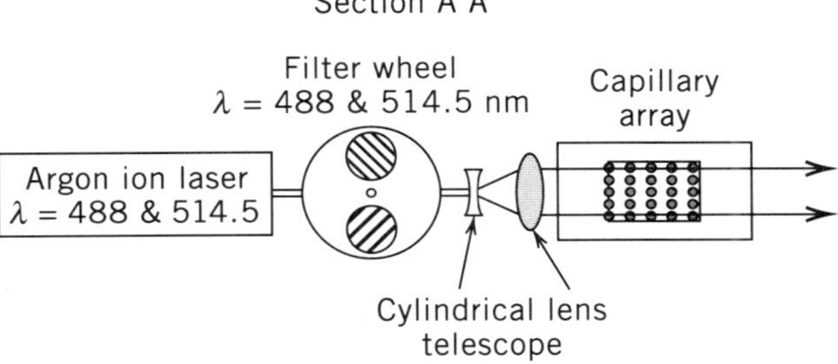

Fig. 7.9 Two-dimensional array of capillaries. A two-dimensional array of capillaries is held snugly in a square or rectangular flow chamber. Sheath fluid passes through the interstitial spaces between capillaries, drawing DNA as a two-dimensional array of sample streams, one stream per capillary. Fluorescence is excited simultaneously with an elliptic-ally shaped argon ion laser beam (a line focus). By offsetting successive rows of capillaries like the steps in a staircase and by tilting the major axis of the laser beam to be parallel with the capillary tips, the entire array of capillaries may be viewed through the wall of the flow chamber. An aberration correction prism is included in the optical train to ensure that the optical path between the capillary tip and the collection optic is of equal length. Section AA' presents a detailed view of the excitation optics.

with 6 microtiter plates. The instruments have an interesting property: After the system is designed for a single capillary, the incremental cost of an additional capillary is simply the cost of the capillary itself. A 1024-capillary instrument costs about $2000 more to construct than a single-capillary device. A 30-cm long capillary draws 2-μA of current when operated at an electric field of 200 V cm^{-1} ($V = 6$ kV) with a capillary filled with 1M TBE buffer and 6%T noncrosslinked polyacrylamide. An array of 1024 capillaries draws 2 mA of current; 10 W of power will be dissipated in the capillaries. A single 30-W power supply easily can operate all the capillaries simultaneously. Our major limitation in the design and construction of the largest-scale systems is simply the cost of DNA fragments to evaluate the system!

The throughput of these large-scale systems is much larger than that of the present generation of sequencers. We have demonstrated 1000 base separations in 135 minutes with our current generation sequencers. If we are conservative and anticipate 600 bases of raw sequence per capillary in a two-hour separation, then one run of the 576-capillary instrument will generate 345,000 bases of raw sequence. Assuming a one-hour period is required to replace the separation medium and prepare the capillary for a subsequent run, then eight runs can be performed per day, generating 2.7 Mbase of raw sequence. At seven fold redundancy, this throughput corresponds to 400,000 bases of *finished* sequence per day per instrument. Throughput for a 1024 or 4096 capillary device is proportionally higher. Of course sample preparation technology will be severely stressed to support one Genosequencer.

FUTURE PROSPECTS

Capillary electrophoresis has demonstrated fast and accurate separation of long sequencing fragments. Over 700 bases of sequence are routinely generated in a two-hour separation. Occasional runs of 1000 bases in 135 minutes have been obtained. These data are generated at high temperature to minimize compressions and with low viscosity separation medium to simplify regeneration of the capillaries between runs.

A number of capillary arrays have been demonstrated. The use of a two-dimensional array of capillaries results in the efficient generation of sequence from a large number of capillaries. A 576 capillary device (a Genosequencer) is under construction in our lab, and we have plans for 1240 and 4096 capillary instruments. These large-scale devices are capable of generating several million bases of raw sequence per day.

The throughput of these largest-scale instruments will be significantly larger than that of the present generation commercial sequencers. At first, only the largest genome centers will require this high throughput. However, large-scale clinical screening applications are sure to become important in the near future. For example, screening of every female in North America for mutations in BRCA-1 and BRCA-2 would require processing several trillion bases of sequence. Further applications in bacterial disease diagnosis and cancer prognosis will produce increased demand for large-scale sequencing.

ACKNOWLEDGMENTS

This work was supported by the Canadian Bacterial Diseases Network, the Canadian Genetic Diseases Network, the Natural Sciences and Engineering Research Council, and SCIEX. I gratefully acknowledge support from a McCalla professorship from the University of Alberta.

REFERENCES

1. A.M. Maxam, W. Gilbert (1977). "A new method for sequencing DNA." *Proc. Nat. Acad. Sci.* 74:560–564.

2. F. Sanger, S. Nicklen, A.R. Coulson (1977). "DNA sequencing with chain-terminating inhibitors." *Proc. Nat. Acad. Sci.* 74:5463–5467.

3. L.M. Smith, J.Z. Sanders, R.J. Kaiser, P. Hughes, C. Dodd, C.R. Connell, C. Heiner, S.B.H. Kent, L.E. Hood (1986). "Fluorescence detection in automated DNA sequence analysis." *Nature* 321:674–679.

4. W. Ansorge, B.S. Sproat, J. Stegemann, C. Schwager (1986). "A non-radioactive automated method for DNA sequence determination." *J. Biochem. Biophys. Meth.* 13:315–317.

5. J.M. Prober, G.L. Trainor, R.J. Dam, F.W. Hobbs, C.W. Robertson, R.J. Zagursky, A.J. Cocuzza, M.A. Jensen, K. Baumeister (1987). "A system for rapid DNA sequencing with fluorescent chain-terminating dideoxynucleotides." *Science* 238:336–341.

6. J.E. Edström (1953). "Nucleotide analysis on the cyto-scale." *Nature* 172:809.

7. R.J. Zagursky, R.M. McCormick (1990). "DNA sequencing separations in capillary gels on a modified commercial DNA sequencing instrument." *BioTechniques* 9:74–79.

8. H. Swerdlow, R. Gesteland (1990). "Capillary gel electrophoresis for rapid, high resolution DNA sequencing." *Nucleic Acids Res.* 18:1415–1419.

9. H. Drossman, J.A. Luckey, A.J. Kostichka, J. D'Cunha, L.M. Smith (1990). "High-speed separations of DNA sequencing reactions by capillary electrophoresis." *Anal. Chem.* 62:900–903.

10. A.S. Cohen, D.R. Najarian, B.L. Karger (1990). "Separation and analysis of DNA sequence reaction products by capillary gel electrophoresis." *J. Chromatogr.* 516:49–60.

11. H. Swerdlow, S. Wu, H. Harke, N.J. Dovichi (1990). "Capillary gel electrophoresis for DNA sequencing: Laser-induced fluorescence detection with the sheath flow cuvette." *Chromatogr.* 516:61–67.

12. J.A. Luckey, H. Drossman, A.J. Kostichka, D.A. Mead, J. D'Cunha, T.B. Norris, L.M. Smith (1990). "High speed DNA sequencing by capillary electrophoresis." *Nucleic Acids Res.* 18:4417–4421.

13. H. Swerdlow, J.Z. Zhang, D.Y. Chen, H.R. Harke, R. Grey, S. Wu, C. Fuller, N.J. Dovichi (1991). "Three DNA sequence methods using capillary gel electrophoresis and laser-induced fluorescence." *Anal. Chem.* 63:2835–41.

14. A.E. Karger, J.M. Harris, R.F. Gesteland (1991). "Multiwavelength fluorescence detection for DNA sequencing using capillary electrophoresis." *Nucleic Acids. Res.* 19:4955–4962, 1991.

15. D.Y. Chen, H.R. Harke, N.J. Dovichi (1992). "Two-label peak-height encoded DNA sequencing by capillary gel electrophoresis: three examples." *Nucleic Acids. Res.* 20:4873–4880.

16. S. Bay, H.R. Harke, J. Elliott, N.J. Dovichi (1993). "Accuracy of Two-Color Peak Height Encoded DNA Sequencing by Capillary Gel Electrophoresis and Laser-Induced Fluorescence." *Proceedings of the Society of Photo-Optical Instrumentation Engineers* 1891:8–11.

17. H.R. Starke, J.Y. Yan, J.Z. Zhang, K. Muhlegger, K. Effgen, N.J. Dovichi (1994). "Internal fluorescence labeling with fluorescent deoxynucleotides in two-label peak-height encoded DNA sequencing by capillary electrophoresis." *Nucleic Acids Res.* 22:3997–4001.

18. N. Best, E. Arriaga, D.Y. Chen, N.J. Dovichi (1994). "Separation of fragments up to 570 bases in length by use of 6%T non-crosslinked polyacrylamide for DNA sequencing in capillary electrophoresis." *Anal. Chem.* 66:4063–7.

19. S.L. Pentoney, K.D. Konrad, W. Kaye (1992). "A single-fluor approach to DNA sequence determination using high performance capillary electrophoresis." *Electrophoresis* 13:467–74.

20. X.C. Huang, M.A. Quesada, R.A. Mathies (1992). "DNA sequencing using capillary array electrophoresis." *Anal. Chem.* 64:2149–54.

21. M. Nelson, J.L. Van Etten, R. Grabherr (1992). "DNA sequencing of four bases using three lanes." *Nucleic Acids Res.* 20:1345–8.

22. D.C. Williams, S.A. Soper (1995). "Ultrasensitive near-IR fluorescence detection for capillary gel electrophoresis and DNA sequencing applications." *Anal. Chem.* 67:3427–3432.

23. J.L. Viovy, T. Duke (1993). "DNA electrophoresis in polymer solutions: Ogston sieving, reptation and constraint release." *Electrophoresis* 14:322–329.

24. G.W. Slater, P. Mayer, S.J. Hubert, G. Drouin (1994). "The biased reptation model of DNA gel electrophoresis: A user guide for constant field mobilities." *Appl. Theoret. Electrophoresis* 4:71–79.

25. P. Mayer, G.W. Slater, G. Drouin (1994). "Exact behaviour of single-stranded DNA electrophoretic mobilities in polyacrylamide gels." *Appl. Theoret. Electrophoresis* 3:147–155.

26. J.Y. Yan, N. Best, J.Z. Zhang, H.J. Ren, J.Z. Jiang, J. Hou, N.J. Dovichi (1996). "The limiting mobility of DNA sequencing fragments for both cross-linked and noncross-linked polymers in capillary electrophoresis: DNA sequencing at 1200 Vcm^{-1}." *Electrophoresis* 17:1037–1045.

27. H. Lu, E. Arriaga, D.Y. Chen, N.J. Dovichi (1994). "High-speed and high-accuracy DNA sequencing by capillary gel electrophoresis in a simple, low cost instrument: Two-color peak-height encoded sequencing at 40°C" *J. Chromatogr.* 680:497–501.

28. H. Lu, E. Arriaga, D.Y. Chen, D. Figeys, N.J. Dovichi (1994). "Activation energy of single stranded DNA moving through cross-linked polyacrylamide gels at 300 V/cm: Effect of temperature on sequencing rate in high-electric-field capillary gel electrophoresis." *J. Chromatogr.* 680:503–510.

29. J.Z. Zhang, Y. Fang, J.Y. Hou, H.J. Ren, R. Jiang, P. Roos, N.J. Dovichi (1995). "Use of non-cross-linked polyacrylamide for four-color DNA sequencing by capillary electrophoresis separation of Fragments up to 640 bases in length in two hours." *Anal. Chem.* 67:4589–4593.

30. H.J. Bode (1977). "The use of liquid polyacrylamide in electrophoresis. I. Mixed gels composed of agar-agar and liquid polyacrylamide." *Anal. Biochem.* 83:364–371.

31. D. Tietz, M.H. Gottleib, J.S. Fawcett, A. Crambach (1986). "Electrophoresis on uncross-linked polyacrylamide: Molecular seiving and its potential applications." *Electrophoresis* 7:217–220.

32. H. Pulyaeva, D. Wheeler, M.M. Garner, A. Crambach (1992). "Molecular sieving of lambda phage DNA in polyacrylamide solutions as a function of the molecular weight of the polymer." *Electrophoresis* 13:608–614.

33. D. Tietz, A. Aldroubi, H. Pulyaeva, T. Guszczynski, M.M. Garner, A. Crambach (1992). "Advances in DNA electrophoresis in polymer solutions." *Electrophoresis* 13:614–616.

34. D.N. Heiger, A.S. Cohen, B.L. Karger (1990). "Separation of DNA restriction fragments by high performance capillary electrophoresis with low and zero crosslinked polyacrylamide using continuous and pulsed electric fields." *J. Chromatogr.* 516:33–48.

35. J. Sudor, F. Foret, P. Bocek (1991). "Pressure refilled polyacrylamide columns for the separation of oligonucleotides by capillary electrophoresis." *Electrophoresis* 12:1056–1058.

36. M. Chiari, M. Nesi, M. Fazio, P.G. Righetti (1992). "Capillary electrophoresis of macromolecules in 'syrupy' solutions: Facts and misfacts." *Electrophoresis* 13:690–697.

37. A. Guttman, B. Wanders, N. Cooke (1992). "Enhanced separation of DNA restriction fragments by capillary gel electrophoresis using field strength gradients." *Anal. Chem.* 64:2348–2351.

38. D. Figeys, N.J. Dovichi (1995). "Multiple separations of DNA sequencing fragments with a non-cross-linked polyacrylamide-filled capillary: Capillary electrophoresis at 300 V/cm." *J. Chromatogr.* 717:113–126.

39. E.N. Fung, E.S. Yeung (1995). "High-speed DNA sequencing by using mixed poly(ethylene oxide) solutions in uncoated capillary columns." *Anal. Chem.* 67:1913–1919.

40. S. Carson, A.S. Cohen, A. Belenkii, M.C. Ruiz-Martinez, J. Berka, B.L. Karger (1993). "DNA sequencing by capillary electrophoresis: Use of a two-laser-two-window intensified diode array detection system." *Anal. Chem.* 65:3219–26.

41. M.C. Ruiz-Martinez, J. Berka, A. Belenkii, F. Foret, A.W. Miller, B.L. Karger (1993). "DNA sequencing by capillary electrophoresis with replaceable linear polyacrylamide and laser-induced fluorescence detection." *Anal. Chem.* 65:2851–2858.

42. N. Chen, T. Manabe, S. Terabe, M. Yohda, I. Endo (1994). "Effects of linear polyacrylamide concentrations on the separation of oligonucleotides and DNA sequencing fragments by capillary electrophoresis." *J. Microcolumn Sep.* 6:539–543.

43. T. Manabe, N. Chen, S. Terabe, M. Yohda, I. Endo (1994). "Effects of linear polyacrylamide concentrations and applied voltages on the separation of oligonucleotides and DNA sequencing fragments by capillary electrophoresis." *Anal. Chem.* 66:4243–52.

44. M.J. Rocheleau, R.J. Grey, D.Y. Chen, H.R. Harke, N.J. Dovichi (1992). "Formamide modified polyacrylamide gels for DNA sequencing by capillary gel electrophoresis." *Electrophoresis* 13:484–486.

45. L. Hernandez, J. Secalona, N. Joshi, N. Guzman (1991). "Laser-induced fluorescence and fluorescence microscopy for capillary electrophoresis zone detection." *Chromatogr.* 559:183–196.

46. J.A. Taylor, E.A. Yeung (1993). "Multiplexed fluorescence detector for capillary electrophoresis using axial optical fiber illumination." *Anal. Chem.* 65:956–960.

47. K. Ueno, E.S. Yeung (1994). "Simultaneous monitoring of DNA fragments separated by electrophoresis in a multiplexed array of 100 capillaries." *Anal. Chem.* 66:1424–1431.

48. H. Kambara, S. Takahashi (1993). "Multiple-sheathflow capillary array DNA analyser." *Nature* 361:565–566.

49. S. Takahashi, K. Murakami, T. Anazawa, H. Kambara (1994). "Separation of long DNA fragments by capillary gel electrophoresis with laser-induced fluorescence detection." *Anal. Chem.* 66:1021–1026.

50. N.J. Dovichi, J.Z. Zhang. "Multiple capillary biochemical analyzer." *US Patent* 5,439,578; 8 August 1995.

8

Scanning Probe Microscopy in Genomic Research

DAVE P. ALLISON, THOMAS G. THUNDAT,
AND ROBERT J. WARMACK

CONTENTS

INTRODUCTION

The international human genome project has served as a catalyst to focus resources
and attention on genomic research. With the goal of sequencing the 3-billion nucleotide
bases of the 22 autosomal and 2 sex chromosomes contained within human DNA,

Automation Technologies for Genome Characterization, Edited by Tony J. Beugelsdijk.
ISBN 0-471-12806-6 © 1997 John Wiley & Sons, Inc.

this project will rank as one of twentieth century's most ambitious scientific efforts. This monumental task has involved all scientific disciplines and has stimulated serious dialogue on the ethical and legal use of genetic information. Already a number of genes have been found that are responsible for inherited human diseases, including Duchenne muscular dystrophy, myotonic dystrophy, retinoblastoma, and cystic fibrosis [1]. Multigene interactions and genes responding to environmental insults have been identified and implicated in heart disease, strokes, and some forms of cancer. As more of the genome is mapped and sequenced, the information gained should contribute significantly to our understanding of gene organization and regulation, perhaps helping to solve the mysteries of human development and aging from a fertilized egg to the adult. Concurrently, as more genes and multigene interactions are discovered, screening methodologies will be developed that will impact genetic counseling and both diagnostic and therapeutic medicine in the twenty-first century.

Prior to the human genome initiative, mapping and sequencing technologies were already well established. Human and mouse programs along with other model organisms including *Escherichia coli* (bacterium), *Saccharomyces cerevisiae* (yeast), and *Drosophila melanogaster* (fruit fly) were underway. Physical mapping was and still is accomplished primarily by identifying restriction-enzyme sites on cloned molecules by partial and multiple enzyme digests followed by gel analysis [2, 3]. Likewise sequencing projects using either the Maxam-Gilbert [4] or Sanger methods [5] had already been applied to sequence a number of plasmid, viral, and bacteriophage DNAs, and the Maxam-Gilbert method is the basis of multiplex sequencing in use today. Nevertheless, although proven technologies exist for mapping and sequencing DNA, both cost-effective and time-efficient improvements must be made, and new technologies must be developed for large genomes. Scanning probe microscopes offer a new technology that could impact the mapping and sequencing of cloned DNAs.

SCANNING PROBE MICROSCOPES

Scanning probe microscopes are instruments that operate by positioning a sensitive probe on or near surfaces for the purpose of recording topographical or other measurable properties of surface features. The first scanning probe microscope was the scanning tunneling microscope (STM), invented in 1982 [6]. It was followed by the atomic force microscope (AFM) in 1986 [7]. Both of these instruments are now commercially available, capable of atomic resolution on flat surfaces, and most of the sample preparation techniques developed and images reported have been accomplished with these two microscopes. Other probe microscopes sensitive to ion conductance [8], photons [9, 10], and magnetic and electric effects have also been reported [11]. In this article we will explore the potential for using scanning tunneling and atomic force microscopes both alone and in combination with other probe sensitivities for the purpose of mapping and sequencing DNA.

OPERATION OF SCANNING TUNNELING AND ATOMIC FORCE MICROSCOPES

Unlike conventional light or electron microscopes, the scanning-tunneling microscope and the atomic force microscope do not require transmitted waves or lenses for their operation. Instead, both instruments use a surface-sensitive probe.

STM

Imaging is accomplished in the STM by moving a sharpened, conductive tip above a sample surface [12]. The tip provides a tiny tunneling current ($\sim 10^{-9}$ a) that is very sensitive to the size of the tiny gap between the tip and sample surface. The tip is precisely positioned by a piezoelectric crystal configured to respond in the XYZ directions. By raster scanning the tip over a conducting sample in the XY plane and regulating the tunneling current to be constant *via* control of the Z position, a topographic map is recorded. Typically 128–512 points per scan in the X direction are recorded with 128–512 of these line scans recorded in the Y direction (Fig. 8.1a). Information in the Z dimension (height) comes from recording the voltage applied to either retract or extend the piezoelectric crystal in order to maintain a constant tunneling current between the tip and a changing surface topography. A tunneling current is produced when the wave functions of the electrons on the sample surface have sufficient overlap with the wave functions of the electrons on the conducting tip. The current is maintained by applying a difference in potential between the tip and the sample.

On the flat crystalline surfaces of conductors and semiconductors, atomic resolution is routinely obtained because the tunneling current has a strong exponential dependence on the distance between the tip and the sample. For example, if the tip is positioned one nanometer above the surface (by the piezoelectric crystal) the tunneling current will typically change by a factor of 10 if the distance between the tip and sample changes by only 0.1 nanometer. The result of this strong gap-dependent current is that the tunneling current is concentrated into a region of the tip only 0.1 to 0.5 nm wide, making atomic resolution possible. However, on surfaces that are not atomically flat, resolution is affected by switching of the tunneling current to other regions of the tip as the tip encounters nanometer-scale deviations from flatness during a scan. This results in degraded resolution and is a factor that should be considered in the selection of samples and in the interpretation of results.

Electrical conductivity is a problem that must be considered in the selection of biological samples for scanning tunneling microscopy. Biomolecules are usually either nonconductors or poor semiconductors, which presents a problem when they are probed by charged particles. If the molecule cannot bleed away excess charge, the image becomes noisy or distorted. In addition electrical insulators typically do not dissipate heat readily such that energy deposited by any probe can cause disintegration. For the highest possible resolution, DNA must be placed directly on a conducting substrate without being stained or coated with evaporated material.

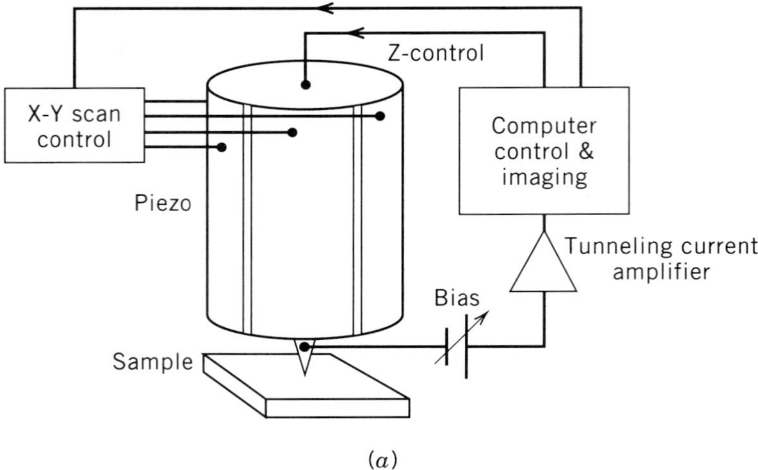

(*a*)

Fig. 8.1 *a* The scanning tunneling microscope depends upon an electrical "tunneling" current that flows between a tip and a nearby surface only a few atomic diameters away or less. Since the current rapidly disappears for larger gaps, only the extreme atoms on the tip participate in the tunneling. This can yield atomic resolution on flat surfaces. The tip is raster scanned by a piezoelectric crystal under computer control that also adjusts the gap with subatomic precision.

AFM

Perhaps the single requirement that serves to differentiate the AFM from the STM is that sample conductivity is irrelevant for AFM imaging [11]. This is because the interaction with the sample is mechanical rather than electrical and therefore even nonconductors can be imaged. This feature is particularly valuable both for imaging marginally conductive biological macromolecules and for the selection of mounting surfaces. The AFM employs a similar cylindrical piezoceramic transducer designed to raster scan the surface (Fig. 8.1*b*). The usual configuration is to mount the sample on the piezoelectric crystal that scans in the *XY* plane beneath a fixed position probe that is in contact or near-contact with the surface. In the AFM the probe is a miniature stylus supported by a micromachined cantilever and height information is gained by monitoring the position of the probe in response to changes in surface topography. Two detection methods are most commonly used in commercial AFMs. One configuration employs optical-beam deflection, in which a laser beam focused on the backside of the cantilever is reflected onto a position-sensitive detector. As the sample is scanned, changes in surface topography cause vertical movement of the probe, detected to sub-Ångstrom sensitivity, by changes in the position where the reflected beam strikes the detector. The other system is an electrical detection scheme that uses a piezoresistive circuit implanted directly into the back of the cantilever. A servocircuit then maintains a constant cantilever deflection or applied force by moving either the sample or the cantilever in response to changes in surface topography.

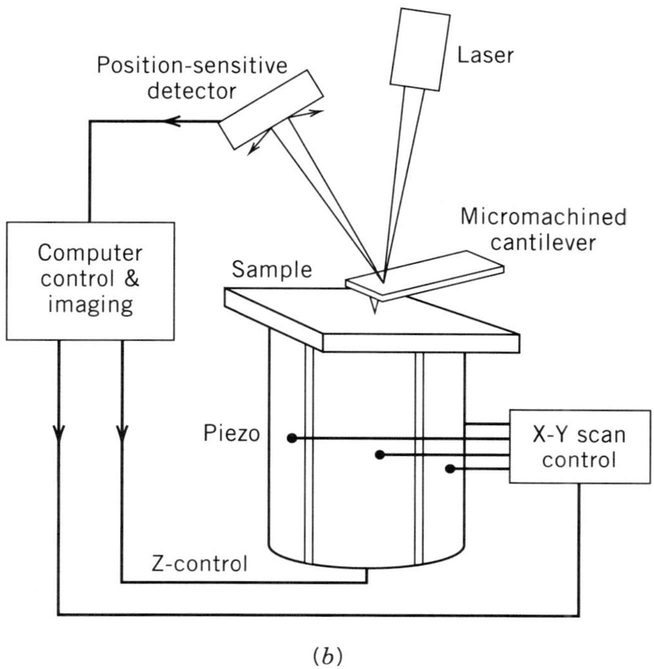

(*b*)

Fig. 8.1 *b* The atomic force microscope uses the same precision control of the STM but touches the surface with a mechanically sensing stylus mounted upon a micromachined cantilever. Vertical motion of the cantilever, replicating surface structure to less than an Ångstrom (10^{-10} m), is monitored in the illustration above with a laser beam that reflects onto an optical position-sensitive detector.

The spring constant of the cantilever is typically chosen to be less than a nanoNewton per meter, which is less than the spring constant of the atomic lattice ($\sim$10 nN/m), thus allowing the tip to elastically glide over the surface without disrupting the atomic crystal structure. Although manufactured tips have limited sharpness, contact between an asperity on the end of the tip with a flat sample can give atomic resolution. Yet, like the STM, the finite curvature of the AFM tip limits the resolution that can be expected for biological samples mounted on atomically flat surfaces, such as freshly cleaved mica. Presently the radius of curvature of commercial tips is the order of 10 nm. This means that for a 2-nm DNA strand, the observed width is 10 nm or more. Athough atomic detail is not expected for biological macromolecules, the resolution achievable under ambient conditions is remarkable, since it is unnecessary to compromise molecular ultrastructure by chemical fixation or staining.

Contact-mode AFM scanning is simple to implement but has some disadvantages. At the point of contact the tip experiences substantial attractive forces that can greatly affect resolution. Primarily these are capillary forces and van der Waals that can frequently amount to tens of nanoNewtons. This can be destructive to the sample and can shear adsorbed structures (e.g., DNA) from the sample surface. Capillary forces result from surface tension created by the liquid bridge that forms between the sample surface and the tip due to the condensation of water vapor and other contaminants upon the surface. For this reason AFMs are frequently operated contained within humidity-controlled chambers.

The noncontact mode is a relatively new concept in AFM imaging and has important advantages for AFM operation on a number of samples. In this mode the cantilever is vibrated near its resonant frequency and, upon approaching the surface, experiences surface forces that cause a shift in the vibration frequency. This shift in frequency is sensed by a servocircuit and is used to maintain tip-surface separation. Since the tip is not in contact with the sample and only experiences nonlocalized forces, the resolution is poorer than contact mode. However, the advantage is that forces that would deform or shear the sample practically vanish. If a compromise between contact and noncontact mode is effected and the excursion of the tip toward the sample is allowed until the tip just barely touches the sample (i.e., briefly experience the repulsive forces), a high-resolution topographic image of the surface can be recorded. This is known as the *cyclical-contact* or *TappingMode*[TM] (Digital Instruments, Inc., Santa Barbara). In this case the amplitude of the vibration, or for greater sensitivity, the phase, is used to regulate the tip-sample separation.

IMAGING DNA WITH SCANNING PROBE MICROSCOPES

After the scanning tunneling microscope was invented and atomic resolution demonstrated in 1982, considerable interest was generated for imaging DNA. The primary rationale was that since this was an instrument capable of atomic resolution, and since imaging could be accomplished on native unstained samples, it might be possible to directly read nucleotide sequences on individual molecules. Although the conduction mechanism in biomolecules is not fully understood, an STM image of the 2-nm diameter DNA molecule can be obtained. The first STM image of DNA [13] was reported in 1984, however, a 1989 image of a portion of a DNA molecule showing the double helix [14] served to awaken the scientific community to the potential for high-resolution imaging of biomolecules with STM. Shortly afterward, other laboratories also reported high-resolution images of DNA [15-26] even showing base stacking within the helix [27] and images of polydeoxyadenylate, a single-stranded synthetic molecule, where adenine could be resolved [28].

Several features common with these early reports were that results were rarely repeatable, the frequency of DNA imaged was far less than the amount of DNA applied, and only parts of DNA molecules either trapped in surface irregularities or salt deposits were imaged. This was not surprising, since most of the images were obtained by air drying a droplet of DNA onto either atomically flat, highly oriented pyrolytic graphite (HOPG) or gold surfaces. HOPG is virtually chemically inert and therefore does not bond biomolecules with sufficient strength to prevent removal by forces exerted by the tip. In addition HOPG has surface features that mimic the helical structure of DNA [29, 30]. Atomically flat gold surfaces must be fabricated, usually by evaporation onto a mica surface at elevated temperatures, and are composed of atomically flat terraces with the largest being on the order of a square micron [31]. Gold and graphite surfaces have been treated both chemically [25, 32, 33] and electrochemically [34, 35] to bond DNA with the result that convincing images of DNA molecules have been produced.

One such treatment, pioneered in our laboratory in collaboration with L. A. Bottomley at the Georgia Institute of Technology, introduces a strong positive charge on the gold surface by reacting the thiol group of 2-dimethylaminoethanethiol with the gold and allowing the positively charged dimethylamino head groups to attract and *electrostatically* immobilize the negatively charged DNA [36]. A curious effect is seen caused by the reduction in tunneling current as the tip moves over the poorly conductive DNA molecule; it is compensated by the servocircuit which moves the tip closer to the sample. This results not only in images of DNA with heights less than the expected 2 nm but also generally in negative images wherein the molecules *appear* to be below the substrate surface [37, 38] (Fig. 8.2). Regardless of the contrast, this method produces fairly routine imaging of DNA molecules at the expected surface coverage [39].

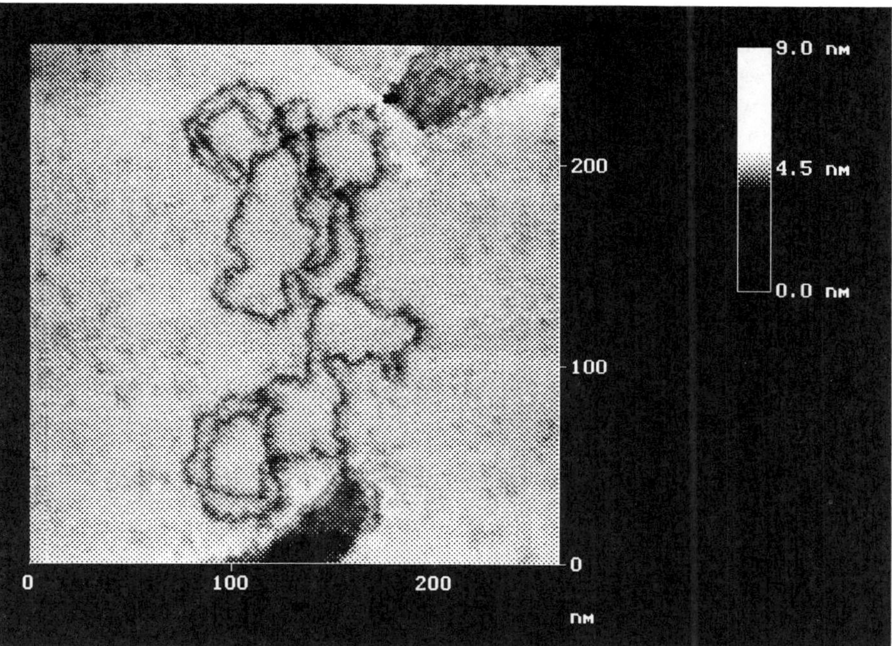

Fig. 8.2 A STM image of a circular plasmid molecule mounted on a gold surface treated with 2-dimethylaminoethanethiol. The positive charge on the dimethylamino groups attract the negatively charged DNA molecules preventing removal by forces exerted by the tip during scanning. To mantain a constant tunnelling current over the poorly conductive DNA molecule, the tip must move downward. This causes the DNA to be imaged in negative contrast appearing to be below the substrate surface.

The conductivity issue is irrelevant to the AFM, which was invented in 1986 [7] and made commercially available in 1989. The significant feature of this instrument, at least from the standpoint of imaging biomolecules, is that both the sample and the mounting surface can be nonconductive. For example, with STM the mounting surface has to be both conductive and atomically flat. This has almost dictated that biomolecules be mounted on atomically flat, conductive substrates such as HOPG or specially produced gold. However, with AFM any atomically flat surface can be used. Mica, being nonconductive, atomically flat over many square microns, and easily prepared by simple cleaving, has proved to be the surface of choice for mounting and imaging DNA molecules.

New sample preparation techniques had to be formulated for biological imaging with the AFM. Early AFM images of DNA on mica were obtained by simply placing a droplet of solution containing DNA onto the surface and allowing the sample to air dry [40]. These images showed definite signs of drying artifacts (Fig. 8.3a). The DNA was obviously moving during drying, and as a result plasmid molecules that should have been imaged as open circles were instead highly tangled. It was also apparent that upon repeated scanning, molecules were being moved due to forces exerted by the AFM tip. In the early 1990s two sample preparation methods were developed for the immobilization of DNA on mica surfaces. One strategy employed the divalent cation Mg^{2+} to interact with the mica surface, apparently replacing K^+ ions in the mica in order to attract and hold the negatively charged DNA [41, 42]. The other method treated mica with 3-aminopropyltriethoxy silane again to attract and hold DNA by electrostatic means [43]. Today the Mg^{2+} method is the most often used for imaging DNA on mica very likely because immobilization can be accomplished by simply adding $MgCl_2$ or $Mg(OAc)_2$ in the mounting solution. Other divalent cations such as Co^{2+} and Ba^{2+} have also been used to bond DNA to mica [44, 45], and Ni^{2+} has been used to complex DNA to mica with sufficient strength that imaging can be accomplished under fluids [46]. In Fig. 8.3b a contact mode AFM image of a field of open circular pBS$^+$ plasmids clearly shows the effect of mounting DNA molecules on mica in the presence of Mg^{2+}. There is no evidence of molecular movement or entanglement that could be attributed to drying effects.

A requirement for genomic work is the ability to image large DNA molecules without spurious noise. Because scanning probe microscopes are topographical imaging systems and thus sense both DNA and other structures on the mounting surface, sample preparation methodology is a key issue. This can be clearly seen from the image of plasmids in Fig. 8.3b where, in addition to DNA molecules, a significant contamination of other structures is also imaged. To ensure high-quality imaging of large DNA molecules necessary for genomic studies, the contribution of buffer salts and other spurious material applied to the sample surface must be minimized. In our laboratory attention is given to ensure that both DNA and mounting buffers are very clean to eliminate background adsorbates on mounting surfaces. Extensive rinsing of mica-mounted DNA samples with both distilled water and 100% ethanol, followed by critical-point drying, in which the liquid goes to the gas phase at equal pressure, also contributes toward eliminating spurious background contaminants and helps eliminates artifacts inherent in air-dried preparations [47, 48].

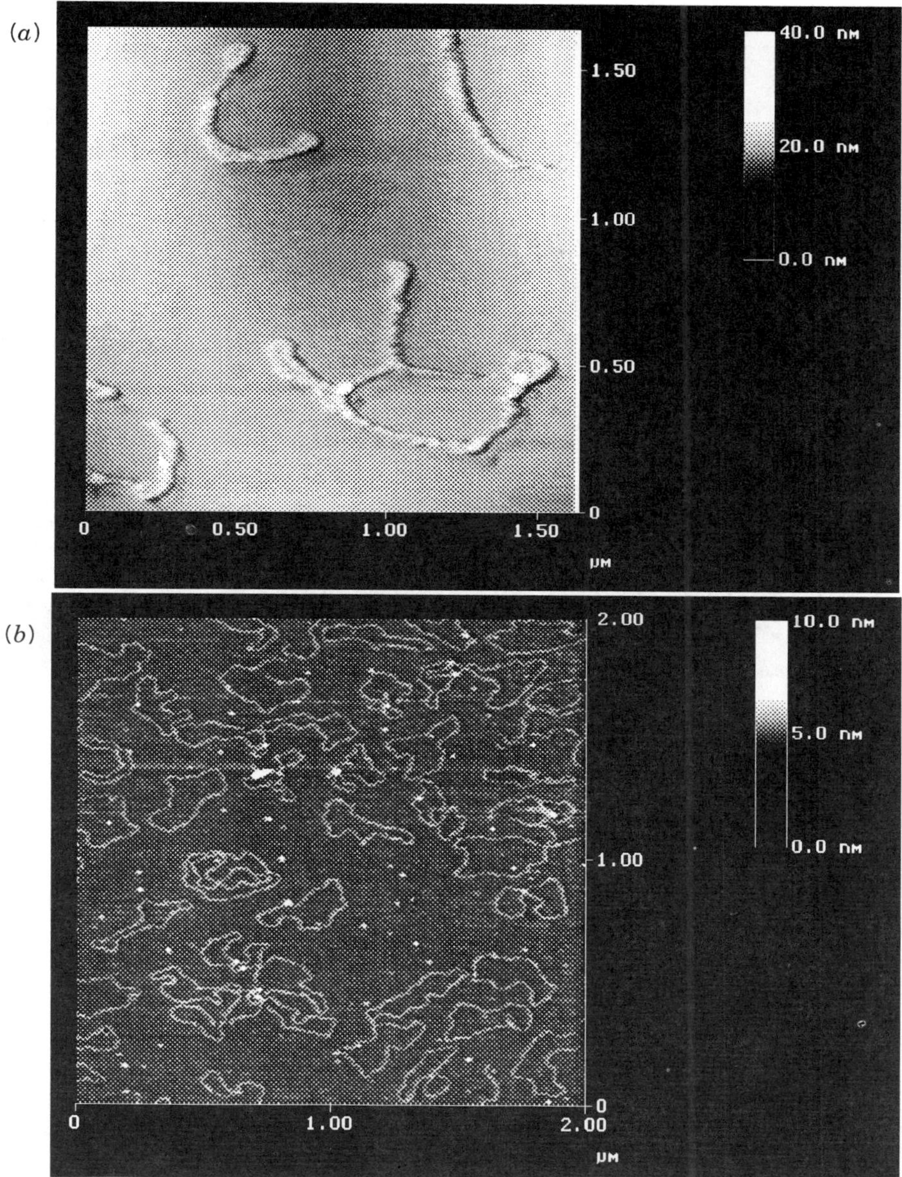

Fig. 8.3 *Atomic force microscope (AFM) images of relaxed pBS[+] plasmid DNA. The early AFM images (a) were taken after a drop of solution containing DNA was applied to a mica surface and allowed to air dry. Although imaging was possible, there was a serious problem of molecular entanglement due to drying, and clear images of open circular molecules were not possible. By either treating mica with Mg^{2+} or adding Mg^{2+} to the DNA solution prior to mounting (b), it was possible to image fields of open circular plasmids at the expected surface coverage. However, imaging problems caused by surfaces being contaminated with buffer salts and other spurious debris were still in evidence.*

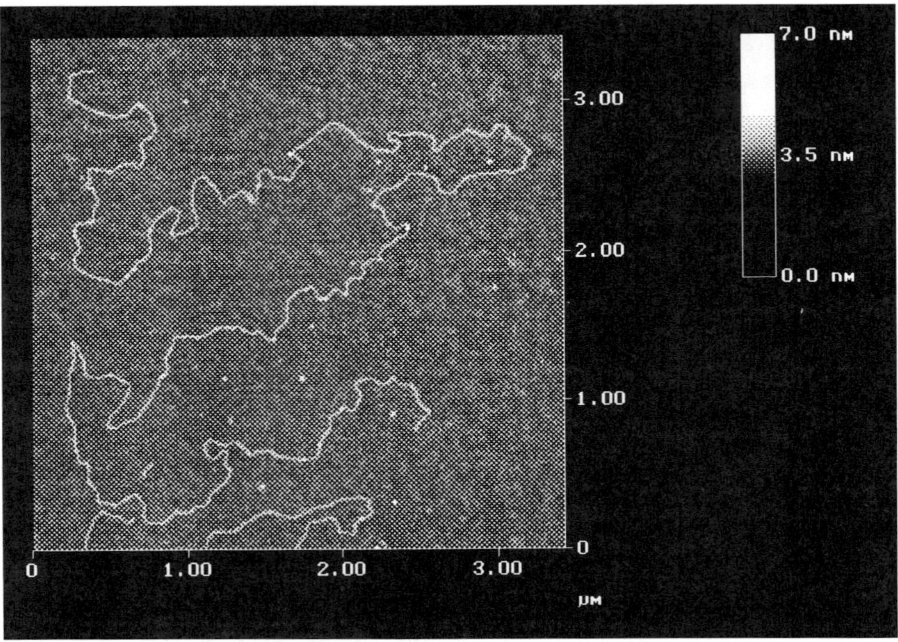

Fig. 8.4 *AFM imaging has benefited from improvements in both sample preparation techniques and instrumentation. Today images of large DNA molecules, such as this single-scan image of lambda phage DNA (48 kb), are routine.*

The result of careful sample preparation can be seen in Fig. 8.4, where a large DNA molecule, in this case lambda bacteriophage DNA (48kb), is imaged virtually free from background contamination. Further improvements in both instrumentation and sample preparation methodologies are desirable for imaging very large molecules, such as BACs (~200 kb). Improvements in instrumentation would include increasing the number of data points per scan line to enlarge the maximum effective scan area to more than 100 μm square, and improvements in flattening the image background over an entire 100-μm scan area. The identity of the contour path of individual molecules could be clarified greatly by straightening the molecules as they are laid upon the mounting surface, thereby reducing entanglement of these such long molecules.

RESTRICTION MAPPING GENOMIC DNA

Physical mapping of genomic DNA is a continuum spanning several levels of resolution. At one extreme, bands on chromosomes serve to identify and discriminate between the 22 autosomal and 2 sex chromosomes, while at the other extreme, the sequence of nucleotide bases in a DNA molecule are identified. The ultimate goal of mapping is to correctly order sequenced fragments to their respective chromosomes. Obviously this is a complex task considering that a sequenced fragment, on the order of 500 to 2000 bases, must be exactly located to a human chromosome that contains 50–250 million nucleotide base pairs. In genomic research the continuum of mapping is accomplished by cutting individual chromosomes into smaller pieces using different rare-cutting restriction endonucleases that site-specifically cleave DNA. The fragments are united with vectors, such as plasmids, virus, and bacteriophage, and propagated in bacterial or yeast cells. Conventional mapping strategies for cloned DNAs rely upon restriction mapping using gel-based techniques that involve multiple partial restriction endonuclease digestions, gel fractionation, Southern blotting, and hybridization with labeled end probes, or fingerprinting that requires multiple overlapping clones. The end result of mapping a cosmid (~50 kb) clone determines the position of all of the sites for one or more restriction enzymes. From this point, mapped cosmids from a chromosome are pieced together to form contigs by identifying overlapping regions. Overlapping contigs are identified to create linked libraries and eventually the entire chromosome is ordered [1–3].

Obviously this is a process that is also very time-consuming and further complicated by DNA fragments that cannot be cloned, thereby making closure in contigs a real problem. Also, as the need arises to restriction map larger clones such as BACs, the number of restriction sites also increases, significantly increasing the difficulty of mapping these large molecules by conventional methods. One alternative to gel-based mapping may be to use scanning probe microscopes to identify restriction sites directly on intact molecules by imaging enzyme-DNA complexes. Such a new technology is currently being developed in our laboratory.

The goal of mapping DNA molecules, using direct AFM imaging, is to identify the site-specific attachment of proteins, such as restriction enzymes or transcription factors, on intact DNA molecules. Under normal conditions restriction enzymes cleave DNA at the active site. This function can be modified either by withholding cofactors such as Mg^{2+} or by engineering mutants that will site-specifically bind but not cleave DNA. *Eco*RI is one of the more frequently used endonucleases for genomic mapping. The enzyme is a dimeric globular protein with a molecular weight of 62,000 Daltons that recognizes the nucleotide sequence GAATTC on DNA molecules. The genetically engineered mutant of *Eco*RI, having Gln substituted for Glu at position 111 in the amino-acid sequence, has a high specific affinity for *Eco*RI binding sites but with rate constants for first- and second-strand cleavage reduced by a factor of 10^4 [49]. Using this mutant endonuclease, it is possible to identify *Eco*RI sites on *intact* DNA molecules by direct AFM imaging.

The feasibility of mapping, *Eco*RI site-specifically bound to DNA by direct AFM imaging, was first determined by using well characterized linearized plasmids, with one or two binding sites [50, 51], and bacteriophage lambda DNA, with five binding sites. These results showed that restriction sites on a large molecule, such as lambda DNA (50 kb), could be very accurately resolved. By imaging and measuring ten molecules or fragment lengths (distance between bound endonuclease), the lengths determined were within 1% of the known values [52]. Also, since this is a direct imaging technique

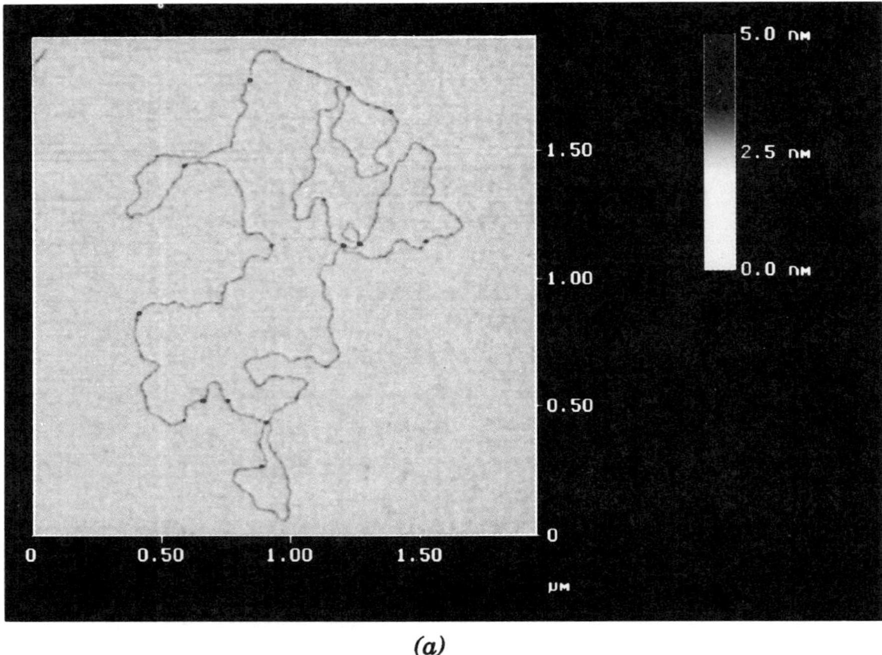

(a)

Fig. 8.5 AFM restriction mapping of individual DNA molecules is a reality. In this top view image (*a*) all of the *Eco*RI restriction endonuclease sites, on a 35 kb cosmid isolated from mouse chromosome 7, are labeled with bound enzyme. The *Eco*RI endonuclease has been genetically engineered to bind but not cleave DNA. Bound endonuclease can easily be identified by exploiting differences in height. The height scale in this image goes from light (*low*) to dark (*high*) such that increased height due to bound enzyme can easily be identified as a black dot. By tilting the image, as in this surface plot (*b*), the differences in height due to bound restriction enzyme can be dramatically illustrated. Perhaps future experiments will use this form of image presentation to identify two or more site-specifically bound proteins by differences in molecular weight and size.

where both the DNA and bound enzyme are clearly resolved, it is reasonable that restriction sites 20–30 base-pairs apart could be resolved with present technology.

In Fig. 8.5*a* an AFM image of a 35-kb cosmid clearly shows *Eco*RI sites labeled with the mutant endonuclease [52]. In the AFM the height information is exploited to document protein binding because imaged width is highly dependent on tip shape. For example, crystallographic analysis of *Eco*RI bound to DNA shows that the two subunits of the dimeric enzyme form a globular structure 5-nm wide at the binding site on the otherwise 2-nm wide DNA molecule [53]. However, the imaged width of DNA alone is on the order of 14 nm, assuming that the AFM tip has a radius of curvature of 12 nm. This can be calculated as follows: The observed width, $w = 4\sqrt{Rr}$, where R is the tip radius and r is the radius of the DNA molecule [54]. When

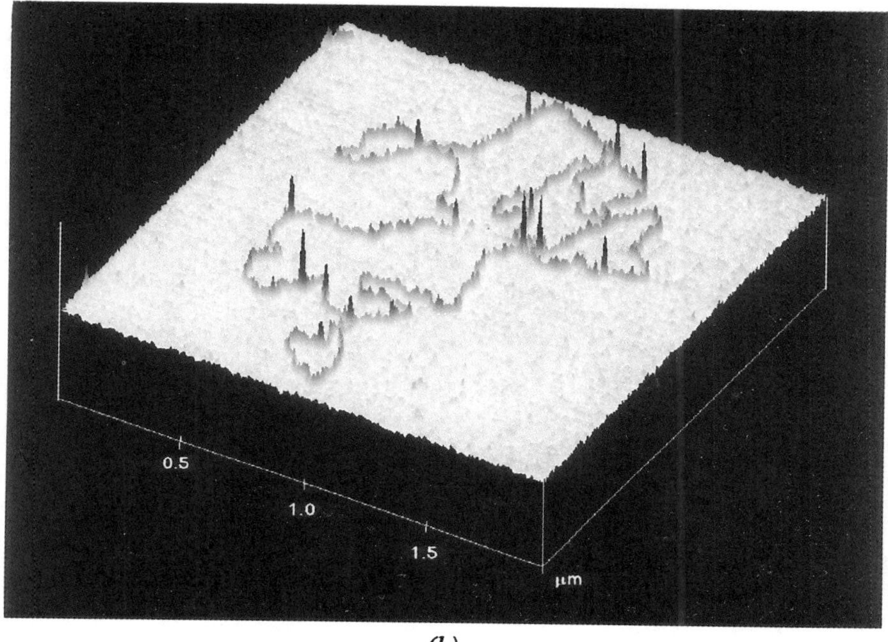

(b)

Fig. 8.5 *(Continued)*

this formula is applied to the enzyme-DNA complex, the calculated width is 22 nm—precisely what is observed. The point here is that even though the crystallographic data shows the enzyme-DNA complex to be two and one-half times the width of DNA alone, the observed ratio in the image is much less. However, the height information is preserved. By setting the vertical range to 4 nm, as in the top view image of Fig. 8.5*a*, small differences in height can be easily identified by the change in color contrast. By tilting the image, as in Fig. 8.5*b*, the contribution to height made by the bound enzyme is even more dramatic. It may even be possible to differentiate between bound endonucleases of different molecular weights to make double or multiple-label experiments possible. Once AFM mapping of genomic molecules using the mutant *Eco*RI system becomes routine, other important endonucleases currently used in genomic mapping could be similarly developed.

MAPPING OLIGONUCLEOTIDE PROBES TO GENOMIC DNA

AFM imaging and site-specific mapping of oligonucleotide gene probes to genomic DNAs is another important technique that can be developed using probe microscopes. RecA-protein-mediated homologous recombination of oligonucleotide sequences as

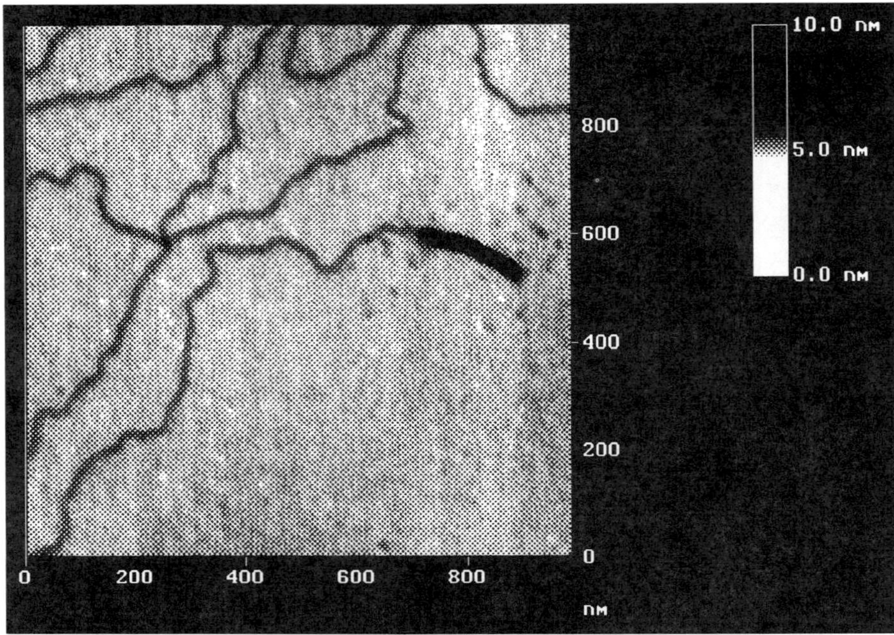

Fig. 8.6 AFM can be used to identify the RecA mediated site-specific recombination of oligonucleotides to double-stranded DNA molecules. In this image a 415-base sequence, homologous to one end of lambda DNA, is site-specifically recombined to a lambda DNA molecule and clearly resolved by AFM. When developed this technology could be used to map exons or other gene probes to important cloned genomic DNAs.

small as 15 bases and as large as several hundred bases have been shown to form stable complexes with duplex DNA [55] and these complexes have been imaged in the electron microscope [56]. In Fig. 8.6 a 400 base fragment isolated from lambda DNA by restriction digest and agarose gel electrophoresis can be seen recombined to its homologous sequence on one end of a lambda DNA molecule. Since there are three strands of DNA involved in the complex and the area of homology is also coated with RecA protein, image contrast is not a problem. Development of this methodology for mapping purposes would allow relatively short sequences, such as promoter elements, as well as larger probes, such as exons, to be located at high resolution on cloned DNAs. This technique could also be combined with restriction mapping, so AFM images of individual molecules would show both endonuclease bound to restriction sites and gene probes on cloned DNAs.

SEQUENCING DNA WITH PROBE MICROSCOPES

At this juncture it seems very likely that sequencing the human genome will be accomplished using conventional methods such as the Maxam/Gilbert or Sanger methods. However, next-generation methodologies will require faster more cost-effective techniques if other genomes or if individual human genomes are to be sequenced. Although there are a number of new technologies on the horizon, including scanning probes, at this juncture probe microscopes have yet to identify their first nucleotide base on an intact DNA molecule. Following the invention of STM, speculation arose as to how rapidly DNA might be sequenced using this instrument. The human genome project was already committed to sequencing the 3-billion nucleotide bases in the human genome and calculations were being made as to how fast the genome could be sequenced. For example, if bases could be sequenced at a rate of one per second, it would still take 100 years to sequence the genome. However, there were predictions made that STM had both the resolution to identify individual nucleotides on DNA molecules and could easily scan at 10 to 100 bases per second. In the following we speculate about the potential for scanning probe microscopes to sequence DNA, even though the first nucleotide is still waiting to be sequenced by this new technology.

As a starting point we have already established that images of DNA uncompromised by chemical fixation or staining can be accomplished with both the STM and AFM. With the STM, imaging is difficult primarily because of the poor conductivity of biomolecules. Nevertheless, if a biomolecule is mounted on a conductive surface as the STM tip passes over the molecule, either conductivity or the lack thereof can be sensed by the tip. Figure 8.7*a* shows the detail in an STM image of DNA where the molecule is imaged as a depression in the surface, as discussed above. The molecular ultrastructure within the molecule is more easily seen by inverting the height scale, as in Fig. 8.7*b*. Here bumps appear on the surface of the molecule that *may* correspond to helical repeats. Although the resolution is not yet sufficient to image individual nucleotide bases, perhaps future efforts may be successful in accomplishing that goal. Isolated bases, as imaged by STM (see Fig. 8.8), have sufficient resolution to identify and discriminate between the single ring pyrimidine bases (thymine, cytosine) and the double ring structure of the purine bases (guanine, adenine) [24, 57].

However encouraging these results are, it would still require much better resolution to differentiate topographically between the two purine and pyrimidine bases. For example, identifying cytosine from thymine would require imaging the subtle differences in groups present or absent on ring position 5 and 6. Thymine has a methyl group at position 5 and a carbonyl oxygen at position 6, while cytosine has an amino group at position 6. Perhaps by either physically or chemically tagging one purine and one pyrimidine of the four bases, resolution would not have to be improved and all four bases could be identified. Although DNA has not yet been sequenced with probe microscopes, the accomplishments to date have been sufficiently encouraging to continue the effort. Future development may depend on gaining a better understanding of imaging biomolecules with probe microscopes and in combining topography with other sensitivities of these instruments.

One such approach might employ secondary techniques along with topographic imaging to identify nucleotide base sequences. For example, topographic and differential conductivity images can be generated simultaneously to differentiate between poorly conductive DNA molecules and surface artifacts [20]. For topographic imaging a dc-bias voltage is applied between the tip and sample to maintain a constant tunneling current. By superimposing an ac-bias voltage between the tip and sample, the derivative

(a)

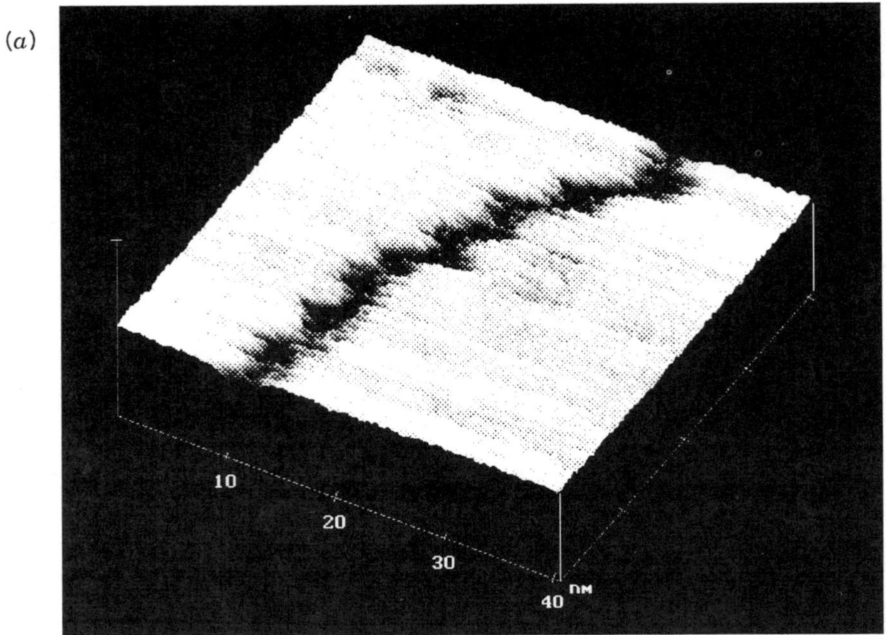

(b)

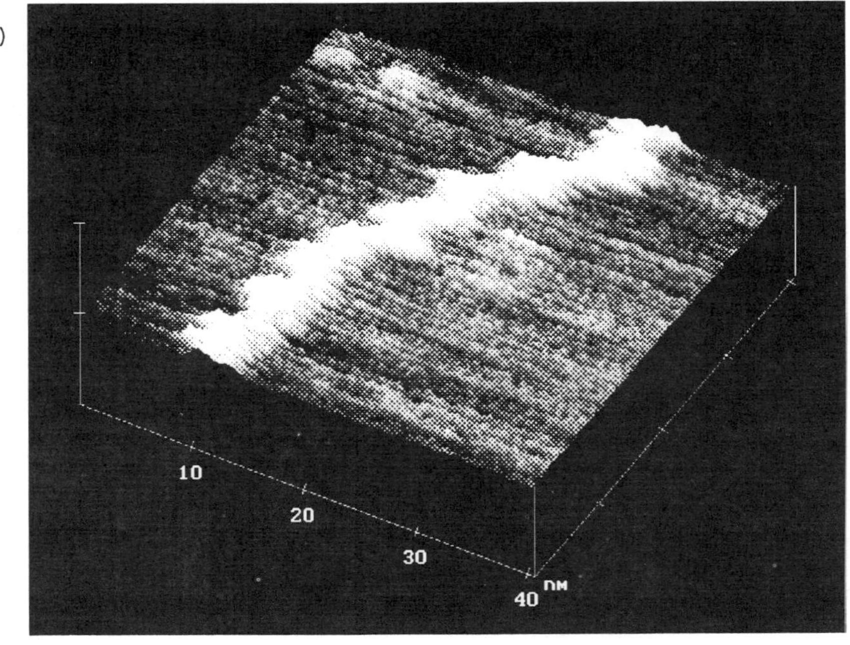

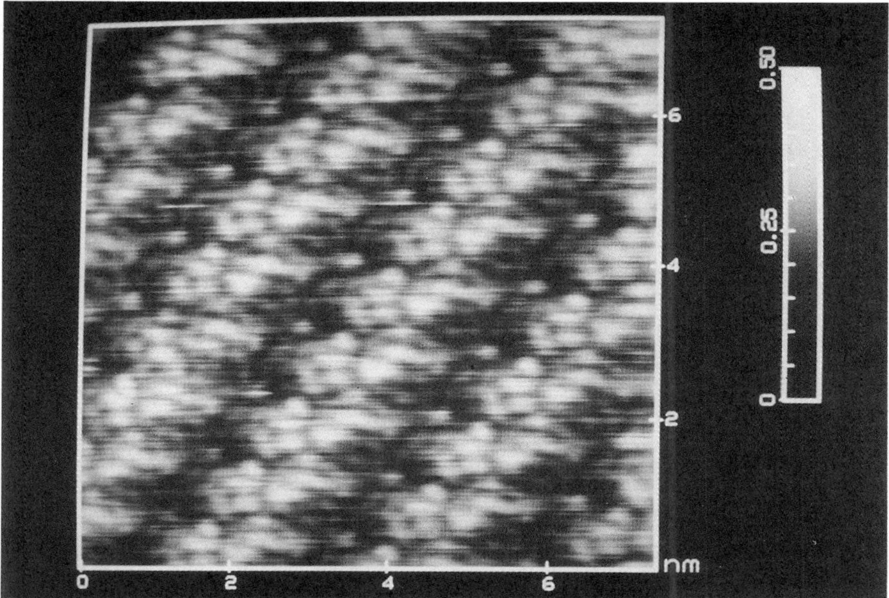

Fig. 8.8 *The high-resolution STM image of adenine molecules oriented into rows shows the double-ring structure of this nucleotide. By simply applying a droplet of solution containing one of the four nucleotide bases to a graphite surface and allowing the surface to air dry, images of all of the nucleotides have been accomplished. Topographic STM imaging of single-stranded DNA may be developed to resolve the double-ring structure of the purine bases (adenine, guanine) from the single-ring pyrimidine bases (thymine, cytosine). Image provided by Dr. Rod Balhorn , Biology and Biotechnology Research Division, Lawrence Livemore National Laboratory, Livermore, CA 94550.*

of the tunneling current with respect to the applied bias voltage is simultaneously recorded. This produces an image of the local electronic conductance for the structure directly beneath the probe tip. By this means large contrasts have been noted between DNA and intrinsic surface artifacts. This technique might be refined and extended to the level where topographic images could be combined with high-resolution electronic

Fig. 8.7 *A surface plot, high-resolution STM image of a portion of a DNA molecule. In (a) the raw data image is presented and clearly shows a trench like structure of the negatively contrasted DNA. Since AFM images are generated by computer analysis of many xyz voltage points the data can be presented in many ways. By simply inverting the image (b), the DNA molecule can be viewed in a more traditional fashion. In this presentation bumps along the DNA fiber may represent the helical repeats of the molecule.*

spectroscopy detecting conductivity differences in the bases so that the sequence could be visualized through contrast changes.

The development of other scanning probe instruments can pave the way toward alternative approachs that may prove feasible for sequencing DNA. For example, the photon scanning tunneling microscope (PSTM) [10] (and the near-field scanning optical microscope (NSOM) [58] are instruments that sense and utilize *optical* information provided by optically conducting probes. Although the resolution of these optical probe instruments is roughly tens of nanometers for visible light, not as good as the STM, the spectroscopic resolution is better than the STM electronic spectroscopy by a factor of 10^6. Perhaps a hybrid instrument may be possible that is capable of integrating STM or AFM high-resolution imaging capabilities with optical spectroscopy or fluorescence within one probe. Already the NSOM has been improved to follow surface topography and record optical data at a resolution of better than 20 nm [59].

Recently DNA has been imaged on mica, a nonconductive surface, using an STM specially equipped with a low-current preamplifier. Instead of using nanoamp tunneling currents, common with most conventional STMs, imaging was accomplished with picoamp tunneling currents [60].

Additionally the images were positive images, meaning that the tip did not exert excessive pressure on the DNA. Nevertheless, this report raises the interesting question as to how poorly conductive DNA molecules, mounted on a nonconductive surface, can be imaged by an instrument that operates by electronically sensing the surface. If imaging biomolecules can be accomplished on mica surfaces with a low-current STM, one ready advantage will be having a surface that is atomically flat over several hundreds of microns with the technology for mounting DNA in place. This will be a great improvement over the conductive gold surface, broken up with terraces, which is the preferred mounting surface for nanoamp STM imaging.

Independent of microscopy, the AFM cantilever has been demonstrated to act as a versatile, sensitive detector for species absorbed upon its surface [61]. This could be combined with sequencing-by-hybridization techniques (SBH). In SBH, all 65,536 possible single-stranded DNA 8-mers, for example, are disposed in an ordered matrix on a silicon surface. This chip is then allowed to react with target single-stranded DNA where complementary regions hybridize. Detection of the hybridized regions allows the sequence of the target DNA to be assessed. One detection approach involves these microcantilevers. The concept involves an array of microcantilevers covered with 8-mers on their surfaces and set into resonance. Hybridization of DNA in the picogram range will cause a detectable change in the resonance frequency of a cantilever, while the unhybridized members of the array remain unaffected. Using a surface coverage of 10^6 molecules/μm^2 determined by Moskos and Southern for glass, approximately 10^{-9} g of hybridized DNA could be expected for AFM-sized cantilevers, sensitivity is favorable for SBH detection.

An alternative approach involves recording the forces of hybridization. The cantilever *tip* would be coated with the DNA to be sequenced. The tip is then brought into contact with an array of all 8-mers immobilized on a hybridization chip immersed in a fluid medium. When the cantilever tip is withdrawn from the surface, the adhesion forces increase if the 8-mer on the cantilever is hybridized with the complimentary

8-mer on the chip. The advantage of this approach is that it could be readily used on hybridization chips that are already developed.

GLOSSARY

Sequencing   Identifying the order of nucleotide bases along the length of a DNA molecule.

Plasmid   Small circular self-replicating DNA molecules found in bacteria.

Restriction endonucleases   Enzymes that recognize and cleave DNA at sites with specific nucleotide sequences.

Oligonucleotide   A short polymer of nucleotides.

Tunneling   The penetration of a barrier by a wave under conditions in which a classical particle of the same energy would not penetrate the barrier.

Piezoelectric crystal   A crystal that changes its volume in response to an electric field applied between electrodes attached to the crystal.

REFERENCES

1. C.R. Cantor, S. Spengler (1992). Human genome 1991–92 program report. *DOE/ER-0544P*, U.S. Department of Energy, Washington, D.C.

2. J.D. Watson, N.H. Hopkins, J.W. Roberts, J.A. Seitz, A.M. Weiner (1988). *Molecular Biology of the Gene.* Benjamin/Cummings Menlo Park, CA, 249.

3. J. Sambrook, E.F. Fritsch, T. Maniatis (1989). *Molecular Cloning: A Laboratory Manual.* Cold Spring Harbor Lab., Cold Spring Harbor, NY.

4. A.M. Maxam, W. Gilbert (1977). A new method for sequencing DNA. *Proc. Natl. Acad. Sci.* 74:560–564.

5. F. Sanger, S. Nilken, A.R. Coulson (1980). DNA sequencing with chain-terminating inhibitors. *Proc. Natl. Acad. Sci.* 74:5463–5468.

6. G. Binnig, H. Rohrer (1982). Scanning tunneling microscopy. *Helv. Phys. Acta.* 55:726–735.

7. G. Binnig, C.F. Quate, C. Gerber (1986). Atomic force microscope. *Phys. Rev. Lett.* 56:930–933.

8. P.K. Hansma, B. Drake, O. Marti, S.A.C. Gould, C.B. Prater (1989). Scanning ion-conductance microscope. *Science* 243:641–643.

9. D.W. Pohl, W. Denk, M. Lanz (1984). Optical stethoscopy: Image recording with resolution lambda/20. *Appl. Phys. Lett.* 44, 651–653.

10. R.C. Reddick, R.J. Warmack, T.L. Ferrell (1989). New form of scanning optical microscopy. *Phys. Rev. B* 39:767–770.

11. D. Sarid (1991). *Scanning Force Microscopy with Applications to Electric, Magnetic, and Atomic Forces*, Oxford University Press, Oxford.

12. C.J. Chen (1993). *Introduction to Scanning Tunneling Microscopy.* Oxford University Press, New York.

13. G. Binnig, H. Rohrer (1984). *Scanning Tunneling Microscopy.* European Physical Society, The Hague.

14. T.P. Beebe, Jr., T.E. Wilson, D.F. Ogletree, J.E. Katz, R. Balhorn, M.B. Salmeron, W.J. Siekhaus (1989). Direct observation of native DNA structures with the scanning tunneling microscope. *Science* 243:370–372.

15. G. Travaglini, H. Rohrer, M. Amrein, H. Gross (1987). Scanning tunneling microscopy on biological matter. *Surf. Sci.* 181:380–390.

16. G. Lee, P.G. Arscott, V.A. Bloomfield, D.F. Evans (1989). Scanning tunneling microscopy of nucleic acids. *Science* 244:475–477.

17. D. Keller, C. Bustamante, R.W. Keller (1989). Imaging of single uncoated DNA molecules by scanning tunneling microscopy. *Proc. Natl. Acad. Sci.* 86:5356–5360.

18. P.G. Arscott, G. Lee, V.A. Bloomfield, D.F. Evans (1989). Scanning tunneling microscopy of Z-DNA. *Nature* 339:484–486.

19. C. Bendixen, F. Besenbacher, E. Laegsgaard, I. Stensgaard, B. Thomsen, O. Westergaard (1990). Deoxyribonucleic acid structures visualized by scanning tunneling microscopy, *J. Vac. Sci. Technol. A* 8:703–705.

20. D.P. Allison, J.R. Thompson, K.B. Jacobson, R.J. Warmack, T.L. Ferrell (1990). Scanning tunneling microscopy and spectroscopy of plasmid DNA. *Scanning Microsc.* 4:517–522.

21. M.-Q. Li, J.-D. Zhu, J.-Q. Zhu, J. Hu, M.-M. Gu, Y.-L. Xu, L.-P. Zhang, Z.-Q. Huang, L.-Z. Xu, X.-W. Yao (1991). Direct observation of B-form and Z-form DNA by scanning tunneling microscopy. *J. Vac. Sci. Technol. B* 9:1298–1303.

22. A. Cricenti, S. Selci, A.C. Felici, R. Generosi, E. Gori, W. Djaczenko, G. Chiarotti (1989). Molecular structure of DNA by scanning tunneling microscopy. *Science* 245:1226–1227.

23. S.M. Lindsay, T. Thundat, L. Nagahara, U. Knipping, and R.L. Rill (1989). Images of the DNA double helix in water. *Science* 244:1063–1064.

24. W.H. Heckl, D.P.E. Smith, G. Binnig, H. Klagges, T.W. Hansch, and J. Maddocks (1991). Two-dimensional ordering of the DNA base guanine observed by scanning tunneling microscopy. *Proc. Natl. Acad. Sci.* 88:8003–8005.

25. Y.L. Lyubchenko, S.M. Lindsay, J.A. DeRose, T. Thundat (1991). A technique for stable adhesion of DNA to a modified graphite surface for imaging by scanning tunneling microscopy. *J. Vac. Sci. Technol. B* 9:1288–1290.

26. S. Selci, A. Cricenti, A.C. Felici, R. Generosi, E. Gori, W. Djaczenko, G. Chiarotti (1990). Molecular structure of organic compounds observed by high resolution scanning tunneling microscopy. *J. Vac. Sci. Technol. A, Vac. Surf. Films* 8:642–644.

27. R.J. Driscoll, M.G. Youngquist, J.D. Baldeschwieler (1990). Atomic-scale imaging of DNA using scanning tunnelling microscopy. *Nature (UK)* 346:294–296.

28. D.D. Dunlap, C. Bustamante (1989). Images of single-stranded nucleic acids by scanning tunnelling microscopy. *Nature (UK)* 342:204–206.

29. W.M. Heckl, G. Binnig (1992). Domain walls on graphite mimic DNA. *Ultramicroscopy* 42–44:1073–1078.

30. C.R. Clemmer, T.P. Beebe, Jr. (1991). Graphite: A mimic for DNA and other biomolecules in scanning tunneling microscope studies. *Science* 251:640–642.

31. J.A. DeRose, T. Thundat, L.A. Nagahara, S. M. Lindsay (1991). Gold grown epitaxially on mica: conditions for large area flat faces. *Surf. Sci.* 256:102–108.

32. W.M. Heckl, K.M.R. Kallury, M. Thompson, C. Gerber, H.J.K. Hoerber, G. Binnig (1989). Characterization of a covalently bound phospholipid on a graphite substrate by X-ray photoelectron spectroscopy and scanning tunneling microscopy. *Langmuir* 5:1433–1435.

33. A. Cricenti, S. Selci, G. Chiarotti, F. Amaldi (1991). Imaging of single-stranded DNA with the scanning tunneling microscope *J. Vac. Sci. Technol. B* 9:1285–1287.

34. S.M. Lindsay, B. Barris (1988). Imaging deoxyribose nucleic acid molecules on a metal surface under water by scanning tunneling microscopy. *J. Vac. Sci. Technol. A, Vac. Surf. Films* 6:544–547.

35. G.M. Brown, D. P. Allison, R.J. Warmack, B.K. Jacobson, F.W. Larimer, R.P. Woychik, and W.L. Carrier (1991). Electrochemically induced adsorption of radio-labeled DNA on gold and HOPG substrates for STM investigations. *Ultramicroscopy* 38:253–264.

36. D.P. Allison, R.J. Warmack, L.A. Bottomley, T. Thundat, G.M. Brown, R.P. Woychik, J.J. Schrick, K.B. Jacobson, and T.L. Ferrell (1992). Scanning tunneling microscopy of DNA: A novel technique using radiolabeled DNA to evaluate chemically mediated attachment of DNA to surfaces. *Ultramicroscopy* 42–44:1088–1094.

37. D.P. Allison, T. Thundat, K.B. Jacobson, L.A. Bottomley, and R.J. Warmack (1993). Imaging entire genetically functional DNA molecules with the scanning tunneling microscope. *J. Vac. Sci. Technol. A* 11:816–819.

38. D.P. Allison, L.A. Bottomley, T. Thundat, G.M. Brown, R.P. Woychik, J.J. Schrick, K.B. Jacobson, R.J. Warmack (1992). Immobilization of DNA for scanning probe microscopy. *Proc. Natl. Acad. Sci.* 89:10129–10133.

39. L.A. Bottomley, J.N. Haseltine, D.P. Allison, R.J. Warmack, T. Thundat , R.A. Sachleben, G.M. Brown, R.P. Woychik, K.B. Jacobson, T.L. Ferrell (1992). Scanning tunneling microscopy of DNA: The chemical modification of gold surfaces for immobilization of DNA. *J. Vac. Sci. Technol. A* 10(pt.1):591–595.

40. T. Thundat, D.P. Allison, R.J. Warmack, and T.L. Ferrell (1992). Imaging isolated strands of DNA molecules by atomic force microscopy. *Ultramicroscopy* 42–44:1101–1106.

41. J. Vesenka, M. Guthold, C.L. Tang, D. Keller, E. Delaine, C. Bustamante (1992). Substrate preparation for reliable imaging of DNA molecules with the scanning force microscope. *Ultramicroscopy* 42–44:1243–1249.

42. C. Bustamante, J. Vesenka, C.L. Tang, W. Rees, M. Guthold, R. Keller (1992). Circular DNA molecules imaged in air by scanning force microscopy. *Biochem.* 31:22–26.

43. Y.L. Lyubchenko, A.A. Gall, L.S. Shlyakhtenko, R.E. Harrington, B.L. Jacobs, P.I. Oden, and S.M. Lindsay (1992). Atomic force microscopy imaging of double stranded DNA and RNA. *J. Biomol. Struct. Dyn.* 10:589–606.

44. T. Thundat, D.P. Allison, R.J. Warmack, M.J. Doktycz, K.B. Jacobson, and G.M. Brown (1993). Atomic force microscopy of single- and double-stranded deoxyribonucleic acid. *J. Vac. Sci. Technol. A* 11:824–828.

45. T. Thundat, D.P. Allison, R.J. Warmack, G.M. Brown, K.B. Jacobson, J.J. Schrick, T.L. Ferrel (1992). Atomic force microscopy of DNA on mica and chemically modified mica. *Scanning Microsc.* 6:911–918.

46. M. Bezanilla, B. Drake, E. Nudler, M. Kashlev, P.K. Hansma, H. Hansma (1994). Motion and enzymatic degradation of DNA in the atomic force microscope. *Biophys. J.* 67:2454–2459.

47. T. Thundat, R.J. Warmack, D.P. Allison, K.B. Jacobson (1994). Critical point mounting of kinetoplast DNA for atomic force microscopy. *Scanning Microsc.* 8:23–30.

48. T. Thundat, D.P. Allison, R.J. Warmack (1994). Stretched DNA structures observed with atomic force microscopy. *Nuc. Acid. Res.* 22:4224–4228.

49. D.J. Wright, K. King, P. Modrich (1989). The negative charge of Glu-111 is required to activate the cleavage center of *Eco*RI endonuclease. *J. Bio. Chem.* 264:11816–11821.

50. D.P. Allison, T. Thundat, P. Modrich, R.J. Isfort, M.J. Doktycz, P.S. Kerper, R.J. Warmack (1995). Mapping site-specific endonuclease binding to DNA by direct imaging with AFM. *SPIE* 2386:24–29.

51. D.P. Allison, P.S. Kerper, M.J. Doktycz, J.A. Spain, P. Modrich, F.W. Larimer, T. Thundat, R.J. Warmack (1996). Direct AFM imaging of *Eco*RI endonuclease site specifically bound to plasmid DNA molecules. *Proc. Natl. Acad. Sci.,* in press.

52. D.P. Allison, P.S. Kerper, M.J. Doktycz, T. Thundat, P. Modrich, F.W. Larimer, L.J. Stubbs, D.K. Johnson, M.L. Mucenski, R.J. Warmack (1996). Restriction mapping individual cosmid DNAs by direct AFM imaging of bound *Eco*RI endonuclease. *Science,* submitted.

53. J.A. McClarin, C.A. Frederick, B.-C. Wang, P. Greene, H.W. Boyer, J. Grable, J.M. Rosenberg (1986). Structure of the DNA-*Eco*RI endonuclease recognition complex at 3A resolution. *Science* 234:1526–1441.

54. T. Thundat, X.-Z. Zheng, S.L. Sharp, D.P. Allison, R.J. Warmack, D.C. Joy, T.L. Ferrell (1992). Calibration of atomic force microscope tips using biomolecules. *Scanning Microsc.* 6:903–910.

55. P. Hsieh, C.S. Camerini-Otero, R.D. Camerini-Otero (1992). The synapsis event in the homologous pairing of DNAs: RecA recognizes and pairs less that one helical repeat of DNA. *Proc. Natl. Acad. Sci. USA* 89:6492–6496.

56. B.M.J. Revet, E.P. Sena, D.A. Zarling (1993). Homologous DNA targeting with RecA protein-coated short DNA probes and electron microscope mapping on linear duplex molecules. *J. Mol. Bio.* 232:779–791.

57. M.J. Allen, M. Balooch, S. Subbiah, R.J. Tench, W. Siekhaus, R. Balhorn (1991). Scanning tunneling microscope images of adenine and thymine at atomic resolution, *Scanning Microsc.* 5:625–630.

58. D.W. Pohl, U.C. Fischer, U.T. Duerig (1988). Scanning near-field optical microscopy (SNOM). *J. Microsc.* 152:853–862.

59. E. Betzig, P.L. Finn, and J.S. Weiner (1992). Combined shear force and near-field scanning optical microscopy. *Appl. Phys. Lett.* 60:2484–2486.

60. R. Guckenberger, M. Heim, G. Cevc, H.F. Knapp, W. Wiegrabe, A. Hillebrand (1994). Scanning tunneling microscopy of insulators and biological specimens based on lateral conductivity of ultrathin water films. *Science* 266:1538–1540.

61. E.A. Wachter, T. Thundat (1995). Micromechanical sensors for chemical and physical measurements. *Rev. Sci. Instrum.* 66:3662–3667.

A Miniature Integrated Nucleic Acid Analysis System

M. ALLEN NORTHRUP,* BART BEEMAN, BILL BENETT,
DEAN HADLEY, PHOEBE LANDRE, AND STACY LEHEW

CONTENTS

---

* Current address: Cepheid, 3410 Garrett Dr., Santa Clara, CA 95054

Automation Technologies for Genome Characterization, Edited by Tony J. Beugelsdijk.
ISBN 0-471-12806-6   © 1997 John Wiley & Sons, Inc.

INTRODUCTION

In the past few years there has been a growth of interest in using traditional electronic integrated circuit (IC) manufacturing technology to perform micromachining of silicon-based materials. These have been converted into micron-sized electrical and mechanical devices for a variety of applications. The utility of this technology as a means to build structural components was pointed out by Petersen in 1982 [1]. One of the most dramatic extensions of micromachining is in the fabrication of electrostatically driven micromotors. These motors have been built with feature sizes in the tens of microns, with the ability to attain very high rotation rates. One example of many is the work of Lee [2] (Fig. 9.1). This silicon-based microvibromotor uses electrostatic forces to drive a piston that impinges on a 60-micron diameter rotor, sending it to rotation rates of over 60,000 rpm. A large variety of microstructures, actuators, sensors, and the like, have been built. This growing enabling technology area is referred to as microelectromechanical systems or MEMS. Advancement in other areas ranging from imaging science, diode technology, circuits, plastics manufacturing, and dye chemistries, for example, strongly compliment MEMS and is allowing new capabilities in analytical instrument components.

Concurrent with the growth of interest of MEMS or micromachining has been the growth and advancement of biotechnology. The ability to perform recombinant genetic techniques, for example, has allowed the explosion of a whole new industry that is

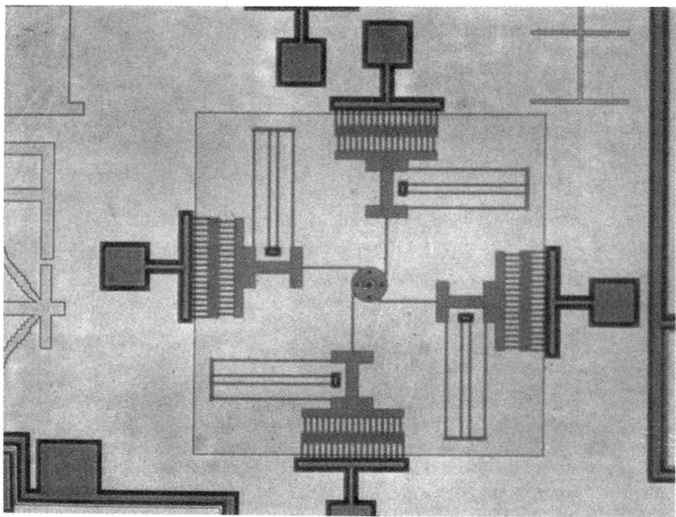

Fig. 9.1 Photograph of microvibromotor (courtesy of A. Lee). Microvibromotor in center is 60 μm in diameter and is impact driven by the surrounding four resonant comb drives with resonant frequencies as high as 15 KHz.

involved with applications as broad as agriculture, medicine, basic biology, and food science, to name a few. Other techniques, such as the invention of the polymerase chain reaction (PCR) in 1985 at Cetus Corporation in Emeryville California, have added new fuel to the momentum of the biotechnology industry. The PCR technique [3–5] is a way to make billions of copies of a specific sequence of DNA with simple thermal control of the reagents and led to the 1993 Nobel Prize in Chemistry. PCR has revolutionized biotechnology and biological science. Its use is so pervasive in the biotechnology industry that it would be impossible to work in the area effectively without it. The instrumentation required to perform this enzyme-mediated synthesis is based on precise control of thermocycling of the reaction, typically between 45°C and 95°C.

The combination of MEMS and biotechnological advancements for instrumentation shows great promise. It has been shown that analytical performance is improved as a result of the miniaturization. Electrophoresis, for example, benefits through the ability to precisely control channel dimensions and electric field strength. Examples of applications of this kind include nucleic acids [6–9] and immunoassays [10]. The restriction to glass substrates in those examples is due to breakdown voltage problems with silicon. Such limitations reflect the fact that materials properties are very important in performing real analyses with these devices. The miniaturization of electrophoresis is a good example of a technology that requires instrument capabilities augmented by MEMS. Micromachining as an enabling technology also applies, in a broad sense, to manufacturing of all possible materials including polymers with micron-sized dimensions. In nearly all cases in the current literature, a variety of materials are used and most still rely on macrosized support equipment.

Issues such as compatibility of the surfaces with the reaction chemistry, fluidic interconnects, and ease of use still plague the acceptance of the micromachined devices for commercial use. As with any leading-edge technology, the ultimate usefulness will provide some momentum for transfer to industry. A recent review [11] provides a summary of what the merger of the technologies has achieved. For example, some groups have developed plain silicon reaction chambers for PCR [12, 13], but without the integration of devices such as heaters. Recently another group [14] has discussed highly integrated DNA analysis systems but have not integrated the components together on a microscale level. Micromachined electrophoresis channels as individual devices, however, have been highly developed and tested [15–20]. The next step is to truly integrate several miniaturized components and perform real analyses.

Micromachined or MEMS-based PCR reaction chambers benefit from miniaturization for fundamental reasons. Reduction in thermal dimensions, conduction, response times, and power consumption been attained with MEMS-based reaction chambers [21–23]. In addition, due to their simplicity, batch fabrication is realistic. Results from the use of MEMS-based reaction chambers to perform the PCR technique, along with real-time, diode-based optical reaction monitoring and integration with microelectrophoretic separation has been shown to exemplify the convergence of the two technologies. Reaction optimization, decreased analysis time and portability have also been shown to result from the combination of technologies. The complete miniature analytical thermocycler instrument (MATCI) has shown equivalence and improvements over the state-of-the-art commercial systems.

METHODS

MEMS-Based Reaction Chambers

Micromachining of silicon using IC fabrication technologies to make MEMS has been divided into two types of processes: bulk and surface. In general, bulk, micromachining involves removing or modifying the single-crystal silicon wafer or other substrate. On the other hand, surface micromachining is based on adding layers to the substrate and creating structures through subsequent modification of those layers. Typical MEMS-based devices have aspects of both, as in our reaction chambers. The PCR reaction chambers (Fig. 9.2) were fabricated in three-inch round, 40-mil thick double-polished 100-oriented silicon wafers. After RCA cleaning and low-pressure chemical vapor deposition (LPCVD) of low stress (200–300 MPa) silicon nitride (1.0–2.0-μm thick), the nitride was photolithographically patterned and reactive ion etched (RIE) through to the silicon (defining the chamber surface area). The photoresist was then removed, the bulk substrate silicon was etched to a depth of 850 μm with KOH to form the chambers, and a repeat RCA cleaning was performed. LPCVD polysilicon (3000 Å thick) was then doped at high temperature with boron to a sheet resistance of 400 ohms/square. The polysilicon was photolithographically patterned, etched, and the photoresist was again removed. Photolithographic definition of the heater contacts was performed with subsequent E-beam deposition of 2500 Å of gold (5 Å/s) on top of titanium (2 Å/s). After lift-off of the undesired metal, the wafer was sawed into individual reaction chamber halves. The complete chambers were formed by bonding two resistivity-matched halves together with polyimide (Epo-Tek 600, Epoxy Technology) followed by over eight hours of curing at room temperature. A 30 AWG Teflon insulated thermocouple (type K Omega, Stamford, CT) was glued with thermal conducting epoxy (Tra-bond BB2151, Tra-Con, Medford, MA) to the outside wall of the reaction chambers to monitor bulk silicon temperatures.

Extrusion-molded, medical grade, thin-walled polypropylene reaction chamber liners (inserts) were fabricated at the Lawrence Livermore National Laboratory (LLNL) plastics manufacturing facility. These inserts are 2 cm long with an internal diameter of 1.0 mm and a 1.7-mm outside diameter that tapers down to 1.5 mm. The top expanded to a 4.0-mm diameter cup-shaped top. The reaction volume was typically 20 μl.

The silicon chambers without inserts have been shown to be reusable with cleaning; however, overall reproducibility is less than with the inserts. Thermal cycling with 0.5 N HCl for a few cycles and rinsing with water was adequate to clean the bare chambers. The cleaned and dried chambers were given a final treatment with 10% dichloro-dimethylsilane in chloroform when reactions were performed without liners.

PCR Reactions

Most of the PCR reagents were supplied by Roche Molecular Systems (Alameda, CA). The beta actin *Taq*man Assay was purchased from Perkin Elmer/Applied Biosystems

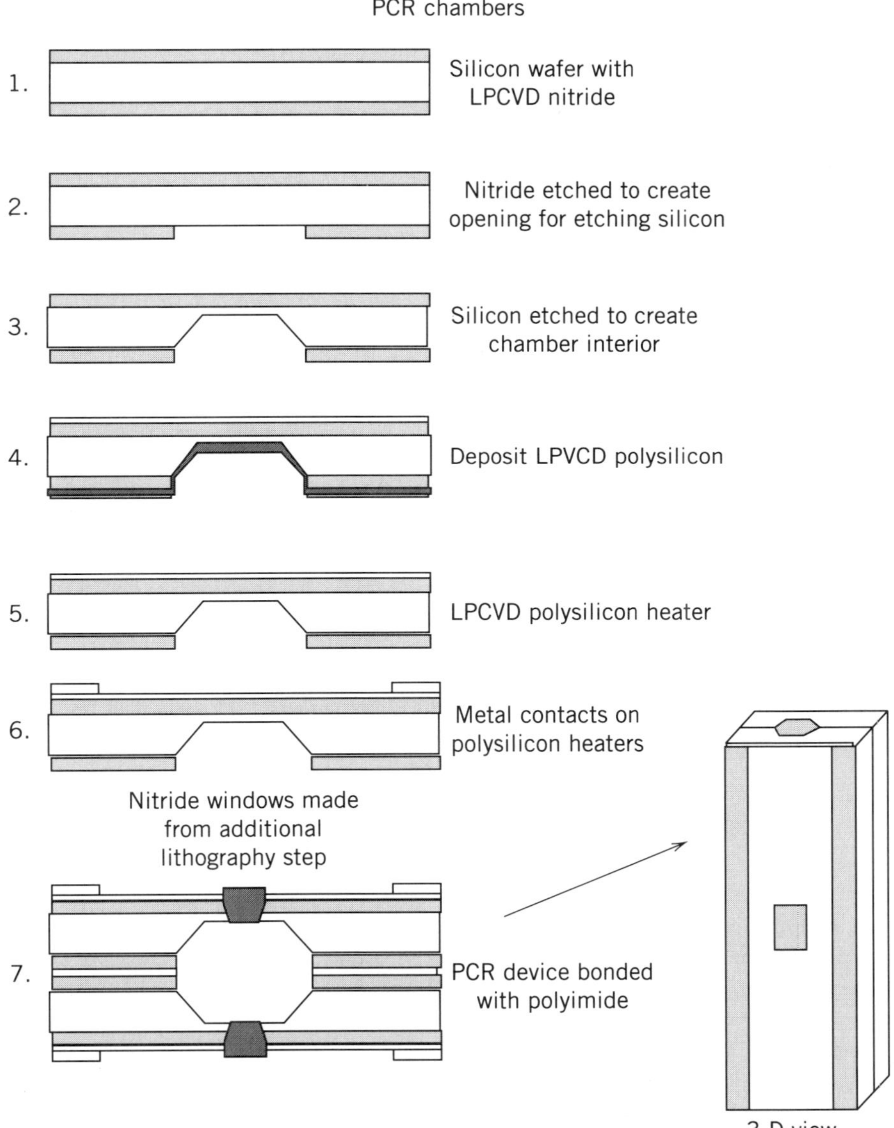

Fig. 9.2 Schematic of the silicon-based PCR reaction chamber fabrication.

Division (Foster City, CA). Master PCR mixes typically contained 50-mM KCl, 10-mM Tris-HCl pH 8.3, 1.5–3.0-mM $MgCl_2$, 200 μM each of deoxynucleotide, 0.5 μM each of two oligonucleotide primers, 2.5 units per 50-μl AmpliTaq® DNA polymerase and target template at a specified copy number per 50-μl reaction. The human genomic multiplex PCR for the detection of cystic fibrosis (CF) used double-stranded DNA derived from the cultured cell line, HL60, as the template. In some reagent mixes dUTP was substituted for dTTP and uracil glycosylase was added at 50 units/ml to eliminate carryover contamination as well as enhance reaction specificity and detection in reactions with template at low initial copy number. Master mix was added to the micromachined silicon chambers and overlaid with mineral oil to seal the top of the chambers. A control reaction was aliquoted from the master mix and thermocycled in a Perkin-Elmer GeneAmp® 9600 Thermocycler. The GeneAmp® 9600 reactions were typically cycled at 93°C, 60°C, and 72°C each for 30 seconds. Product verification was done by agarose gel electrophoresis in 1X TBE buffer (89-mM Tris, 89-mM borate, 2-mM EDTA), stained with ethidium bromide and photographed using UV light (302 nm) illumination.

The surface immobilized probe, reverse dot-blot assay (provided by Roche Molecular Systems) uses PCR product-complimentary oligonucleotides immobilized onto a nylon membrane. Hybridization to the PCR products is detected with a biotin-streptavidin, horseradish peroxidase, colorimetric assay. The substrate is added, and the colorimetric change associated with the enzyme-mediated reaction is detected visually.

For the kinetic monitoring system, ethidium bromide (2 μg/ml) was added to the PCR reaction mixtures prior to amplification. The fluorescence-energy (Taqman) procedure followed the instructions for the Perkin Elmer/Applied Biosystems beta actin control.

MATCI Control Electronics and Software

The system to control temperature within the chamber and for thermal cycling uses Pulse Width Modulation (PWM) heating. First, a temperature setpoint is implemented by the computer. This turns the PWM "full on," and the PCR reaction chamber starts to heat while reading the thermocouple bonded to the reaction chamber. As the temperature of the reaction chamber reaches the setpoint, the PWM starts to control, and the setpoint temperature is maintained to an accuracy of $+/- 0.5$°C. The temperature is sent to the external computer "real time," and a graph is updated to display temperature. After the setpoint is set to cool the reaction chamber, the PWM turns off, and the reaction chamber starts to cool. At this time the fan circuit turns and draws air across the exposed chamber surfaces. The PWM will start to control again as the temperature reaches the new setpoint. The fan circuit is enabled when over-temperature is indicated. Average operating power was 1.2 W. Reaction (20-μL) heating rates of 15–20°C/s with polypropylene liners were achieved. At 15 VDC input voltage, 3–6°C/s heating rates were typical. Cooling rates ranged from 4–8°C/s. An automated 5-VDC, 1.5-inch diameter cooling fan was used to augment the cooling. Fast heating rates (20–30°C/s) were achieved by increasing the input voltage to 25–35 VDC. Figure 9.3 shows an example of an ultrafast thermal cycle series (35 cycles in 10 minutes).

The optical detection apparatus for real-time detection consisted of low-cost components. The excitation source was a "Superbrite" blue light emitting diode (LED) (Ledtronics, Inc.) with its emission centered at 450 nm, $+/- 30$ nm at half max. The power input was 4 VDC, 80 mA. Both photo diodes and CCD imagers were used to collect the fluorescence emission. The photo diode used was a low-light-level detector

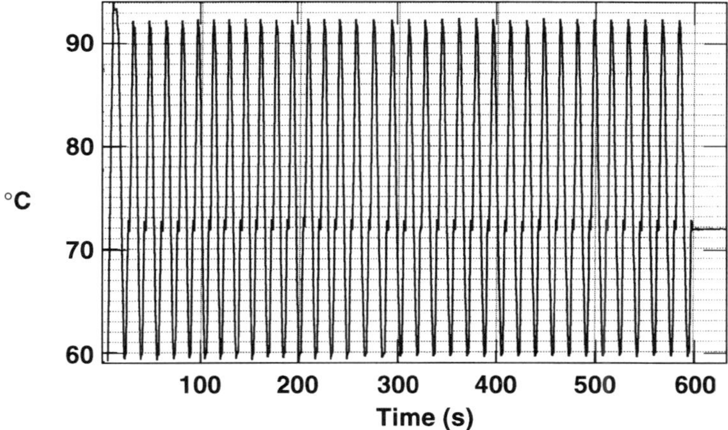

Fig. 9.3 Plot of the temperature of the PCR chamber as a function of time during a fast PCR on the MATCI. Thirty-five cycles were done in 10 minutes (92°C for 1 s, 60°C, for 1 s, 72°C for 2 s) to successfully amplify a 268 base-pair product from a β-globin target cloned in M13.

(Centronic, p/n OSD35-LR-A). Excitation wavelength filter for ethidium bromide monitoring was a 520-nm short pass and the detection filter was a 597-nm long pass. Taqman detection filters were band-pass and centered at 540 +/− 40 nm and 580 +/− 40 nm for the fluorescein and rhodamine labels, respectively. A 480 +/−40-nm short-pass filter was used as the excitation filter. A 0.5-inch diameter collumnating lens was used to focus the LED light into the top of the reaction chamber. Either plastic molded caps or a mineral oil layer (detection experiments) over the PCR reaction fluid prevented evaporation during thermal cycling.

A Macintosh Quadra 650 using IGOR Software (Wavemetrics Inc., Lake Oswego, OR) was used for input/output. The software performed all the instrument timing functions and data collection. A National Instruments board (Model MI0-16, Austin, TX) was used as the I/O interface.

Each reaction chamber was initially calibrated prior to use by correlating reaction fluid temperatures with thermocouple readings, and the power and software were adjusted to compensate for differences.

RESULTS AND DISCUSSION

Multiplex PCR Reactions

A seven primer pair multiplex PCR system (provided by R. Saiki, Roche Molecular Systems, Alameda, CA) has been developed for the detection of cystic fibrosis (CF) that amplified approximately 2558 base pairs (7 separate amplicons) of the coding region. The primer pairs were chosen to result in equivalent amplification of discrete amplicons ranging from 263 to 482 base pairs. Efficient and nearly equivalent amplification of each of the seven amplicons was demonstrated in both the miniature and commercial instruments as shown by agarose gel electrophoresis analysis (Fig. 9.4). Since multiplex amplification and the use of genomic DNA as a template for PCR have proved more challenging to amplify than single amplicon systems, these results demonstrated the versatility and biocompatibility of the new reaction chambers. These results are from PCRs performed directly in cleaned and silanized silicon chambers, without polypropylene liners.

Surface-Immobilized DNA Probes

Figure 9.5 represents reverse dot-plot analysis of products of the CF multiplex amplification generated on both the MATCI and GeneAmp thermocyclers. The results are equivalent. On the far left of the test strips, the letter *F* represents the normal condition (no mutation), and the Δ represents a CF mutation. All three possible allelic conditions of having the disease were determined (i.e., homozygous normal and CF positive, and heterozygous carrier). Fourteen additional mutations that can be detected are represented by the letters *a–n* on the test strip and were not tested here.

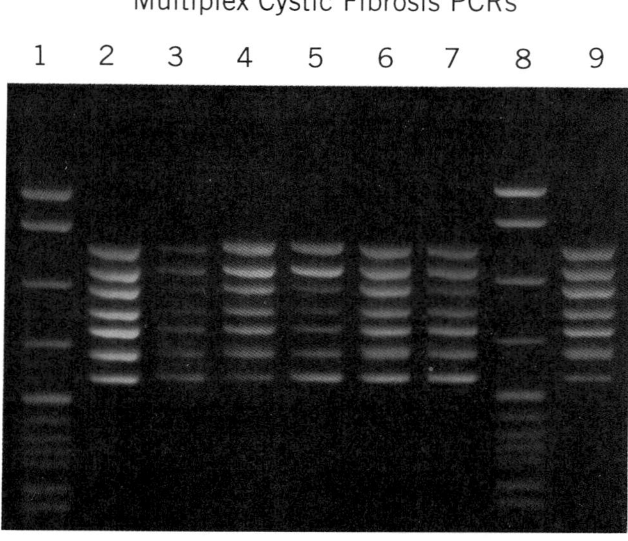

Fig. 9.4 Agarose gel electrophoresis (3% NuSeive, 1% Seakem agarose, FMC Corp.) results of a 7 amplicon multiplex amplification of a cystic fibrosis target to compare amplification in the Perkin-Elmer Gene Amp® 9600 to that in the MATCI. Lanes 2, 4, and 6 show the results of amplification in the GeneAmp® 9600 of the cystic fibrosis multiplex PCR system (263 to 482 base pairs) from three human genomic DNA targets. Lanes 3, 5, and 7 show the results of the same targets amplified in the MATCI.

Cystic Fibrosis Detection

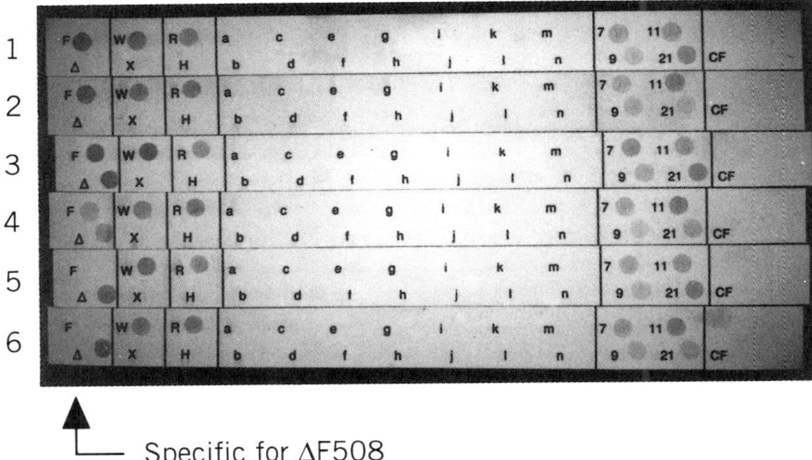

Fig. 9.5 Reverse dot blot results from the cystic fibrosis multiplex amplification with the GeneAmp® 9600 (rows 2, 4, and 6) and the MATCI (rows 1, 3, and 5).

Real-Timing Monitoring

Double-Stranded DNA Production   The kinetic, real-time, detection of amplification was demonstrated in the MATCI using ethidium bromide. These results are similar to those achieved on a large table-top system [24]. Figure 9.6 depicts the miniaturized real-time detection system module used in the instrument. Figure 9.7 shows the results obtained by the MATCI monitoring of a betaglobin target (268 bp) using genomic DNA. Four different starting concentrations (10 ng–10 μg) were used to demonstrate the quantitative nature of the information gathered. Calculation of the expected reaction productivity demonstrates 85% efficiency in this instrument. In addition to the demonstration of kinetic monitoring in these devices, two additional conclusions were reached. First, comparison of the fluorescence profiles of reactions (cycle number when the first positive signal was observed and slope of the subsequent fluorescence increase) carried out in standard microfuge tubes and GeneAmp 9600 to those carried out in silicon devices demonstrated that the amplification efficiency per cycle was equivalent. This indicates that the biocompatibility and thermal parameters of the silicon devices matches that of the more typically used reaction tubes and thermocyclers. Second, kinetic PCR will greatly facilitate future studies in these devices because of the opportunity to independently examine amplification efficiency and parameters that effect plateau.

Real-time, Specific Probe Detection   Sequence-specific, fluorescently labeled probes can be used to exploit the exonuclease activity of Taq DNA polymerase to signal the accumulation of specific product [25–27]. This has been designed for the

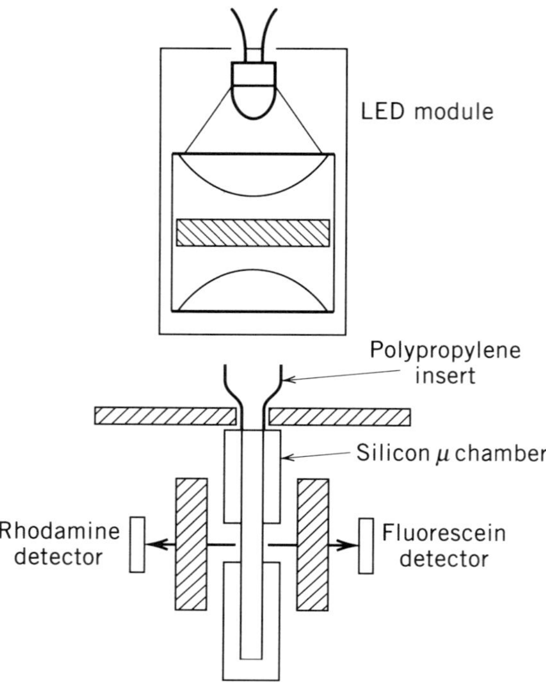

Fig. 9.6 *Schematic of the real-time optical detection module in cross section.*

detection of a PCR product from the amplification of the beta-actin gene from human genomic DNA and has also been performed successfully in the MATCI. It is unique in its ability to obtain and display the fluorescence real-time during the reaction. This greatly speeds up the assay, allows optimization, and real-time quantitation. Figure 9.8 shows the Taqman detection of the beta-actin gene in the MATCI. In this case the fluorescein (540-nm) fluorescence started to deviate from background between 24 and 26 cycles. This agrees with the expected results based on the starting concentration (10 ng/50 μl) of target DNA. This assay has given consistent quantitative results in the MATCI. Comparisons with the commercial system (ATC 7700, Perkin-Elmer, Applied Biosystems) have been performed as has the analysis of a large series of NIH standards and patient tissue biopsies for HIV and Hepatitis C viral RNA [28]. The commercial system does not provide the results real-time, and one must wait for the thermal cycling to be completed ($\sim$2.0 hours for 40 cycles) to get results. The MATCI has completed Taqman detection in less than 40 minutes (50 cycles) of HIV RNA (2.2 $\times$ 10^5 starting copies)—total reaction time was 45 minutes. This was after 16 cycles and an initial 30-minute reverse transcriptase (RT) step (converts RNA to DNA) at 40°C. In direct DNA PCR, such as the beta-actin Taqman, the MATCI has performed 40 cycles in 25 minutes.

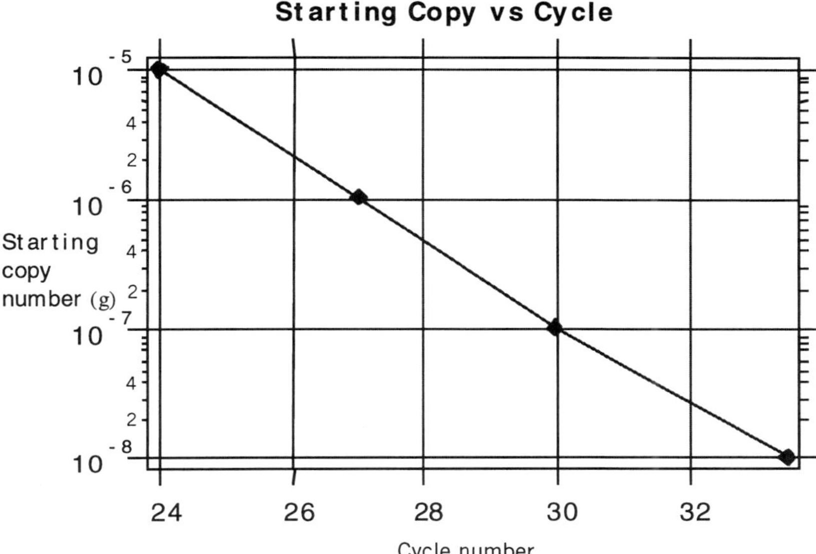

Fig. 9.7 Plot of the starting copy number as a function of cycle number when the first fluorescent signal is obtained. The amplification of the 268 base-pair product from four concentrations of β-globin genomic target (10 ng–10 μg) was monitored with ethidium bromide using the real-time monitoring system as described.

Fast Thermal Cycling and Reaction Optimization

Wittwer et al. [29] have shown that thermal cycling parameters affect PCR productivity. In the MATCI, anneal temperatures and times and extension times were also shown to have such effects. The micromachined silicon-based reaction chambers were typically thermal cycled with the controller at low power (15 Vdc, 0.5 amps [max], 50% duty cycle). We have also performed rapid thermal cycling to determine how fast the silicon-reaction-chamber instrument can cycle and what the effects will be on PCR productivity. For example, we have increased the power input to the chamber to 30 V and 0.8 amps (peak) to obtain heating rates that approach 30°C/s. for a reaction volume of 50 μl. At the same time the denaturation time at 96°C and annealing periods at 55°C were reduced to minimal times (1 second). Extension times at 72°C were reduced to 2 seconds. These fast PCR thermal cycling conditions produced nearly equivalent amounts of PCR product (betaglobin, 268 bp target on genomic DNA) as did the commercial instrument, which has a maximum heating and cooling rate of 1°C/s. The commercial system took over one hour to perform the standard 30 cycle profile (10^8 starting copies of betaglobin target cloned onto M13 bacteriophage), whereas the MATCI required only 8 minutes. We studied the effect of anneal temperature on fast PCR product and showed that increasing anneal temperatures decreased productivity (Fig. 9.9). As a result we were able to optimize the betaglobin PCR to be 1 second each at the denature

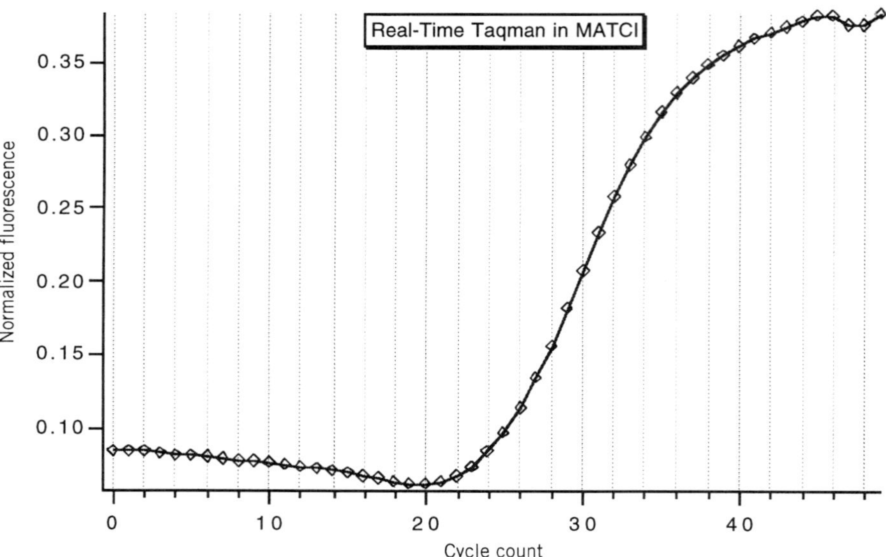

Fig. 9.8 Plot of normalized fluorescence as a function of cycle count. The Taqman amplification and real-time detection of the β-actin control (Perkin-Elmer) was performed using the MATCI with integrated real-time optical detection.

(96°C) and anneal temperature (50°C), with a 10-second extension time. However, it is important to note that the 2-second extension time was able to produce sufficient product for standard gel electrophoresis detection, this indicates a greater than 100 bp/s extension rate for the enzyme. The same betaglobin product was amplified from 100 ng of human genomic DNA (40 cycles, 14 minutes). Other optimization studies include the amplification of an eight amplicon CF multiplex system (provided by Roche Molecular Systems) employing polypropylene reaction tube liners in 30 minutes (45 cycles) at low-power settings. In addition we showed the augmented productivity afforded by the MATCI of a 5-amplicon multiplex PCR for the identification of the pathogenic bacteria *Clostridium perfringens,* where the optimal PCR profile was 8.8, and 5 seconds each for denature, extend, and anneal times, or a total of 19 minutes to perform 40 cycles [30]. These results demonstrate that DNA amplification can be optimized through faster thermal cycling and with low power.

The combination of extremely fast thermal cycling and real-time monitoring will allow for detailed studies of the kinetics and limiting parameters for reactions such as PCR, thereby permitting new capabilities for reaction optimization. Real-time detection and quantitation of the amplification results and more recent rapid thermal cycling times permitted by the silicon reaction device will now permit analytical investigation of extremely rapid thermal cycling. It will also cut the cost and time required for important nucleic acid based analytical methods. The system is also highly portable and battery operated for field studies.

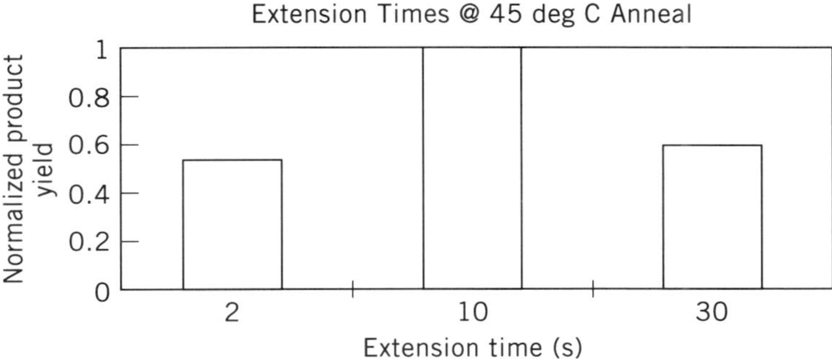

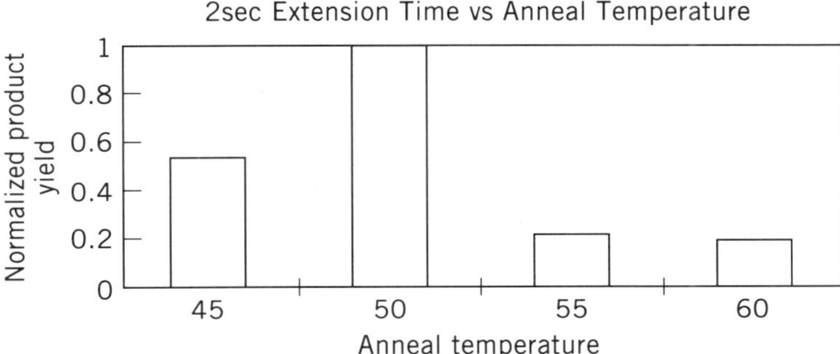

Fig. 9.9 *Plot of normalized product yield as a function of extension times or anneal temperatures in fast PCR in the MATCI. PCR product yield was determined by densitometer scanning of ethidium bromide stained agarose gels. Comparison of extension times (96°C for 1 s, 45°C for 1 s, 72°C for 2, 10, or 30 s) is shown in the upper panel. Comparison of anneal temperatures (96°C for 1 s, 45°, 50°, 55°, or 60°C for 2 s) is shown in the lower panel.*

SUMMARY

Development of a miniature, reaction-based analytical instrument has been achieved. Employing a microfabricated silicon-based, reaction chamber and simple feedback control circuitry, we have demonstrated (1) multiplex PCR amplification on genomic DNA, (2) confirmation of amplification specificity by using probe immobilized strips, (3) ultrafast thermal cycling and determination of key reaction parameters, (4) incorporation of a diode-based fluorescence monitoring of DNA production, and (5) real-time fluorescence detection of the generation of specific PCR products during amplification. The combined detection and fast thermal cycling of the low-cost, miniaturized, portable system has facilitated analytical studies. The integration of amplification and detection in the same device allows for the rapid assessment of samples with minimal handling as well. The coupling of the MATCI thermocycler reaction chamber module with a microchannel electrophoresis system has also been demonstrated [31]. That system uses an electroosmotic valve to electrokinetically inject the PCR reaction mixture into the channel directly for on-line detection of the product. In those results real-time PCR monitoring was accomplished by injecting at a series of intermittent cycles during the reaction.

Due to the typically small volumes, forces, and energy input of bioanalytical instrumentation, MEMS technology is particularly well suited for integration into such systems. True integration and maximal optimization of the reaction parameters require control at the "point" (microscopic) level. Such control will allow more precise experimental verification of reaction optima, limits, and requirements. In addition, due to increased efficacy, they become inexpensive alternatives to current commercial instruments.

ACKNOWLEDGMENTS

This work was performed under the auspices of the U.S. Department of Energy by Lawrence, Livermore National Laboratory, contract number W-7405-ENG-48. The authors acknowledge the support of Dr. Ken Gabriel of the MEMS program of the Advanced Research Projects Agency and the collaboration of Robert Watson Jr., John J. Sninsky, and Robert Watson, Sr. of Roche Molecular Systems.

ADDITIONAL READINGS

Transducers 91, 93, 95; 6th, 7th, and 8th, International Conferences on Solid State Sensors and Actuators, Available from IEEE, New York.

Journal of Microelectromechanical Systems (J. MEMS); A joint ASME and IEEE publication, New York.

REFERENCES

1. K.E. Petersen (1982). Silicon as a mechanical material. *Proc. IEEE* 70:420–457.

2. A.P. Lee, A.P. Pisano (1992). *J Microelectromechanical Systems* 1:70–76.

3. K. Mullis, F. Falonna (1987). Specific synthesis of DNA *in vitro* via a polymerase-catalyzed chain reaction. *Meth. Enzymol.* 155:335–350.

4. R.K. Saiki, D.H. Gelfand, S. Stoeffel, S.J. Scharf, R. Higuchi, G.T. Horn, K.B. Mullis, H.A. Erlich (1988). Primer-directed enzymatic amplification of DNA with a thermostable DNA polymerase. *Science* 239:487–491.

5. R.A. Gibbs (1990). DNA amplification by the polymerase chain reaction. *Anal. Chem.* 62:1202–1214.

6. S.C. Jacobson, R. Hergenroder, L.B. Koutny, R.J. Warmack, J.M. Ramsey (1994). Effects of injection schemes and column geometry on the performance of microelectrophoresis devices. *Anal. Chem.* 66:1107–1113.

7. S.C. Jacobson, J.M. Ramsey (1996). *Anal. Chem.* 68:720–723.

8. A.T. Woolley, R.A. Mathies (1994). Ultra-high-speed DNA sequencing using microfabricated capillary array electrophoresis chips. *Proc. Natl. Acad. Sci., USA* 91:11348–11352.

9. A.T. Wolley, R.A. Mathies (1995). Ultra-high-speed DNA fragment separations using microfabricated capillary electrophoresis chips. *Anal. Chem.* 67:3676–3680.

10. J.E. Harrison (1996). Presented at the *Microfabrication and Microfluidics Conference,* August 11–13. Sponsored by International Business Communications Inc., Southborough, MA.

11. G.A. Kovacs, K. Petersen, M. Albin (1996). *Anal. Chem, News and Features* (July 1): 407A–412A.

12. P. Wilding, M.A. Shiffner, L.J. Kricka (1994): "PCRina silicon microstructure" *Clin. Chem.* 40:1815–1818.

13. J. Cheng, M.A. Shoffner, G.E. Hvichia, L.J. Kricka, "Investigation of different PCR amplification systems in microfabriated silicon-glass chips" *Nuc. Acid. Res.* 24:380–385.

14. M.A. Burns, C.H. Mastrangelo, T.S. Sammarco, F.P. Man, J.R. Webster, B.N. Foerster, D, Jones, Y. Fields, A.R. Kaiser, D.T. Burke (1996). *Proc. Nat. Acad. Sci.* 93:5556–5561.

15. A. Manz, N. Gaber, H.M. Widmer (1990). Sensors and actuators. B1: "Miniaturized total chemical analysis systems: a novel concept for chemical sensing" 244–248.

16. A. Manz, D.J. Harrison, E.M.J. Verpoorte, J.C. Fettinger, A. Paulus, H. Ludi, H.M. Widmer (1992). Planar chips technology for miniaturization and integration of separation techniques into monitoring systems: Capillary electrophoresis on a chip. *J. Chromato.* 593:253–256.

17. D.J. Harrison, K. Fluri, K. Seiler, Z. Fan, C.S. Effenhauser, A. Manz (1993). Micromachining a miniaturized capillary electrophoresis-based chemical analysis system on a chip. *Science* 261:895–897.

18. C.S. Effenhauser, A. Manz, H.M. Widmer (1993). Glass chips for high-speed capilary electrophoresis separations with submicrometer plate heights. *Anal. Chem.* 65:2637–2642.

19. A. Manz, C.S. Effenhauser, N. Burggraf, D.J. Harrison, K. Sieler, K. Fluri (1994). "Electroosmotic pumping and electrophoretic separation for miniaturized chemical-analysis systems" *J. Micromechan. Microeng.* 4:257–265.

20. K. Seiler, Z.H. Fan, K. Fluri, D.J. Harrison (1994). "Electroosmotic pumping and valveless control of a fluid-flow within a manifold of capillaries on a glass chip" *Anal. Chem.* 66:3485–3491.

21. M.A. Northrup, M.T. Ching, R.M. White, R.T. Watson (1993). DNA amplification in a microfabricated reaction chamber. In *Transducers '93, Seventh International Conference on Solid State Sensors and Actuators,* Yokohama, Japan. IEEE, New York, 924–927.

22. M.A. Northrup, R.F. Hills, P. Landre, S. Lehew, D. Hadley, R. Watson (1995). A MEMS-based DNA analysis system. In *Transducers '95, Eighth International Conference on Solid State Sensors and Actuators,* Stockholm. IEEE, New York, 764–767.

23. M.A. Northrup, B. Beeman, P. Landre, S. Lehew, D. Hadley. A DNA-based miniature instrument based on micromachined silicon reaction chambers. Submitted to *Anal. Chem.,* 1997.

24. R. Higuchi, C. Fockler, G. Davinger, R. Watson (1993). Kinetic PCR Analysis: Real-timing Monitoring of DNA Amplification Reactions. *Bio/Techn.* 11:1026–1030.

25. P. Holland, R.D. Abramson, R. Watson, D.H. Gelfand (1991). Detection of specific polymerase chain reaction product by utilizing the 5′ to 3′ exonuclease activity of *Thermus acquaticus* DNA polymerase. *Proc. Natl. Acad. Sci.* 88:7276–7280.

26. L.G. Lee, C.R. Connelly, W. Bloch (1993). Allelic discrimination by nick-translation PCR with Fluorogenic Probes. *Nuc. Acid. Res.* 21:3761–3766.

27. K.J. Livak, S.J.A. Flood, J. Marmaro, W. Giusti, K. Deetz (1995). Oligonucelotides with fluorescent dyes at opposite ends provide a quenched probe system useful for detecting PCR product and nucleic acid hybridization. *PCR Methods and Apps.* 22:357–362.

28. M.A. Northrup, B. Beeman, P. Landre, D. Hadley, J.J. Sninsky, R. Watson, M. Fisher, N. Constantine, D. Oldach (1997). Real-time, quantitative detection of HIV and HCV RNA in human plasma and tissue. In preparation.

29. C.T. Wittwer, G.C. Fillmore, D.J. Garling (1990). Minimizing the time required for DNA amplification by efficient heat transfer to small samples. *Anal. Biochem.* 186:328–331.

30. J. Kidd, J. Stilwell, M. Northrup, M. Segraves, J. Lamerdin, C. Strout, Carrano, A. Carrano, (1996). Application of a Micro-PCR instrument for fast, reproducible and field-portable detection of *Clostridium perfringens.* May 19–23. American Society of Microbiology, New Orleans, LA.

31. A.T. Woolley, D. Hadley, P. Landre, A.J. deMello, R.A. Mathies, M.A. Northrup (1996). Functional integration of PCR amplification and capillary electrophoresis in a microfabricated DNA analysis device. *Anal. Chem.,* 68:1040–1046.

10

Genosensors and Model Hybridization Studies

MITCHEL J. DOKTYCZ AND KENNETH L. BEATTIE

CONTENTS

INTRODUCTION

During the late 1980s several research groups independently proposed the concept of sequencing by hybridization (SBH), whereby a complete set of 4^n oligonucleotide probes (representing all sequences of length n, such as all 65,536 octamers) are independently hybridized to a target sequence; then the hybridizing n-mer sequences are arranged (by an appropriate computer algorithm) according to unique $(n - 1)$ overlaps to generate the complete base sequence of the target strand [1–4]. Drmanac and Crkvenjakov

Automation Technologies for Genome Characterization, Edited by Tony J. Beugelsdijk.
ISBN 0-471-12806-6 © 1997 John Wiley & Sons, Inc.

originally pursued the so-called *format 1* version of SBH, wherein the complete set of oligonucleotide probes are sequentially hybridized to high-density filter arrays of cloned genomic segments [3, 5–8]. However, a significant fraction of the SBH effort has focused on the "DNA chip" approach (format 2), wherein an individual fragment is hybridized to a compete set of oligonucleotides arrayed on a surface [2, 4, 9–17]. Although the SBH concept is extremely elegant and could potentially result in a 100- to 1000-fold increase in the speed of sequence determination, there are a number of inherent problems that must be overcome before SBH is useful for *de novo* sequence determination. First, the natural occurrence of short-tandem repeats and the chance multiple occurrence of a given oligonucleotide sequence within the target create sequence reconstruction ambiguities (more than one sequence consistent with the hybridization pattern) which must be resolved by additional experimental strategies. A more serious issue is the imperfect specificity of hybridization. A hybridization signal may be due to a unique match of the probe within the target, but it may occasionally be due to an imperfect match or the additive contribution of several mismatched bases. Thus additional ambiguity in sequence reconstruction is introduced by lack of absolute +/− specificity of hybridization, and until a great deal more information is available about the precise nature of hybridization ambiguity (i.e., which mismatches are poorly discriminated against in certain nearest-neighbor contexts), or unless hybridization specificity can be made absolutely unambiguous, it is unlikely that SBH will be implemented for *de novo* sequencing.

The above challenges have motivated most researchers in the SBH field to focus on simpler, but as important, applications of oligonucleotide arrays for *comparative* sequence analyses rather than for *de novo* sequence determinations [15, 18–20]. The predominant focus in DNA chip research is currently the commercially relevant applications of DNA diagnostics. Although comparative SBH for DNA diagnostics has recently received a great deal of interest and funding, there remain significant applications of array hybridization for the human genome project [21–23].

The term "genosensors" has been coined to describe miniature devices containing arrays of surface-tethered oligonucleotide probes, individually addressable for detection of hybridization across the array [24–27]. This term may be more broadly interpreted to include a variety of hybridization array strategies (employing longer fragments as well as short probes immobilized at each test site). When implemented in the miniaturized chip format, array hybridization may greatly enhance the speed, economy, and throughput of genome mapping and sequencing.

CONSTRUCTION OF OLIGONUCLEOTIDE ARRAYS

Two approaches are being used to create arrays of synthetic oligonucleotides for the DNA chip or genosensor: (1) the *in situ* synthesis of numerous sequences onto a support surface and (2) the attachment of presynthesized oligonucleotides at each site in the array. Both strategies employ the standard phosphoramidite method of solid phase chemical synthesis of oligonucleotides [28]. One *in situ* approach, under development by Affymetrix and others [29–33], utilizes photolithographic deprotection of 5'-OH groups at specific sites on the chip to direct specific base addition at those sites during chemical synthesis by the phosphoramidite method. The Southern group, in partnership with Beckman Instruments, is employing physical masking techniques to direct the coupling of the four phosphoramidites within parallel channels formed between two flat-glass synthesis supports. By orienting the channels perpendicularly in two successive coupling cycles, a 4×4 matrix of the 4^2 dinucleotides is synthesized; then by continuing this process and dividing the channel width by four after each pair of coupling cycles, a matrix of 4^n sequences is eventually built up [2, 18, 19]. The Southern method of *in situ* synthesis has also been adapted to polypropylene sheets [34]. In another *in situ* strategy, being developed by Combion, Inc. [35], miniature "ink jets" are used to deliver tiny droplets of phosphoramidite solutions to precise locations on a surface, permitting synthesis of numerous sequences on a single surface.

In the postsynthesis attachment strategy of oligonucleotide array preparation the desired set of oligonucleotides is first synthesized, then chemically immobilized at specific sites on the surface of the chip [4, 9, 24–27, 36, 37]. Synthesis of thousands of oligonucleotides, in quantities sufficient for millions of chips, is readily accomplished by the segmented synthesis strategy [38–41] of Genosys Biotechnologies, Inc. (The Woodlands, TX), which enables chemical synthesis of about a thousand short oligonucleotides per working day on a single machine. Three Genosys machines could synthesize all 65,536 octamers in about a month. Several methods are currently available for delivery of tiny droplets to the genosensor surface. Using a Hamilton MicroLab 2200 fluid-dispensing system, up to eight droplets of as little as ten nanoliters can be simultaneously applied to a flat surface [37]. Using rows of tiny piezoelectric "ink jet" tips, Microfab Technologies, Inc. (Dallas, TX) can deliver tiny droplets (tens to hundreds of picoliter volumes) to precise locations onto a genosensor chip [27]. Two-dimensional arrays of microfabricated piezoelectric microjets are under development at Accelerator Technologies Corp. (Bryan, TX), which are designed to simultaneously "print" large numbers of probes onto the genosensor surface [42]. The Mirzabekov group has developed a miniature robotic multi-pin printing device [43] to simultaneously deliver numerous minute droplets of probe solutions to a thin polymer gel employed in their "sequencing by hybridization to oligonucleotide matrix" (SHOM) version of SBH [9, 15, 16]. In Mirzabekov's gel matrix genosensor, hybridization occurs in three dimensions, to oligonucleotide probes covalently immobilized within thin [10–20 μm] slabs of polyacrylamide gel arrayed on a glass support [9, 15, 16].

SIMPLIFIED PROBE ATTACHMENT CHEMISTRY

A relatively simple, reliable procedure for covalent linkage of oligonucleotide probes to silicon dioxide surfaces, involving specific condensation of a primary amine on the 5′- or 3′-terminus of the oligonucleotide with an epoxysilane group on the glass, has been reported [26, 36, 37]. We subsequently discovered a much simpler procedure, involving direct coupling of 3′-propanolamine-derivatized oligonucleotides to unmodified SiO_2 surfaces [44]. The presumed reaction is shown below in Fig. 10.1.

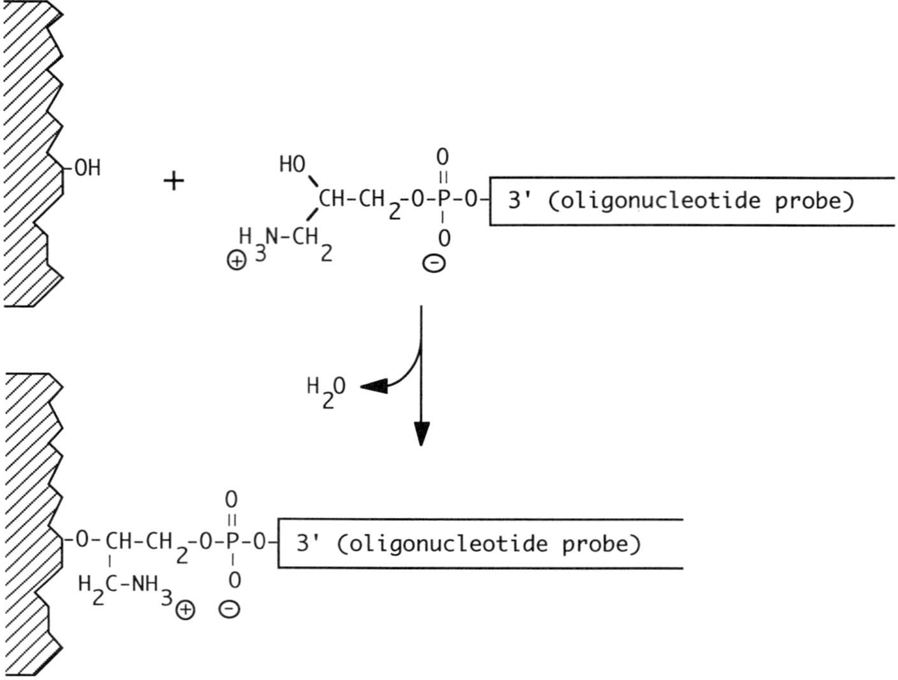

Fig. 10.1 *Proposed reaction scheme for covalent attachment of 3′-propanolamine-derivatized oligonucleotides to underivatized glass surfaces.*

This reaction scheme is supported by the following results [44]: The linkage is (1) stable in hot water, enabling multiple cycles of hybridization, (2) stable in mild acid but labile in mild base (favoring the ester linkage over the amide linkage), (3) not formed with 5′-hexylamine-derivatized oligonucleotides (primary amine alone is insufficient), (4) inhibited by pretreatment of glass with propanolamine but not propylamine, and (5) blocked by acetylation of primary amine on oligonucleotide suggesting that the linkage reaction is assisted by the amine function. The attachment reaction proceeds rapidly in aqueous solution at room temperature and gives a lower background of nonspecific binding of target DNA to the surface, compared with the previous epoxy-amine linkage method. The attachment density (10^{10}–10^{11} molecules/mm^2) is similar for both methods.

Although the *in situ* approaches may seemingly have an advantage over the postsynthesis attachment option in the preparation of extensive arrays needed for *de novo* SBH, the *in situ* synthesis strategy has a potentially serious limitation in quality control. Unless the coupling efficiency is quantitative throughout the entire *in situ* array synthesis, there will be significant differences in the quantity of full-length product at different positions in the array, which would lead to unpredictable effects on hybridization across the array. In the postsynthesis attachment strategy, quality control of each oligonucleotide is carried out prior to attachment to verify the length and quantity of product, but it is still critical to achieve uniform tethering of probes at different positions across the array. Another consideration is the eventual need for mass production of genosensors. If rapid, parallel microfluidic delivery of oligonucleotides to individual sites on the chip can be achieved, mass production of genosensors should be feasible in the postsynthesis attachment scenario. It is less obvious how the *in situ* synthesis strategies can achieve the ultrahigh throughput required for mass production of genosensors, given that the multistep *in situ* probe synthesis would need to be part of the mass production process. For mass production of miniature arrays of genomic fragments, highly parallelized immobilization of DNA fragments across the surface will have to be provided in any case.

HYBRIDIZATION

Assuming that complex arrays of DNA fragments can be achieved, a second major hurdle exists in the use of the genosensor chip. Predictable hybridization between the immobilized probes and the target DNA is necessary. This hybridization is characterized by a rich variety of interactions that lead to the formation of perfect Watson-Crick (W-C) duplexes as well as imperfectly matched hybrids. These interactions lead to a sequence-dependent stability that has been elusive to predict with the accuracy demanded by genosensor applications. An attempt at improving upon these stability predictions can begin by characterizing the thermodynamics of the various contributions. Table 10.1 lists some potentially important interactions and suggests the large number of interactions that are possible. If only W-C base pairing contributed to duplex stability then understanding only two types of interactions would suffice. However, taking account of mismatched pairs adds 8 more interactions. Nearest-neighbor effects of W-C base pairs may need to be considered, adding 10 more interactions. If nearest-neighbor effects that include a single mismatch are accounted for, then an additional 48 interactions would be appended. The stability of these interactions is compounded by unpaired nucleotides that can occur at the ends of duplexed regions (dangling ends) or internally as hairpin loops and bulges. The length and sequence of the target and probe DNA may be influential. Furthermore the position of the bases involved in these interactions, relative to the ends of duplexed regions, may be significant as well as the solution environment and the position and type of pendant labels, attachment linkers, and unusual bases.

The large number of interactions makes stability analysis by conventional methods a laborious process. Characterization of these interactions may best be assessed by parallel analysis methods such as genosensor chips [45]. Fortunately many higher-order, intramolecular interactions would not be expected to occur on the short, immobilized probe sequences but may occur on the longer target strand. Procedures aimed at limiting the length of the target strand can reduce intramolecular folding. The short length of the hybridization probes, however, emphasizes short-range interactions due to the lower total number of interactions that would be present. For example, if base pairing were the only factor involved in duplex stability, then a single base pair would contribute

TABLE 10.1 Interaction List

Interaction	Example
Watson-Crick base pairs	$A \cdot T$, $G \cdot C$
Mismatched base pairs	$A \cdot A$, $A \cdot C$, $A \cdot G$, $T \cdot T$, $T \cdot C$, $T \cdot G$, $C \cdot C$, $G \cdot G$
Watson-Crick base-pair doublets	AT/AT, TA/TA, AA/TT, AC/GT, AG/CT, TC/GA, TG/CA, GG/CC, GC/GC, CG/CG
Mismatched base-pair doublets	AA/TA, AT/AA, AG/CA, AC/GA, AA/TG, AT/AG, AG/CG, AC/GG, AA/TC, AT/AC, AG/CC, AC/GC, TA/TT, TT/AT, TG/CT, TC/GT, TA/TG, TT/AG, TG/CG, TC/GG, TA/TC, TT/AC, TG/CC, TC/GC, GA/TA, GT/AA, GG/CA, GC/GA, GA/TT, GT/AT, GG/CT, GC/GT, GA/TG, GT/AG, GG/CG, GC/GG, CA/TA, CT/AA, CG/CA, CC/GA, CA/TT, CT/AT, CG/CT, CC/GT, CA/TC, CT/AC, CG/CC, CC/GC
Dangling ends	$5'$TCGCGTTAGCATG$3'$ GCAATCGT
Secondary structures	Bulge loops T AC CT T GCTCCG GA GAG T CGAGGC CT CTC T TT End loop
Other effects	Positional dependence Solution environment Pendant labels Attachment linkers Unusual bases Unusual backbones

12.5% to the total stability of an eight base-pair duplex, while only contributing 2% to the total stability of a 50 base-pair duplex. Consequently the overall stability of short DNA duplexes is sensitive to influences that do not significantly affect longer sequences. These influences could include alterations of a single-base, nearest-neighbor stacking interactions, mismatching, and end effects. This greater sensitivity to local interactions can be advantageous for genosensor applications. For example, a single mismatched base pair would be more disruptive to the total stability of a short duplex than to a longer duplex. Also short-sequence probes will suffer more from end effects. End effects arise from the "fraying" at the ends of duplexed regions. Destabilization due to end fraying can propagate several base pairs into the duplex, resulting in less discrimination against mismatches. For short probes, end effects can influence a significant fraction of the molecule.

One method of characterizing hybrid stability is to determine the T_m by optical melting. Typically the absorbance, at an appropriate wavelength, of a mixture of two complementary oligonucleotides is monitored as a function of temperature. Upon increasing temperature, a hyperchromicity over a narrow temperature range is observed due to the unpairing and unstacking of the aromatic nucleotide bases. This transition is referred to as melting, and the midpoint of the transition as the T_m. At the melting temperature, approximately half of the DNA bases are paired; the actual structure differs between long and short duplexes. For long DNA molecules, this usually accounts for partially melted structures interspersed among duplexed regions. Short molecules, however, typically melt by an all or none process, and the T_m corresponds to the fraction of molecules melted. In either case the T_m can be defined as the point at which one-half of the potential base pairs are present. This value serves as a convenient reference point with which to compare sequence- and structure-dependent stability provided that the experimental conditions are the same or corrected for.

An example of a melting curve of a short DNA duplex is shown in Fig. 10.2. The T_m can be evaluated by several methods [46]. Commonly the curve is normalized by fitting base lines to the approximately linear regions surrounding the transition. The absorbance at each temperature (A) is then corrected to determine the fraction of broken base pairs (Θ_B):

$$\Theta_B = \frac{A - A_l}{A_u - A_l}$$

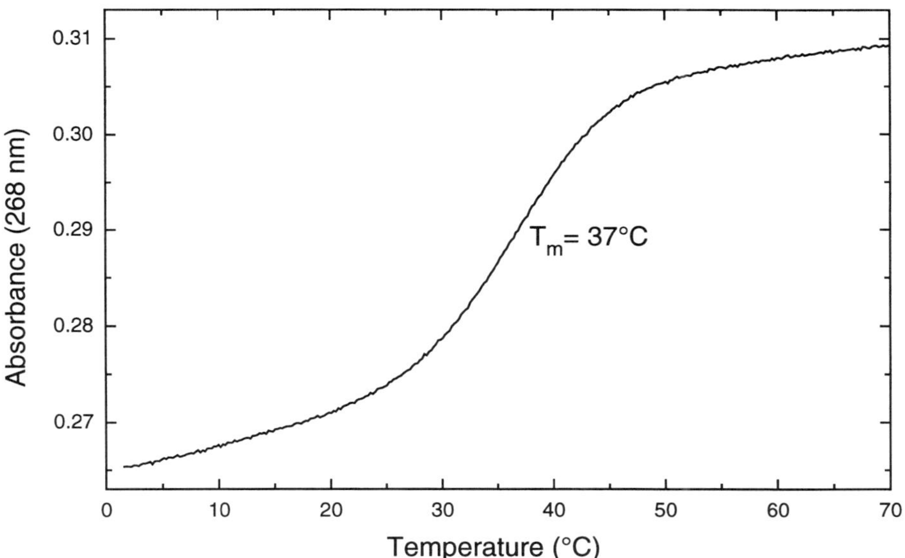

Fig. 10.2 Raw melting data of the duplex sequence GGATGAGC/GCTCATCC in 10-mM sodium phosphate buffer with 1-M sodium chloride. The absorbance was monitored at 268 nm as the temperature was ramped up at 0.5°C/min. The data were collected at every 0.2°C.

where A_l is the absorbance value of the lower base line and A_u is the absorbance value of the upper base line at the specified temperature. The T_m can then be determined by identifying the temperature corresponding to $\Theta_B = 0.5$. Utilization of the derivative of the melting curve eases the identification of multiple transitions.

If two-state behavior is ensured, then the slope of the melting curve at the T_m can be used to evaluate the van't Hoff enthalpy. For a bimolecular reaction, the van't Hoff enthalpy is

$$\Delta H_{van't\,Hoff} = -6RT_m^2 \left(\frac{d\Theta}{dT}\right)_{T=T_m}$$

However, the enthalpy obtained in such a manner is subject to the limits chosen for the baselines. These limits can influence the transition slope, altering the calculated enthalpy, although the T_m is usually not affected. Further complications arise with melting transitions that occur at either high- or low-temperature extremes. For example, the low-melting temperatures of short duplexes often prevent the establishment of a lower baseline. Methods for extracting the enthalpy in these situations have been discussed [46]. Alternatively, the van't Hoff enthalpy can be determined from the concentration dependence of the T_m. This latter method is less sensitive to base-line choice but requires the evaluation of multiple melting experiments over a wide range of concentrations.

Short duplex stability is currently being evaluated to determine the influence of the various interactions listed in Table 10.1. Optical melting experiments can identify the significant interactions and eventually lead to a general algorithm that describes short oligonucleotide duplex stability. An extensive set of oligonucleotides has been designed and synthesized to highlight possible positional and sequence dependent stability of base pairs in short duplexes [47]. This set contains 256 octadeoxyribonucleotides and is based on four general sequences. Two of these sequences are $^{5'}$XYZTGGAC$^{3'}$ and $^{5'}$GTCCAXYZ$^{3'}$ where X, Y, and Z may be A, T, G, or C. These molecules allow the formation of 64 perfectly matching duplexes with every base-pair type in every nearest-neighbor environment occurring in the end and penultimate positions. The other two sequences, $^{5'}$GCXYZGAC$^{3'}$ and $^{5'}$GTCXYZGC$^{3'}$, form 64 perfectly matching duplexes to evaluate the sequence dependent stability of internal base pairs. Among the 256 sequences are molecules that can be mispaired to evaluate the positional and sequence-dependent stability of mismatching.

Melting experiments performed on the matched duplexes of the above set yielded T_m's ranging from 25.1°C to 50.2°C in a buffer of 10-mM sodium phosphate and 1-M NaCl [47]. Sequence variations led to T_m changes of as much as 8.2°C for molecules of similar base-pair content. To evaluate the sequence and positional dependence of the melting temperature, and associated thermodynamic parameters, various models were created to describe and predict sequence-dependent stability. These models tallied the number and type of interactions present, such as base-pair type and position as well as nearest-neighbor type and position, and allowed their influence on the melting temperature to be determined. The importance of the various interactions were determined by statistical analysis of the models by refitting the data set of 128 duplexes. The predictive usefulness of these parameters was demonstrated with a set of arbitrarily chosen sequences. The results show that the base-pair and the nearest-neighbor relationships are significant in characterizing the duplex stability. From the positional dependent stability of these interactions, the effect of end fraying was found to be minor; accounting for positional effects gives predictions that are improved only slightly and are within the variance due to errors associated with the measurements. The resultant set of coefficients that account for base-pair and nearest-neighbor type is shown in Table 10.2. These values can be used to calculate the T_m, or the ΔG^0, of any eight base-pair sequence by simply summing the various contributions. Since the 12 individual parameters cannot be solved uniquely, due to interrelation among the parameters, physical significance cannot be assigned to any of the individual values. However, the calculated, total stability of the duplex can be used to predict melting temperatures. These values would be useful for comparing the stability of 8-mer duplexes and determining experimental hybridization parameters. The melting temperatures of different length sequences can also be predicted by multiplying the sum of the interactions by the length of the sequence and dividing by eight. Application of these parameters to longer sequences has not been thoroughly investigated, although examination of a series of 9-mers shows promising results.

Besides characterizing the stability of perfectly matched duplexes, an important component of hybridization analysis is the relative stability of mismatched molecules. Mismatching interferes with the interpretation of hybridization patterns, and determination of conditions that either avoid or predict their formation would be useful. Several sets of mismatch data exist [48–53], for which positional dependent information does not exist. When selected, 8-mer molecules were used to evaluate the effect of single base-pair mismatches in various positions. End effects were shown to be significant. Table 10.3 displays the changes in melting temperature and enthalpy for eight different

TABLE 10.2   Nonunique Coefficients for Base-Pair and Nearest-Neighbor Interactions

Parameter	T_m Estimate (°C)	ΔG^0 Estimate (kcal/mole)
A/T	1.9	0.30
G/C	4.7	−0.33
AA/TT	−0.1	−0.01
GG/CC	1.1	−0.17
AT	0.9	−0.25
TA	−2.8	0.53
CG	2.7	−0.53
GC	3.6	−0.72
AC/GT	1.7	−0.37
CA/TG	1.0	−0.24
AG/CT	0.4	−0.11
GA/TC	0.0	0.00

TABLE 10.3 Effect of End and Internal Mismatch

Mismatch	Perfect Match	End Mismatch $\Delta T_m{}^a$	End Mismatch $\Delta\Delta H^b$ (kcal/mole)	Internal Mismatch $\Delta T_m{}^a$
A · C	A · T	−5.1	8.9	−18.3
	G · C	−9.5	15.9	−24.2
T · G	T · A	−3.1	3.2	−14.6
	C · G	−7.5	4.6	−19.6
G · A	G · C	−9.0	5.7	−21.5
	T · A	−3.9	10.9	−14.4
C · T	C · G	−8.9	3.5	−21.5
	A · T	−5.2	0.9	−17.7
A · A	A · T	−3.7	4.6	−20.6
	T · A	−3.0	5.8	−19.4
T · T	T · A	−3.8	4.3	−14.4
	A · T	−4.5	3.1	−15.6
C · C	C · G	−8.7	11.0	−25.9
	G · C	−9.4	15.4	−28.0
G · G	G · C	−8.9	11.0	−17.7
	C · G	−8.2	6.6	−15.6

[a] ΔT_m corresponds to the change in melting temperature resulting from the perfectly matched molecule subtracted from the mismatched molecule.

[b] $\Delta\Delta H$ corresponds to the enthalpy change of the perfectly matched molecule subtracted from the mismatched molecule.

single-base end mismatches when they are compared to their perfectly matched counterparts. A mismatch in an interior position was also examined, and these data are also shown in Table 10.3. Calculations of thermodynamic changes are unreliable for many of these centrally mismatched molecules due to the low melting temperatures. A large change in overall stability for centrally mismatched molecules, compared to end-mismatched molecules, is clearly apparent. As a mismatch is moved to the internal positions, a trend of increased destabilizing effects is observed. This is shown in Fig. 10.3, where the change in melting temperature as a function of position is shown for a G · T mismatch. Destabilizing effects appear to increase as the mismatch occurs in the second or third position from the end for a variety of mismatches examined.

Another interaction that is likely to occur during hybridization of short oligonucleotides is unpaired dangling ends. The length and sequence of these ends may lead to altered stabilities of perfectly matched or mismatched molecules. In general, the stabilizing effects of dangling ends originate from end stacking onto the duplex [54–56], although the effect of this stacking with increased dangling end length has not been fully investigated. Considering the importance of stacking hyteractions on the stability of base pairing, a set of oligonucleotides duplexes designed to have a single dangling base on either the 5′ or 3′ end were melted. A summary of the melting data, relative to the perfectly matched 8-mer duplex, is shown in Table 10.4. These dangling ends

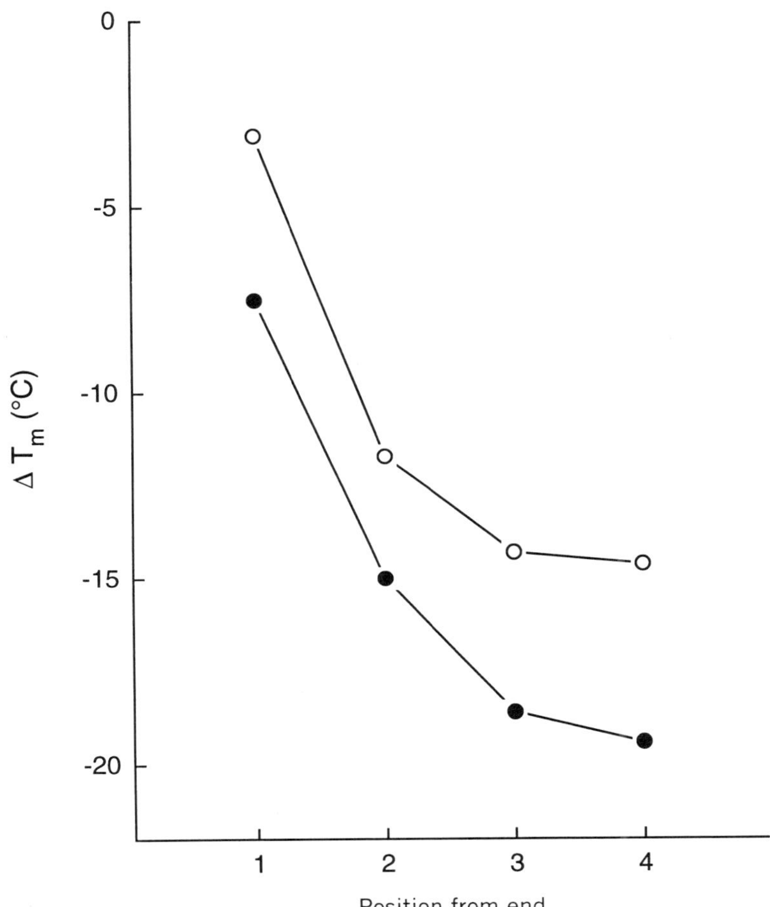

Fig. 10.3 The relative change in melting temperature for a T · G mismatch type as a function of position from the end of an 8 base-pair duplex. The reference duplexes contain either a T · A (*open circles*) or a C · G base pair (*closed circles*).

were adjacent to a terminal A · T base pair and produce a range of stabilizing effects. From this limited set of oligonucleotide pairs, all the 5′ dangling ends, stacked on an adenine base, stabilize the duplex, while the complementary set, stacked on a thymine base, barely alters the T_m. The 3′ dangling end set actually has lower enthalpy values compared to the reference molecules. Apparently the identities of both the dangling base and the adjacent base influence overall stability.

Currently there is ongoing application of equilibrium melting data to genosensor hybridization data. The correlation between the two techniques is dependent on the experimental method of capturing and discriminating between the perfectly matched and mismatched targets. A detailed review connecting the thermodynamics and models

TABLE 10.4   Effect of Dangling Bases on a Terminal A-T Base Pair

Dangling Base	ΔT_m^a	$\Delta\Delta H^b$ (kcal/mole)
5'A	2.8	−3.8
5'T	3.3	−5.3
5'G	3.0	−3.2
5'C	2.0	−2.6
3'A	0.8	1.1
3'T	−0.2	2.3
3'G	1.2	0.8
3'C	−0.1	1.9

[a] ΔT_m corresponds to the change in melting temperature obtained by subtracting the T_m of the perfectly matched molecule from the dangling ended molecule.
[b] $\Delta\Delta H$ corresponds to the enthalpy change of the perfectly matched molecule subtracted from the dangling ended molecule.

of nucleic acid hybridization to practical applications in molecular biology has been provided by Wetmur [57]. Highlights concerning the present application will be emphasized here. The method for monitoring the amount of hybrid formation may use equilibrium conditions or posthybridization washing of the target-probe hybrid in a large volume of buffer. The former situation would require both hybridization and melting to be occurring at the same time, while the latter case is dependent on the melting kinetics. Both situations can be described using the thermodynamic parameters obtained from optical melting experiments.

For measurements made under equilibrium conditions, the amount of target molecule captured by the genosensor probe will be dependent on temperature, buffer conditions, and the concentrations of target and probe. Detailed investigations that utilize short immobilized probes and defined length targets are lacking. However, if the nucleation rate constant k_N' is assumed to be independent of sequence variations for short molecules, then only the DNA concentrations, and the buffer conditions will govern the hybridization rate. This rate constant, k_2, (in $M^{-1}s^{-1}$) is independent of temperature and can be expressed as

$$k_2 = \frac{k_N'\sqrt{L_s}}{N}$$

where L_s is the length of the fixed probe strand and N is the complexity or total number of base pairs present in the free target. The value for the nucleation rate constant for a sodium ion concentration between 0.2 and 4.0 M is [57]

$$k_N' = \{4.35 \log_{10} [Na^+] + 3.5\} \times 10^5$$

At temperatures near the melting point, the release kinetics becomes important and allows for discrimination between perfectly matched and mismatched targets. The release kinetics is first order and depends on the activation energy required for melting. This activation energy can be equated to the transition enthalpy plus the activation energy for hybridization. This latter term is independent of sequence and temperature and is 4 kcal/mole [58]. The reverse rate constant can be expressed as

$$k_r = Ae^{(E_d/RT)}$$

The ratio of the rate constants for hybridization and melting yields the equilibrium constant, which can be related to the Gibbs free energy by standard equations. Such values can also be obtained from the optical melting experiments, and they serve as a link between the two techniques.

Alternatively, the target molecule can first be captured and then "washed" from the probe molecules. In this case rehybridization is extremely slow due to the large dilution that results. Consequently there is a low target concentration; only the release kinetics will influence the amount of probe-target hybrids that are present at a given time. An extended washing time will enhance discrimination until the background signal becomes dominant [59]. There is ongoing experimental validation of the relations between optical melting and genosensor analysis.

FLOWTHROUGH GENOSENSORS

There is continuing development of genosensors. We have begun to develop an advanced genosensor configuration in which the hybridization reactions occur within three-dimensional volumes of porous silicon dioxide or channel-array glass rather than on a two-dimensional surface [37]. The flowthrough genosensor concept is illustrated below in Fig. 10.4. In the design shown here an array of square regions of porous silicon is first formed in a silicon wafer; then the porous silicon wafer is bonded to a second wafer in which square holes (aligned with porous patches) have been acid-etched in a layer of silicon or silicon dioxide. The thickness of the silicon wafer and geometry of the porous patches and sample wells can be varied at will. For detection of hybridization using a phosphorimager, 200-μm diameter porous cells, spaced 500 μm apart in 200-μm thick silicon, is a suitable geometry. For optical detection of hybridization (using a CCD camera), 50-μm diameter porous cells at 150-μm pitch would be appropriate.

Shown in Fig. 10.5 are scanning electron micrographs of a regular array of square pores formed in pure silicon using an electrochemical, acid-etch process [60], following an initial photolithography step that defines the position, size, and shape of each pore. The photograph on the left shows the appearance of individual pores (3-μm diameter, 10-μm pitch) viewed from above, and the photograph on the right shows a cross-sectional view.

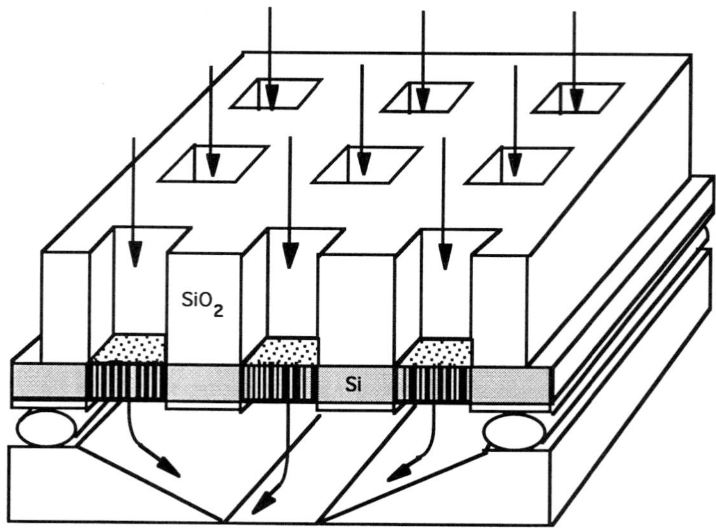

Fig. 10.4 Schematic representation of a flowthrough genosensor. Shown is an array of sample wells, etched in silicon dioxide and bonded to a corresponding array of porous patches formed in a silicon wafer. Each porous patch contains densely packed straight channels of diameter 1–10 μm, oriented perpendicular to and connecting the upper and lower faces of the wafer. The silicon is oxidized to form a lining of silicon dioxide in each channel, to which the DNA probes are covalently tethered. The dimensions of the porous silicon array can be varied to fit the desired application. The porous wafer is typically 50–500 μm thick and individual porous patches are typically 50–500 μm in diameter. Fluidic flow through the porous silicon genosensor is provided by slight pressure from the upper side or slight vacuum from the lower chamber.

Flowthrough Genosensor Advantages and Applications

The anticipated advantages of the porous geometry include (1) a great increase in surface area (on the order of 100-fold) per unit cross section, compared with the flat surface design, thus increasing the binding capacity per unit cross-sectional area and consequently improving detection sensitivity; (2) avoidance of sample drying during the chemical attachment reaction (this has been a difficult problem with the flat surface design); (3) improved accessibility of target strands to surface-tethered probes (DNA strands flowing through the micropores rapidly encounter surface-tethered probes due to the very short distance to the surface); (4) ability to analyze dilute solutions of nucleic acids (by slowly flowing a volume through the hybridization chip); (5) improved hybridization of denatured double-stranded PCR fragments to the porous glass-tethered probes (due to the ability to analyze dilute target strands, slowing their reassociation), which should eliminate the need to prepare single-stranded PCR fragments prior to hybridization; and (6) the ability to recover bound material (from a given hybridization cell) for further analysis such as cloning and sequencing. The 100-fold increase in binding capacity per site enables a higher degree of miniaturization of the array, since

Fig. 10.5 Scanning electron micrographs of a portion of porous silicon formed as described in the text, viewed from above (*left panel*) and in cross section (*right panel*). In the structure shown, pores are 3 µm diameter and spaced at 10 µm pitch. Photograph courtesy of Dr. Peter McIntyre, Accelerator Technology Corp., Bryan, TX.

a given number of bound molecules would occupy a tenfold smaller cross-sectional area in the porous configuration, enabling 100 times as many hybridization reactions to be carried out per unit cross-sectional area, with a concomitant reduction in the required sample volume. Taken together, these anticipated advantages of the porous silicon genosensor over flat surface devices lead to predictions of increased sample throughput and reduced cost.

The advantages of porous glass over flat glass hybridization supports are shown in Fig. 10.6, which summarizes the results of an experiment in which PCR fragments were hybridized to complementary glass-tethered, 9-mer probes [61]. In this experiment three parameters were compared: (1) single-stranded PCR fragments versus denatured double-stranded PCR fragments; (2) concentrated versus diluted (50–100-fold) target strands, and (3) hybridization on flat microscope slides versus in capillary array glass supports. The phosphorimager data at the top represent hybridization signals from ^{32}P-labeled target strands, hybridized to small arrays of 9-mer probes (complementary to a 9mer sequence in the PCR fragment) immobilized within porous glass (right image in each pair) or on flat glass (left image in each pair). Represented below the hybridization images are the corresponding relative hybridization intensities in bar graph form, for hybridization in porous glass (darkly shaded bars) versus on flat glass (lightly shaded bars). The data on the left side of Fig. 10.6 represent hybridization of 20 µL of concentrated (30-nM) target strands to glass-tethered probes, while the data on the right half represent hybridization following 50-fold dilution of target strands (1 mL of 0.6-nM target strands). In the case of 9-mer probes tethered to flat glass, appreciable hybridization occurred only with single-stranded targets hybridized at high concentration. For hybridization of dilute target strands to glass-tethered probes, the porous glass support was dramatically superior to the flat glass support. Finally, hybridization of heat-denatured, double-stranded PCR fragments to support-bound probes was far superior with the porous glass substrates.

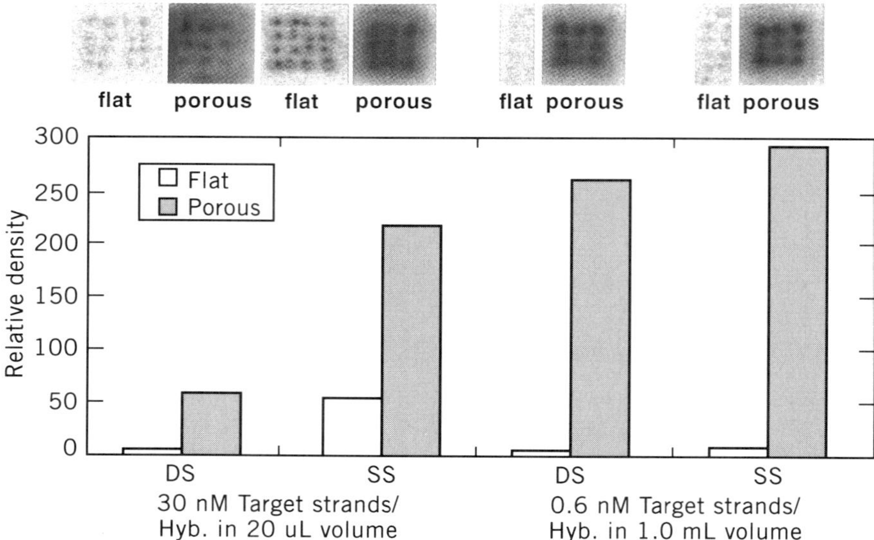

Fig. 10.6 *Comparison of hybridization of single-stranded and denatured double-stranded PCR fragments to 9-mer probes tethered to flat glass (light bars) and porous glass (dark bars). Experimental details are discussed in the text and in [61].*

The ability to analyze dilute solutions of nucleic acids is a major advantage of porous glass over flat glass genosensors. This capability of the porous glass genosensor may enable direct nucleic acid analysis without PCR amplification in certain applications, such as mRNA profiling and analysis of mitochondrial DNA or microbial genomes. In addition the ability to analyze PCR fragments without the cumbersome step of generating single-stranded targets is another important advantage of the porous glass genosensor over its flat glass counterpart. The flowthrough feature of the porous glass genosensor should also enable analysis of large sample volumes without the need to concentrate the sample. Finally, the porous glass geometry should facilitate elution of bound nucleic acids from individual positions in the array, for further manipulation or analysis such as cloning and sequencing.

We anticipate that the high-density gridding of DNA samples within porous silicon would enable further miniaturization of current hybridization schemes, which utilize DNA fragments gridded onto membranes. For example, in the high-density filter hybridizations of Drmanac and Crkvenjakov [8, 23], the DNA samples are spaced approximately 1 mm apart in dots of about 300-μm diameter, and each dot on the membrane contains about 10^8 molecules of PCR fragment or cloned segment [62]. If we conservatively assume that DNA fragments can be immobilized to the cylindrical walls of porous silicon dioxide at an average density of 10^8–10^9 molecules/mm^2 (1/ 100 the density achieved with glass-tethered oligonucleotide probes), then a 100 × 100 × 100 μm cube of porous silicon will contain on the order of 10^8–10^9 molecules of DNA (up to ten times the quantity per dot on the membrane), and about ten times as many hybridization cells per unit cross-sectional area will be achieved within the porous silicon wafer than in the filter format. Thus we anticipate at least a tenfold higher density of arrayed DNA fragments within the porous silicon hybridization array, compared with filter hybridization. This economy of scale should enable higher throughput of analysis at greatly reduced sample volumes.

SUMMARY

Genosensor research is an ongoing and evolving field. The construction of genosensors with improved methods for attaching arrays of different DNA probes to solid surfaces has been achieved as well as a framework for predicting the hybridization properties of the immobilized probes. The flowthrough genosensor is a recent improvement on previous "conventional" designs using two-dimensional surfaces. The advantages of the flowthrough design will have to be coupled to improvements and increased understanding in other aspects of genosensor development. For example, effective fluid transfer mechanisms as well as low-cost, high-throughput detection strategies will have to be worked out. These techniques will need to take advantage of the flow-through capabilities, the increased surface area, and greater number of DNA probes. A better understanding of hybridization reactions will also be necessary. DNA hybridization is an essential feature of genosensors. Optimization of hybridization parameters, including DNA concentrations, hybridization time, and buffer conditions, as well as increasing the predictive powers of theoretical analysis will aid in the effective implementation of genosensors. Genosensor applications are numerous and will soon be realized with the maturation of this exciting technology.

ACKNOWLEDGMENTS

This work was supported in part by NIH Grant 1 P20 HG00665 (to K.L.B.) at the Houston Advanced Research Center and by the Laboratory Director's Research and Development Fund at the Oak Ridge National Laboratory, managed by Lockheed Martin Energy Research Corp. for the U.S. Department of Energy under contract number DE-AC05-96OR22464. The authors thank Dr. Peter McIntyre (Physics Dept., Texas A&M University, and Accelerator Technology Corp.) for providing scanning electron micrographs of porous silicon substrates and Dr. K. B Jacobson and Dr. R. J. Warmack for a critical reading of the manuscript.

REFERENCES

1. W. Bains, G.C. Smith (1988). A novel method for nucleic acid sequence determination. *J. Theor. Biol.* 135:303–307.

2. E.M. Southern (1988). Analyzing polynucleotide sequences. International patent application PCT GB 89/00460.

3. R. Drmanac, I. Labat, I. Brukner, R. Crkvenjakov (1989). Sequencing of megabase-plus DNA by hybridization: Theory of the method. *Genomics* 4:114–128.

4. K.R. Khrapko, Y.P. Lysov, A.A. Khorlyn, V.V. Shick, V.L. Florentiev, A.D. Mirzabekov (1989). An oligonucleotide hybridization approach to DNA sequencing. *FEBS Lett.* 256:118–122.

5. Z. Strezoska, T. Paunesku, D. Radosavlijevic, I. Labat, R. Drmanac, R. Crkvenjakov (1991). Sequencing by hybridization: First 100 bases read by a non gel-based method. *Proc. Natl. Acad. Sci., USA* 88:10089–10093.

6. R. Drmanac, S. Drmanac, I. Labat, R. Crkvenjakov, A. Vicentic, A. Gemmell (1992). Sequencing by hybridization: Towards an automated sequencing of one million M13 clones arrayed on membranes. *Electrophoresis* 13:566–573.

7. R. Drmanac, R. Crkvenjakov (1992). Sequencing by hybridization (SBH) with oligonucleotide probes as an integral approach for the analysis of complex genomes. *Int. J. Genome Res.* 1:59–79.

8. R. Drmanac, S. Drmanac, Z. Strezoska, T. Paunesku, I. Labat, M. Zeremski, J. Snoddy, W.K. Funkhouser, B. Koop, L. Hood (1993). DNA sequence determination by hybridization: A strategy for efficient large-scale sequencing. *Science* 260:1649–1652.

9. K.R. Khrapko, Y.P. Lysov, A.A. Khorlin, I.B. Ivanov, G.M. Yershov, S.K. Vasilenko, V.L. Florentiev, A.D. Mirzabekov (1991). A method for DNA sequencing by hybridization with oligonucleotide matrix. *DNA Seq.* 1:375–388.

10. P.A. Pevzner, Y.P. Lysov, K.R. Khrapko, A.V. Belyavsky, V.L. Florentiev, A.D. Mirzabekov (1991). Improved chips for sequencing by hybridization. *J. Biomol. Struct. Dynam.* 9:399–410.

11. W. Bains (1991). Hybridization methods for DNA sequencing. *Genomics* 11:294–301.

12. C.R. Cantor, A. Mirzabekov, E. Southern (1992). Report on the sequencing by hybridization workshop. *Genomics* 13:1378–1383.

13. W. Bains (1993). Characterizing and sequencing cDNAs using oligonucleotide hybridization. *DNA Seq.* 4:143–150.

14. K.L. Beattie (1994). Report on the 1993 international workshop on sequencing by hybridization. *Human Genome News,* January.

15. A.D. Mirzabekov (1994). DNA sequencing by hybridization—A megasequencing method and a diagnostic tool? *Trends Biotechnol.* 12:27–32.

16. Y.P. Lysov, A.A. Chernyi, A.A. Balaeff, K.L. Beattie, V.L. Florentiev, A.D. Mirzabekov (1994). DNA sequencing by hybridization to oligonucleotide matrix. Calculation of continuous stacking hybridization efficiency. *J. Biomolec. Struct. Dynam.* 11:797–812.

17. N.E. Broude, T. Sano, C.L. Smith, C.R. Cantor (1994). Enhanced DNA sequencing by hybridization. *Proc. Natl. Acad. Sci., USA* 91:3072–3076.

18. E.M. Southern, U. Maskos, J.K. Elder (1992). Analyzing and comparing nucleic acid sequences by hybridization to arrays of oligonucleotides: Evaluation using experimental models. *Genomics* 13:1008–1017.

19. U. Maskos, E.M. Southern (1992). Parallel analysis of oligodeoxyribonucleotide (oligonucleotide) interactions. I. Analysis of factors influencing oligonucleotide duplex formation. *Nuc. Acid. Res.* 20:1675–1678.

20. S.C. Case-Green, J.K. Elder, K.U. Mir, U. Maskos, E.M. Southern, J.C. Williams (1994). Parallel synthesis and analysis: Applications of spatially addressable oligonucleotide arrays. In *Innovation and Perspectives in Solid Phase Synthesis*, R. Epton, ed. Proc. Third International Symposium on Solid Phase Synthesis. Mayflower Worldwide Ltd., Birmingham, U.K., 77–82.

21. S. Meier-Ewert, E. Maier, A. Ahmadi, J. Curtis, H. Lehrach (1993). An automated approach to generating expressed sequence catalogues. *Nature* 361:375–376.

22. J.D. Hoheisel (1994). Application of hybridization techniques to genome mapping and sequencing. *Trends Genet.* 10:79–83.

23. S. Drmanac, R. Drmanac (1994). Processing of cDNA and genomic kilobase-size clones for massive screening, mapping and sequencing by hybridization. *BioTechniq.* 17:328–336.

24. K.L. Beattie, M.D. Eggers, J.M. Shumaker, M.E. Hogan, R.S. Varma, J.B. Lamture, M.A. Hollis, D.J. Ehrlich, D. Rathman (1992). Genosensor technology. *Clin. Chem.* 39:719–722.

25. M.D. Eggers, M.E. Hogan, R.K. Reich, J.B. Lamture, K.L. Beattie, M.A. Hollis, D.J. Ehrlich, B.B. Kosicki, J.M. Shumaker, R.S. Varma, B.E. Burke, A. Murphy, D. Rathman (1992). Genosensors: Microfabricated devices for automated DNA sequence analysis. In *Advances in DNA Sequencing Technology*, R. Keller, ed. SPIE Proceedings, Los Angeles, 1891–2004.

26. J.B. Lamture, K.L. Beattie, B.E. Burke, M.D. Eggers, D.J. Ehrlich, R. Fowler, M.A. Hollis, B.B. Kosicki, R.K. Reich, S.R. Smith, R.S. Varma, M.E. Hogan (1994). Direct detection of nucleic acid hybridization on the surface of a charge coupled device. *Nuc. Acid. Res.* 22:2121–2125.

27. M. Eggers, M. Hogan, R.K. Reich, J. Lamture, D. Ehrlich, M. Hollis, B. Kosicki, T. Powdrill, K. Beattie, S. Smith, R. Varma, R. Gangadharan, A. Mallik, B. Burke, D. Wallace (1994). A microchip for quantitative detection of molecules utilizing luminescent and radioisotope reporter groups. *BioTechniq.* 17:516–525.

28. M.D. Matteucci, M.H. Caruthers (1981). Synthesis of deoxyoligonucleotides on a polymer support. *J. Am. Chem. Soc.* 103:3185–3191.

29. S.P.A. Fodor, J.L. Read, M.C. Pirrung, L. Stryer, A.T. Lu, D. Solas (1991). Light-directed, spatially addressable parallel chemical synthesis. *Science* 251:767–773.

30. J.W. Jacobs, S.P.A. Fodor (1994). Combinatorial chemistry—Applications of light-directed chemical synthesis. *Trends Biotechnol.* 12:19–26.

31. A.C. Pease, D. Solas, E.J. Sullivan, M.T. Cronin, C.P. Holmes, S.P.A. Fodor (1994). Light-generated oligonucleotide arrays for rapid DNA sequence analysis. *Proc. Natl. Acad. Sci., USA* 91:5022–5026.

32. M.C. Pirrung, J.C. Bradley (1995). Comparison of methods for photochemical phosphoramidite-based DNA synthesis. *J. Org. Chem.* 60:6270.

33. M.J. Doktycz, K.B. Jacobson, K.L. Beattie, R.S. Foote (1995). Optical Melting as a tool for optimizing sequencing by hybridization analysis of DNA. In *Ultrasensitive Instrumentation for DNA Sequencing and Biochemical Diagnostics*, G.E. Cohn, J.M. Lerner, K.J. Liddane, A. Scheeline, S.A. Soper, ed. *Proc. SPIE* 2386:30–34.

34. R.S. Matson, J.B. Rampal, P.J. Coassin (1994). Biopolymer synthesis on polypropylene supports. *Anal. Biochem.* 217:306–310.

35. T. Therialt (1996). Synthesis of oligonucleotide arrays using Chem-Jet technology. IBC Conference on Biochip Array Technologies, Marina del Rey, CA.

36. W.G. Beattie, L. Meng, S. Turner, R.S. Varma, D.D. Dao, K.L. Beattie (1995). Hybridization of DNA targets to glass-tethered oligonucleotide probes. *Mol. Biotechnol.* 4:213–225.

37. K. Beattie, W. Beattie, L. Meng, S. Turner, C. Bishop, D. Dao, R. Coral, D. Smith, P. McIntyre (1995). Advances in genosensor research. *Clin. Chem.* 41:700–706.

38. K.L. Beattie, J.D. Frost, III (1992). Porous wafer for segmented synthesis of biopolymers. U.S. Patent #5,175,209.

39. K.L. Beattie, N.J. Logsdon, R.S. Anderson, J.M. Espinosa-Lara, R. Maldonado-Rodriguez, J.D. Frost, III (1988). Gene synthesis technology: Recent developments and future prospects. *Appl. Biochem. Biotechnol.* 10:510–521.

40. K.L. Beattie, R.F. Fowler (1991). Solid phase gene assembly. *Nature* 352:548–549.

41. K.L. Beattie, G.D. Hurst (1994). Synthesis and use of oligonucleotide libraries. In *Innovation and Perspectives in Solid Phase Synthesis. Proc. Third International Symposium on Solid Phase Synthesis,* R. Epton, ed. Mayflower Worldwide Ltd., Birmingham, U.K., 69–76.

42. P. McIntyre, personal communication.

43. G.M. Yershov, A.I. Belgovsky, L.D. Drobyshev, V.N. Sushkov, N.V. Mologina, D. Guschin, J. Steele, A. Gemmel, A. Zaslavsky, D. Naylor, A.D. Mirzabekov (1996). Robot for manufacturing SHOM oligonucleotide microchips. IBC Conference on Biochip Array Technologies, Marina del Rey, CA.

44. W.G. Beattie, G.C. Raia, J.D. Pratt, E.I. Budowsky, S. Kumar, R.S. Varma, K.L. Beattie (1996). Oligonucleotide array hybridization: improvements in probe attachment to glass. *Nuc. Acid. Res.,* submitted.

45. U. Maskos, E.M. Southern (1993). A study of oligonucleotide reassociation using large arrays of oligonucleotides synthesized on a glass support. *Nuc. Acid. Res.* 21:4663–4669.

46. L.A. Markey, K.J. Breslauer (1987). Calculating thermodynamic data for transitions of any molecularity from equilibrium melting curves. *Biopolymers* 26:1601–1620.

47. M.J. Doktycz, M.D. Morris, S.J. Dormady, K.L. Beattie, K.B. Jacobson (1995). Optical melting of 128 octamer DNA duplexes: Effects of base pair location and nearest neighbors on thermal stability. *J. Bio. Chem.* 270:8439–8445.

48. N. Tibanyenda, S.H. De Bruin, C.A.G. Haasnoot, G.A. Van der Marel, J.H. Van Boom, C.W. Hilbers (1984). The effect of single base-pair mismatches on the duplex stability of d(T-A-T-T-A-A-T-A-T-C-A-A-G-T-T-G) · d(C-A-A-C-T-T-G-A-T-A-T-T-A-A-T-A). *Eur. J. Biochem.* 139:19–27.

49. F. Aboul-ela, D. Koh, I. Tinoco, Jr., F.H. Martin (1985). Base-base mismatches: Thermodynamics of double helix formation for $dCA_3XA_3G + dCT_3YT_3G$ (X, Y = A, C, G, T). *Nuc. Acid. Res.* 13:4811–4824.

50. H. Werntges, G. Steger, D. Riesner, H.-J. Fritz (1986). Mismatches in DNA double strands: Thermodynamics parameters and their correlation to repair efficiencies. *Nuc. Acid. Res.* 14:3773–3790.

51. Y. Kawase, S. Iwai, H. Inoue, K. Miura, E. Ohtsuka (1986). Studies of nucleic acid interactions I: Stabilities of mini-duplexes $(dG_2A_4XA_4G_2 \cdot dC_2T_4C_2)$ and self-complementary d(GGGAAXYTTCCC) containing deoxyinosine and other mismatched bases. *Nuc. Acid. Res.* 14:7727–7736.

52. S. Ikuta, K. Takagi, R.B. Wallace, K. Itakura (1987). Dissociation kinetics of 19 base paired oligonucleotide-DNA duplexes containing different single mismatched base pairs. *Nuc. Acid. Res.* 15:797–811.

53. S.-H. Ke, R.M. Wartell (1993). Influence of nearest neighbor sequence on the stability of base pair mismatches in long DNA: Determination by temperature-gradient gel electrophoresis. *Nuc. Acid. Res.* 21:5137–5143.

54. M.J. Doktycz, T.M. Paner, M. Amaratunga, A.S. Benight (1990). Thermodynamic stability of the 5′ dangling-ended DNA hairpins formed from the sequences 5′-(XY)$_2$GGATAC(T)$_4$-GTATCC-3′, where X,Y = A,T,G,C. *Biopolymers* 30:829–845.

55. M. Senior, R.A. Jones, K.J. Breslauer (1988). Influence of dangling thymidine residues on the stability and structure of two DNA duplexes. *Biochem.* 27:3879–3885.

56. J.C. Williams, S.C. Case-Green, K.U. Mir, E.M. Southern (1994). Studies of oligonucleotide interactions by hybridisation to arrays: The influence of dangling ends on duplex yield. *Nuc. Acid. Res.* 22:1365–1367.

57. J.G. Wetmur (1991). DNA Probes: Applications of the principles of nucleic acid hbridization. *Crit. Rev. Biochem. Mol. Bio.* 26:227–259.

58. R.S. Quartin, J.G. Wetmur (1989). Effect of ionic strength on the hybridization of oligodeoxynucleotides with reduced charge due to methylphosphonate linkages to unmodified oligodeoxynucleotides containing the complementary sequence. *Biochem.* 28:1040–1047.

59. R. Drmanac, Z. Strezoska, I. Labat, S. Drmanac, R. Crkvenjakov (1990). Reliable hybridization of oligonucleotides as short as six nucleotides. *DNA Cell Bio.* 9:527–534.

60. V. Lehmann (1993). The physics of macropore formation in low doped n-type silicon. *J. Electrochem. Soc.* 140:2836–2843.

61. K.L. Beattie, B. Zhang, J.D. Pratt, W.G. Beattie (1996). Flowthrough genosensors employing porous glass hybridization substrates. *Nuc. Acid. Res.*, submitted.

62. Radoje Drmanac, personal communication.

11

Mass Spectrometric Methods in DNA Characterization

PETER WILLIAMS, CHAU-WEN CHOU, AND
DAVID M. SCHIELTZ

CONTENTS

INTRODUCTION

Over the past 20 years steady advances have occurred in the capability of mass spectrometry to characterize biomolecular analytes. These advances have resulted from the development of novel methods for volatilization and ionization of these labile materials without excessive fragmentation. In particular, in the late 1980s two methods were developed

Automation Technologies for Genome Characterization, Edited by Tony J. Beugelsdijk.
ISBN 0-471-12806-6   © 1997 John Wiley & Sons, Inc.

that allow transport of intact molecular ions of very large biopolymers into the gas phase. One is electrospray ionization (ESI), which produces ions from evaporating charged droplets electrosprayed from a capillary needle at high potential [1]. The second approach uses pulsed laser ablation of a solid matrix in which the biomolecules have been embedded [2] and has received the unwieldy name "matrix-assisted laser desorption-ionization" (MALDI). Over the last five years both of these techniques have been applied to problems of DNA characterization.

ESI and MALDI have different strengths and weaknesses. Electrospraying occurs near room temperatures, and only moderate heating is necessary to desolvate the molecular ions. The ions are thus produced with low internal energies, and very large molecules can survive the process without fragmentation. Detection of intact T4 DNA molecular ions with a nominal molecular weight $\sim 1 \times 10^8$ Da has been reported [3]. ESI produces ions in very high charge states; electrosprayed DNA molecules, for example, can carry as much as one negative charge (missing proton) for every 3–6 nucleotides. This is a great advantage, since it moves the mass-to-charge ratio (m/z) into the range where relatively inexpensive quadrupole mass spectrometers operate ($m/z \sim 1000$–2000). Thus the ionization technique is readily adapted to this mature mass spectrometric technology and can benefit from other aspects of the technology, in particular, the capability to induce fragmentation, through gas phase collisions, revealing structural or sequence information. The mass resolution of these instruments is typically quite good and accurate determination ($\pm 0.01\%$) of the m/z values is possible. However, the accompanying disadvantage is that analytes acquire a spread in charge states which is typically within a factor of two of the mean charge, so a 20 kDa molecule, for example, may appear with charges ranging from ~ 10–30. The multiplicity of peaks in the mass spectrum resulting from a single molecular species makes ESI unsuitable, in general, for analysis of mixtures, especially mixtures with the complexity of DNA sequencing ladders. Determination of mass from the m/z values requires knowledge of the value of the ion charge z. Where charge states are resolved, z is usually determinable from the progression in spacing of the peaks arising from the different charge states, such as appearing at $M/20$, $M/19$, or $M/18$, for a single molecule of mass M. One of the promising uses of electrospray technology is in concert with Fourier transform ion cyclotron resonance mass spectrometry (FTICR-MS). The extremely high mass resolution of these devices allows m/z determination with very high accuracy and correspondingly accurate identification of DNA fragments, again given that z can be determined.

In MALDI mass spectra, singly charged ions typically dominate. Depending on the size of the molecule and analytical conditions, ions in higher charge states are detectable, but the mass spectra are always considerably simpler than ESI spectra, making this approach a viable analytical technique for mixtures. Due to the low ion charge, the mass/charge ratio is high and typically time-of-flight mass spectrometers, which have no intrinsic mass limit, are used for mass analysis. Mass resolution for large molecules in these devices is typically $\sim 1:500$ or less; even so, mass accuracy as good as 0.01% can be achieved, at least for molecules up to about 10 kDa. The dominant problem in MALDI is that the ablation process is much more energetic than ESI and appears to produce vibrationally excited (hot) biomolecular ions that may fragment on a

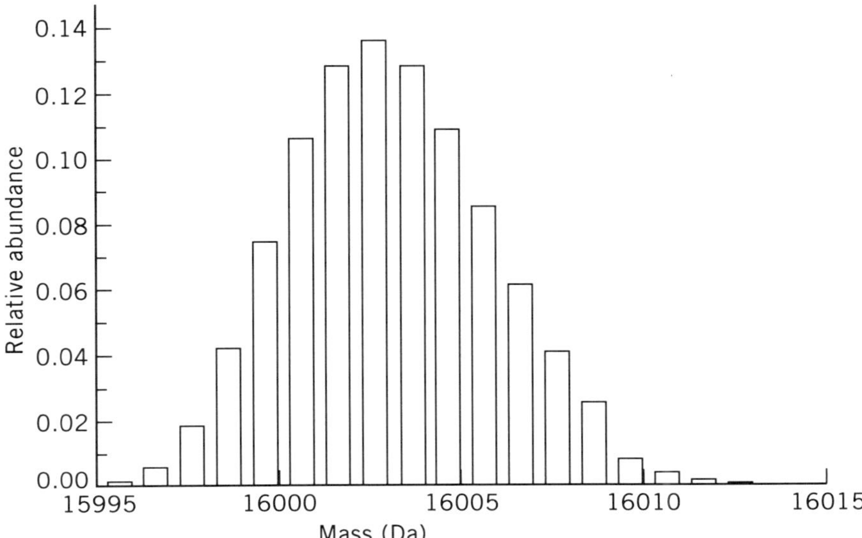

Fig. 11.1   Isotopic distribution calculated for a DNA 52-mer $(C_{13}A_{13}G_{13}T_{13})$.

microsecond time scale, depending on the species, the matrix, and the laser power needed for ablation. Finding matrix compounds that minimize fragmentation of DNA has been a major goal of research groups in this field.

Mass spectrometers are conventionally separated into "low" and "high" resolving power categories. Resolving power may be understood as the mass at which two ion peaks differing by only 1 mass unit can be distinguished from each other. The dividing line is vague, but resolving power less than ~1000 would be considered "low," while usefully high resolving power begins above ~10,000–20,000. Above masses of a few thousand daltons, peaks are intrinsically broadened due to the existence of the minor isotopes of C, N, and O. Figure 11.1 shows the calculated isotopic distribution for a DNA 52-mer. $(C_{507}O_{310}N_{195}H_{638}P_{51})$. Twelve peaks are visible above 10% relative intensity. The pure ^{12}C, ^{14}N, ^{1}H, ^{16}O peak is barely detectable at a mass of 15,995.7 Da, with a relative intensity of only ~0.5% of the most abundant peak. Unless the isotopic peaks can be resolved, which in the present example would require a resolving power ~16,000, this isotopic spread limits mass resolving power to ~1000–2000, and can complicate the interpretation of high-resolution FTICR mass spectra as will be seen later.

MALDI-TOFMS APPLIED TO SEQUENCE LADDER READOUT

MALDI has produced extremely good results for a wide variety of protein analytes. Mass resolving power of several hundred is readily achieved in simple TOF instruments, and detection of molecules up to ~1 MDa has been demonstrated [4]. A number of groups have worked to apply MALDI to DNA analytes. The most widespread approach has used matrix compounds that absorb strongly in the UV, typically around 337–355 nm (the wavelengths of nitrogen lasers and frequency-tripled Nd-YAG lasers) or 266 nm (frequency-quadrupled Nd-YAG), although some success has been achieved also in the IR at 2.9 mm (Er-YAG laser) or ~10 mm ($CO2$ laser). However, the UV matrices that yield good protein mass spectra perform very poorly for DNA. Figure 11.2 *a* and *b* shows a comparison of mass spectra of insulin and a 20-mer DNA strand of comparable mass from the same matrix (sinapinic acid, a protein-friendly matrix) [5]. The DNA spectrum shows evidence of extensive fragmentation. Fragmentation clutters the mass spectrum with peaks that could obscure components of a sequence ladder mixture, and typically for DNA produces a high background signal.

Significant effort has been expended in searches for optimized DNA matrices. The searches have been hampered by a general lack of understanding of the parameters that characterize a "good" matrix, other than the necessity for strong absorption at the wavelength of the laser employed. The best DNA MALDI matrix, discovered by Becker et al. [6], is 3-hydroxy picolinic acid (3-HPA), a UV matrix. Figure 11.2 *c* shows a mass spectrum of the DNA 20-mer ablated from 3-HPA [5]. Fragmentation is very much reduced; residual peak broadening results largely from sodium adduct formation, which can be minimized by ion-exchange removal of sodium from the sample or by adding an excess of ammonium salts to the matrix solution. The ammonium ion is labile under the ablation conditions and protonated phosphate groups are left as

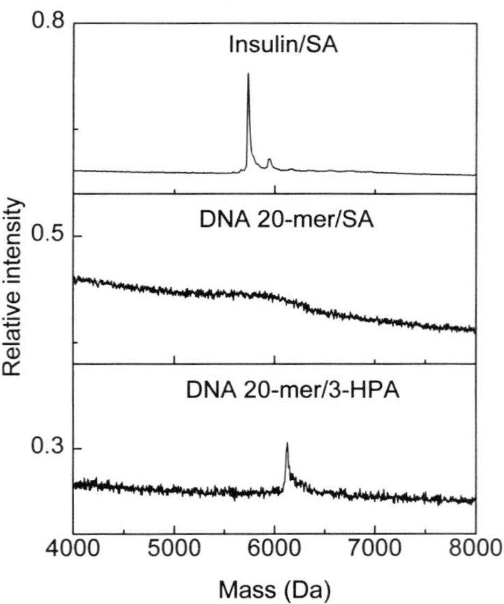

Fig. 11.2 Positive ion time-of-flight mass spectra from different matrices: (*a*) Insulin ablated from a sinapinic acid (SA) matrix (protein-friendly), (*b*) a mixed base DNA 20-mer ablated from sinapinic acid, and (*c*) the DNA 20-mer ablated from a 3-hydroxypicolinic acid (3-HPA) matrix (nucleic acid-friendly). From Chou [5].

ammonia departs. Figure 11.3 is a mass spectrum of a mock sequencing mixture assembled from pmol quantities of separately synthesized DNA strands and ablated from a 3-HPA matrix using 355-nm radiation [7]. The limitations of the UV MALDI approach for DNA are that both sensitivity and mass resolving power degrade significantly with increasing molecular weight as seen in Fig. 11.3. This is true also for protein analytes but to a much smaller extent. The sensitivity decrease results from a variety of factors that are not completely understood but that must include a lower detection sensitivity for larger, slower ions, and significantly increased probability of fragmenta-

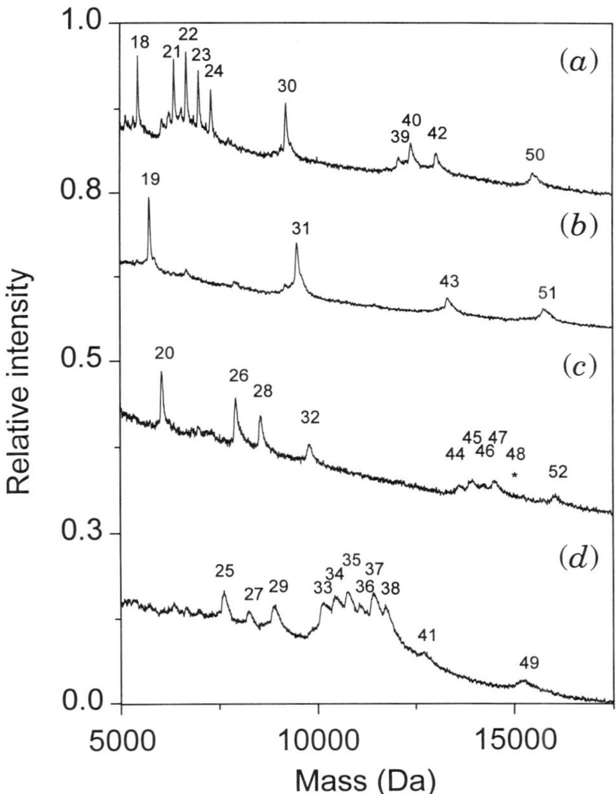

Fig. 11.3   Positive ion time-of-flight mass spectra of a mock sequencing mixture assembled from pure synthetic samples. The samples contained oligonucleotides terminated with (a) adenine, (b) thymine, (c) guanine, and (d) cytosine, ranging from the 18-mer to the 52-mer. The 48-mer (position marked by an *) was intentionally omitted from the mixture to evaluate error detection. The overall sequence was: 5'-(18-mer "primer") CT AAAA GTGT GACT GGGG GGAA GACT TTTT GACT-3'. Multiple base resolution checks are apparent at A_{21-24}, G_{33-38} and T_{44-48}. *Matrix:* 3-hydroxypicolinic acid; *laser:* 355 nm. Concentrations ranged from 0.7 pmol (smaller fragments) to 3.5 pmol (largest fragments) From Chou and Williams [7].

tion. Fragmentation can also contribute to degraded mass resolution, as can the formation of unresolved adducts of alkali metal ions and the matrix itself. Given that the resolution criterion for sequence ladder readout is single nucleotide resolution, the readout length currently accessible to the UV MALDI approach is only about 40 nucleotides. This is noncompetitive with gel techniques for large-scale sequencing but is sufficient to be useful in characterizing short lengths of synthetic DNA, such as antisense DNA and short PCR products.*

Sanger reactions typically produce few-femtomole amounts of individual ladder components. Subfemtomole quantities of analyte are typically consumed to obtain UV MALDI mass spectra, although typically picomole amounts are loaded into the mass spectrometer simply for convenience in sample handling. Becker and colleagues have shown that MALDI mass spectra can be obtained from pure DNA analytes loaded at the few-femtomole level as shown in Fig. 11.4 [8]. However, real samples pose additional problems because they contain impurities that adversely affect the mass spectrum and must be removed. Cleanup procedures tend to be time-consuming and require larger amounts of sample. Becker's group recently demonstrated the first mass spectrometric sequence ladder readout from a real Sanger reaction with a 2-pmol truncated template (45-mer) producing sub-pmol quantities of ladder components [8]. Two cleaning procedures were investigated. One cleanup approach used ultrafiltration through a membrane with a 3000 molecular weight cutoff, and the second used a glass resin to

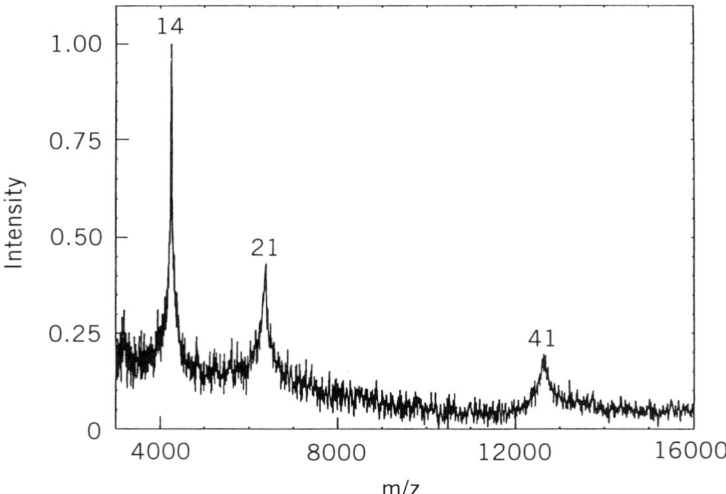

Fig. 11.4 Positive ion mass spectrum of a pure synthetic mixture containing 5 fmol each of a 14-, a 21, and a 41-mer mixed-base single-stranded DNA, ablated from 3-hydroxypicolinic acid, summing 100 laser shots at 355 nm. From Shaler et al. [8].

* Monforte and Becker have recently reported readout to 95 bases, including the primer. (J. A. Monforte and C. H. Becker (1997) "High-throughput DNA analysis by time-of-flight mass spectrometry," *Nature Medicine,* 3:360–362

bind the DNA while buffer salts were washed away. Using fairly high ratios (1 : 4) of dideoxy to normal nucleotide triphosphates for a short sequencing read, they report that sequence can be read out about 19 bases beyond a 12-mer primer. Mass spectra of sequence ladders produced using the 12-mer primer are shown in Fig. 11.5. Sequence ladder peaks are visible with useful intensity out to the 31-mer. A significant false-stop or misincorporation by the enzyme is seen in the ddT reaction at the 14-mer position, possibly due to the short length of the primer, but the sequence is otherwise readable. Using a 21-mer primer, the authors detect peaks out to the 41-mer, but with readable sequence only out to about the 35-mer.

Dominant in the sequence ladder spectra with the 12-mer primer are strong peaks for the singly and doubly ionized template which is not removed in the cleanup. The tail of the intense singly charged template peak may obscure ladder peaks nearby, while the doubly charged peak interferes with sequence readout in the center of the ladder. Chou and Williams have investigated affinity cleanup techniques that bind the ladder fragments to a DNA strand complementary to the primer [7]. In such an approach the template should not be captured and could be washed away together with the buffer components. Two approaches were investigated: one in which the complementary strand was coupled to a solid support (paramagnetic beads) prior to annealing, and the second in which the strand was biotinylated and annealing occurred in the solution phase, followed by capture on streptavidin-coated paramagnetic beads. The latter method was significantly more rapid and efficient with approximately 30% of a synthetic ladder mixture being retained after cleanup [7]. The quality of the sequence ladder mass spectra after this affinity cleanup may be assessed by comparing the mass spectra of Fig. 11.6 with those of the pure mixtures in Fig. 11.3.

An alternative approach to laser ablation produces gas-phase DNA by laser ablation from a frozen aqueous solution using a pulsed laser operating in the visible [9,10]. An advantage of this approach derives from the use of a laser wavelength that is not directly absorbed in the sample, in either the DNA analyte or the water matrix. Instead, the laser power is primarily deposited in the copper substrate on which the frozen film is formed. Volatilization of the ice appears to result from shock heating driven by the formation of a plasma at the metal/ice interface due to the very high laser power density used (up to 10^9 W/cm^2 UV MALDI typically operates at a laser power density $\sim 10^6$ W/cm^2) [11]. In initial experiments, collection of 32P-labeled ablated DNA followed by gel electrophoresis showed that double-stranded DNA up to 622 base pairs (~ 410 kDa) survived the ablation process substantially intact [9]. However, a major disadvantage of the technique for mass spectrometry is that ionization only rarely accompanies the ablation event, so most laser pulses produce no DNA signal in the mass spectrometer, even though copious ions resulting from the plasma at the metal substrate can be seen. Very rarely ions are observed from laser impact near the edges of the deposited ice films. When this occurs the resulting mass spectra are quite good. Figure 11.7 shows mass spectra of two "lanes" of a mock sequence ladder mixture assembled from synthesized fragments of a 90 mer [12]. Picomole quantities of material were loaded in the sample films from which these spectra were obtained, and the sensitivity, measured as the peak integrals, is roughly constant across the mass range. However, to obtain these data, several dozen frozen samples were loaded and scanned

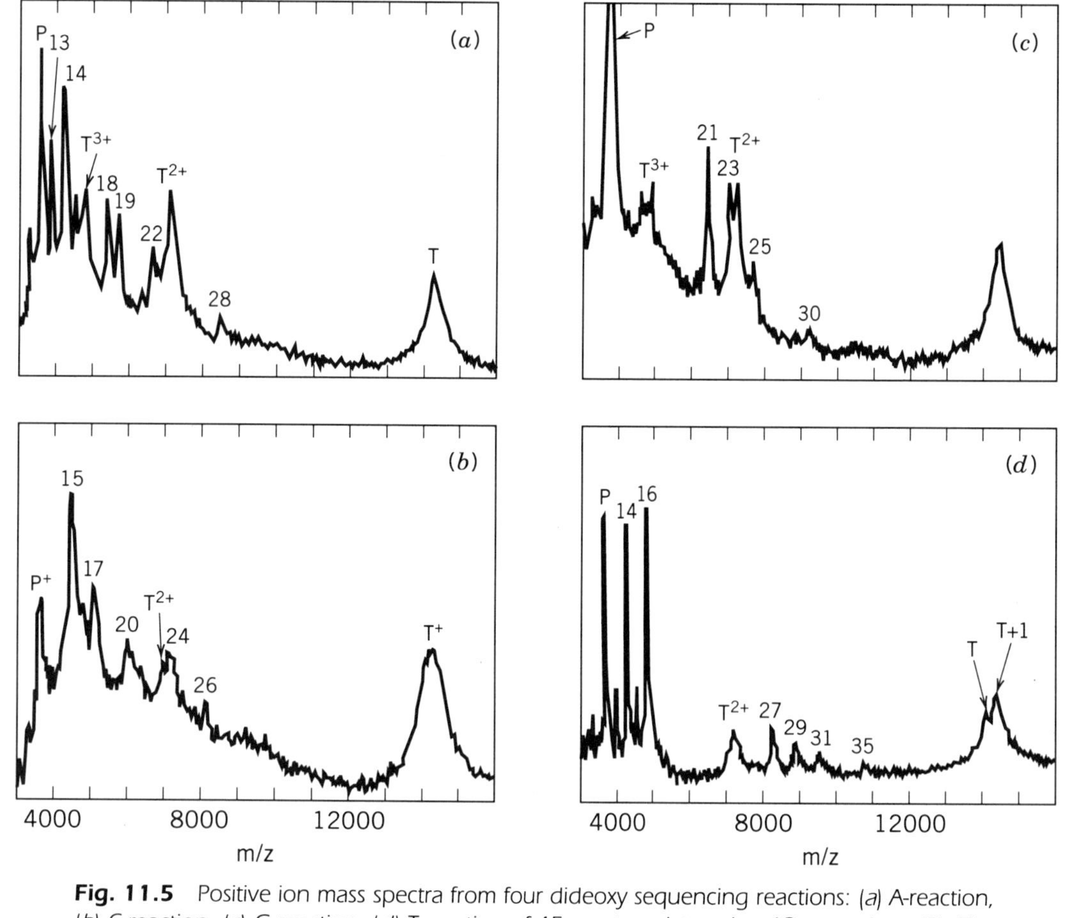

Fig. 11.5 Positive ion mass spectra from four dideoxy sequencing reactions: (*a*) A-reaction, (*b*) C-reaction, (*c*) G-reaction, (*d*) T-reaction of 45-mer template using 12-mer primer (P). The spectra were obtained from 3-hydroxypicolinic acid, summing 100 laser shots at 355 nm. From Shaler et al. [8].

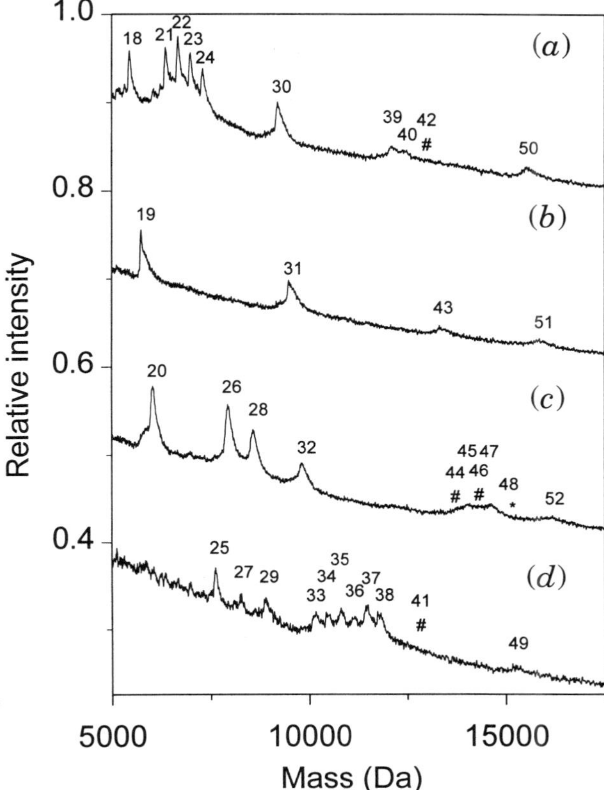

Fig. 11.6 Positive ion mass spectra of the four synthetic mixtures used in Fig. 11.3, after affinity cleanup of added Sanger reaction impurities (template, buffer, deoxy- and dideoxynucleotides). *Matrix:* 3-hydroxypicolinic acid; *laser:* 355 nm. From Chou and Williams [7].

under the laser beam to find the one sample for each mixture that yielded data of the quality shown.

Although neither UV MALDI nor ice ablation have yet achieved the level of performance required for extended sequence ladder readout, the data of Fig. 11.7 allow discussion of the potential advantages of mass spectrometric readout.

Error Flagging

The complementary nature of the two mass spectra in Fig. 11.7 is apparent, and this can be seen more clearly in the expanded views of low and high mass regions of the spectra shown in Fig. 11.8. It demonstrates the first major advantage of mass spectrometry for sequence readout: the mass scale is absolute. Even though these two mass spectra were

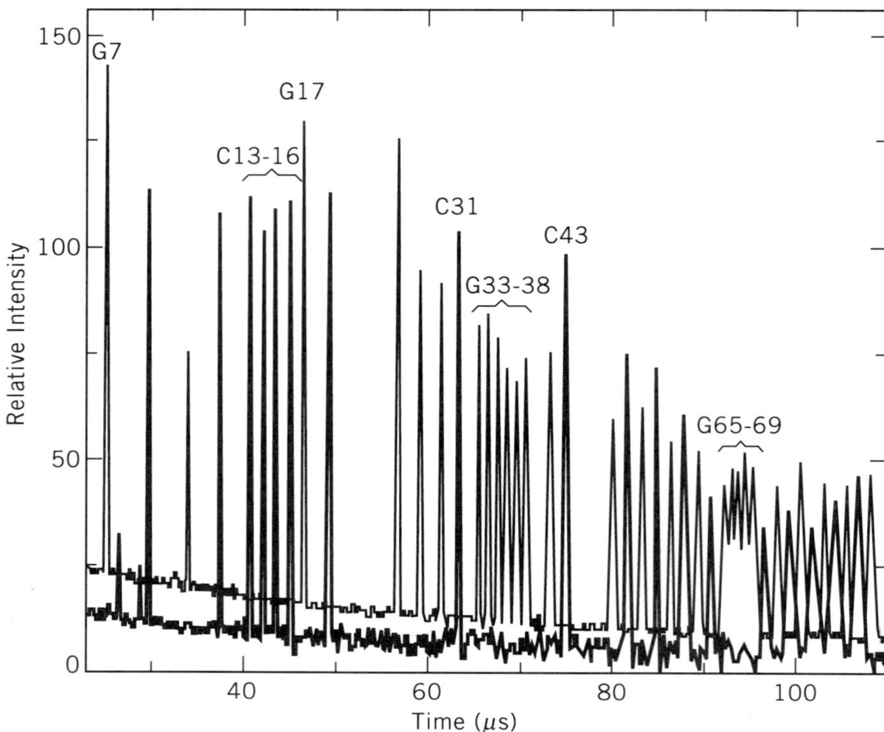

Fig. 11.7 *Positive ion mass spectra of two "lanes" (C- and G-terminated) of a synthetic mixture assembled from synthesized fragments of a 90-mer sequence. The spectra were obtained by visible pulsed laser ablation of thin frozen films of aqueous solutions of the two mixtures. Each spectrum is the sum of 12 laser shots. The overall 90-mer sequence from which the fragments were selected is 5′-GACT GACT GACT CCCC GACT AAAA GTGT GACT GGGG GGAA GACT TTTT GACT GACT GACT GACT GGGG GACT GACT GACT GACT GACT GA-3′. Fine line: G-terminated spectrum; Heavy line: C-terminated spectrum. From Schieltz [12].*

obtained some months apart in time, the reproducibility of the mass spectrometer is such that the peaks appear in the correct sequence, and gaps in the sequence are clearly detectable. This capability makes mass spectrometric readout potentially immune to frame shift errors. Gaps in the sequence are clearly identifiable. Because most uses of sequence data will be tolerant to a moderate incidence of one-letter gaps, usually mass spectrometric readout would obviate the need for multiple re-sequencing.

Error Recovery

The ability to flag errors gives mass spectrometric sequence readout a high degree of error *immunity*. However, mass spectrometry has an additional advantage in that the

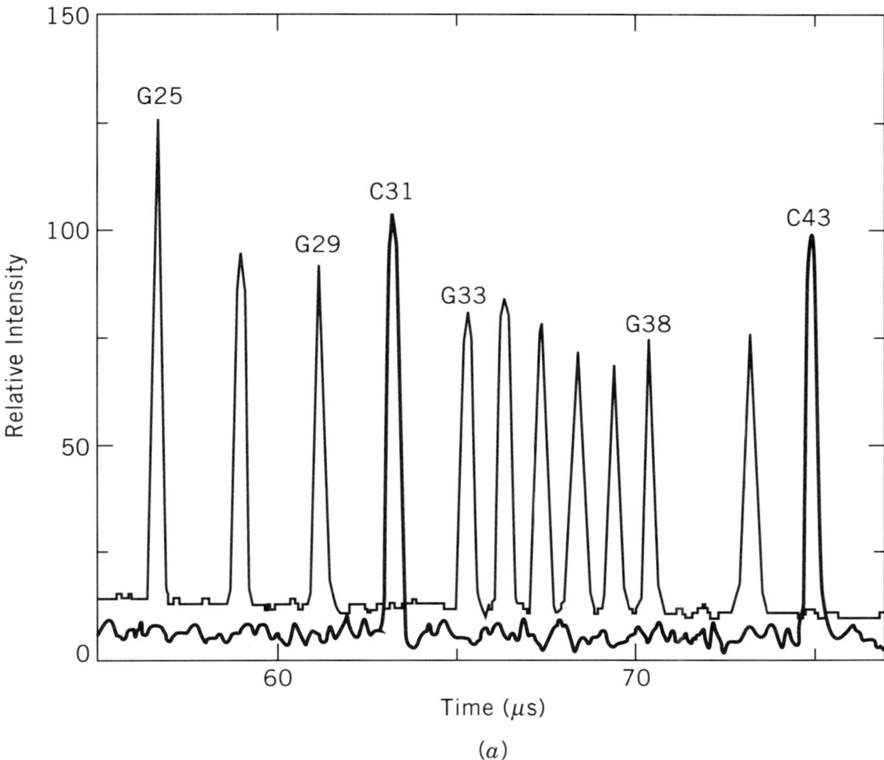

Fig. 11.8 (*a*) Expanded view of middle-mass region (25 mer–43 mer) of the spectra of Fig. 11.6, showing the G-repeat resolution check at $G_{33–38}$. (*b*) Expanded view of high-mass region (49 mer–89 mer) of the spectra of Fig. 11.6, showing the G-repeat resolution check at $G_{65–69}$. *Fine line:* G-terminated spectrum; *Heavy line:* C-terminated spectrum. From Schieltz [12].

ability to perform precise mass measurement, or more exactly precise mass *difference* measurement, allows a degree of error *recovery*. If a peak is missing in the mass spectrum, the information needed to complete the genetic code readout is the identity of the terminal nucleotide on the missing fragment. If the next peak in the sequence is detectable, then the 2-nucleotide mass difference that spans the gap can be measured. Given that the terminal nucleotide on the peak on the high mass side of the gap is identifiable from the "lane" in which it appears (this is a major advantage of retaining the four-lane readout methodology of the classic Sanger approach), precise measurement of the 2-nucleotide mass difference allows the identity of the terminal nucleotide on the missing peak to be deduced.

In Table 11.1, the masses of the four DNA nucleotides are listed. The four have characteristic masses, but in the case of A and T these differ by only 9 Da. To distinguish, with >90% certainty, between A and T on the basis of mass difference requires a mass

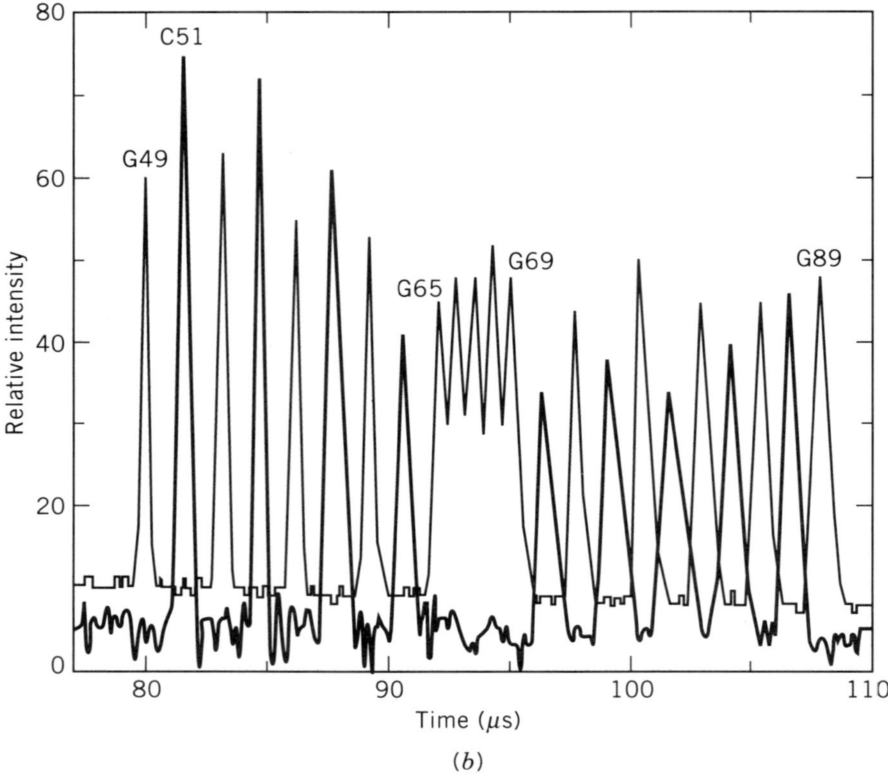

(*b*)

Fig. 11.8   *(Continued)*

accuracy of ± 3 Da (1σ) or better. The mass spectra of Fig. 11.7 constitute overdetermined sets of data to which a best fit mass calibration curve can be applied. Figure 11.9 shows the deviations of the true masses of the G-terminated strands from the masses calculated from the flight times and the best-fit line. Also shown on this plot are lines bounding errors of $\pm 1 : 3000$. It can be seen that somewhat more than two-thirds of the data points lie within these bounds indicating that a relative mass accuracy (1σ) a little better than $\pm 1 : 3000$ was achievable for these data. At this accuracy, errors smaller than ± 3 Da should be achievable up to masses $\sim 10,000$ Da.

Since mass spectra were not obtained for the A- and T-terminated mixtures shown in Fig. 11.7, the mass spectra allow demonstration of the error recovery process. Consider the gaps in the combined spectra at the 26-mer and 28-mer positions, seen clearly in Fig. 11.8*a*. If experimentally these peaks were missing from the entire four-lane mass spectrum, error recovery would use the sequence ladder readout up to the 25-mer to identify the sequence and produce a mass listing for calibration. This calibration can then be extended to calculate the (27-mer–25-mer) and (29-mer–27-mer) mass differences. Experimentally, these mass differences are found to be

TABLE 11.1   Masses of the Individual Nucleotides (Da)

Nucleotide	Mass (Da)	Difference (Da)
Guanidine, G	329.2	
		} 16
Adenine, A	313.2	
		} 9
Thymine, T	304.2	
		} 15
Cytidine, C	289.2	

632.7 and 634.7 Da. The 27-mer and 29-mer are G-terminated, so part of the 2-nucleotide mass difference in each case corresponds to G. Subtracting the G mass (329.2 Da) yields the second nucleotide masses in the 2-nucleotide gaps as 303.5 and 305.5 Da. These are within ~1 Da of the T mass (304.2 Da) and 8 Da or more different from the other nucleotide masses (C = 289.2, A = 313.2 Da). Indeed, the missing nucleotides in this sequence are T's. This is a best-case example: Statistically the error in this region can be as large as ~3 Da for these data.

To extend such error recovery methods to masses in the 100 kDa region (~300 nucleotides), which is a read length routinely accessible to gel electrophoresis, would require a mass measurement accuracy of $3:10^5$. This is significantly better than the best accuracy yet achieved for proteins using time-of-flight mass spectrometry (~ $1:10^4$ at masses below ~10 kDa, degrading to ~$1:10^3$ or worse at ~100 kDa) and may not be achievable for large biomolecules by time-of-flight mass spectrometry. However,

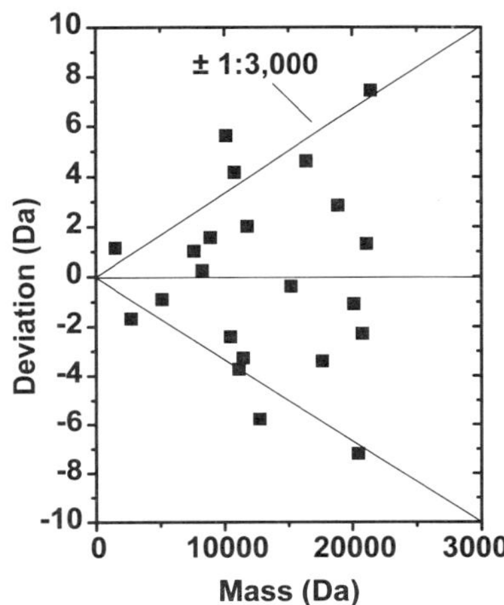

Fig. 11.9 Deviations of peak centroids in G-terminated spectrum of Fig. 11.6 from best-fit mass calibration.

it may well be possible instead to modify the consituent nucleotides of the DNA strands chemically so as to increase the minimum mass differences between the four nucleotides. Nucleotides derivatized with a variety of dye molecules have been shown to be incorporated into growing DNA strands by DNA polymerase. If it is desired only to modify the mass without worrying about any functional utility of the modification, the number of possible derivatives must be quite large and can be selected for compatibility with the enzyme. For example, simply substituting thiophosphate for phosphate in A and G, which makes a negligible change in the chemistry, increases the minimum mass difference between the nucleotides to 15 Da (C–T), relaxing mass accuracy requirements by about 50%. It seems quite feasible to introduce derivatized nucleotides that would have mass differences of 100 Da or more so that a mass accuracy of $3:10^4$ or less would be sufficient to identify missing nucleotides up to the 100 kDa region.

MALDI-TOFMS SEQUENCE ANALYSIS OF SHORT DNA STRANDS

Sequence ladder readout is a potential, but still only a potential, grand-scale application of TOFMS. However, for sequencing short lengths of DNA, TOFMS already is becoming the technique of choice. Major applications here are in sequence confirmation of synthetic DNA, for example, primers and antisense compounds.

The central requirement for sequencing, if only moderate mass resolution is available, is a unidirectional fragment ladder. Then the sequence is read from high to low mass by determining successive fragment mass differences. If nucleotide loss from either end of a strand is allowed, then two overlapping ladders are produced that can give rise to unresolvable peaks. For example, loss of 3′T or 5′A would produce two fragment peaks differing in mass by only 9 Da, whereas in a unidirectional ladder the minimum mass separation is always a single nucleotide (~300 Da). The *maximum* mass difference between 3′ and 5′ single nucleotide losses is 40 Da (G–C). For a mass resolution of, say, 1 : 100, read length in a unidirectional fragment ladder would in principle be 100 nucleotides. However, in a bidirectional ladder unresolved peaks may arise at a parent mass of only ~900 Da (3 nucleotides) and are inevitable by 4,000 Da or ~13 nucleotides.

In Sanger sequencing, unidirectional ladders are produced by polymerase-catalyzed strand extension (Maxam-Gilbert sequencing protocols are unsuitable for mass spectrometric sequencing because they produce bidirectional sequence ladders). For sequence analysis of short DNA strands, methods to produce unidirectional ladders include use of failure sequences in synthetic samples and phosphodiesterase digestion.

Keough and colleagues were the first to show that UV MALDI can be used to confirm short synthetic sequences through the analysis of the failure sequence mixture [13]. Although these authors did not have access to a good DNA matrix (the work was done prior to the introduction of 3-HPA), they were able to obtain sufficiently precise masses for the failure sequence fragments of a synthetic 15-mer antisense DNA to confirm the sequence. A particularly attractive aspect of this approach is that no further chemistry is required. At about the same time Pieles et al. discovered a significantly better DNA matrix compound, 2,4,6-trihydroxy acetophenone, and demonstrated the use of unidirectional exonucleases, snake venom or calf spleen phosphodiesterases, to produce unidirectional sequence ladders by digesting single-stranded DNA exclusively from the 3' or 5' ends, respectively. Uneven digestion produces a sequence ladder from which, again, the sequence can be read from the measured mass differences as shown in Fig. 11.10 (note the improved mass spectral performance from the DNA-friendly matrix) [14]. At present the convenience, speed, and accuracy of these analyses makes them the technique of choice for characterization of short DNA strands, particularly if modified nucleotides are incorporated.

RECENT AND FUTURE DEVELOPMENTS IN MALDI

Efforts to improve performance in MALDI are proceeding along several fronts. Searches continue for improved UV matrices or for approaches that work better at other laser wavelengths, particularly in the infrared. Although surprisingly good results can be obtained on occasion, IR irradiation to date has not produced data better than UV MALDI using the 3-HPA matrix. Furthermore IR MALDI typically has poor pulse-to-pulse reproducibility and so is not currently the technique of choice for robust, high-throughput operation. Matrix searches are hampered by the fact that the parameters that make a good matrix are not yet understood. Clearly a high extinction coefficient at the wavelength used is essential, but beyond that, chemically quite similar compounds can behave quite differently as matrices. Chen and coworkers have investigated mixtures of matrix compounds. Using a mixture of picolinic and 3-hydroxypicolinic acids, Tang et al. reported detection of a molecular ion peak of a DNA 500-mer, albeit with poor signal-to-noise ratio and mass resolution $\sim 1:15$ [15]. This mixture does not appear to be preferable to 3-HPA alone for good mass resolution and sensitivity at lower masses. The difficulty in searches for mixed matrices is that without a basic understanding of matrix operation, the mixture parameter space to be searched is immense.

More promising are methods to improve the resistance of the DNA analyte to photochemically or thermally induced cleavage. A thorough series of studies on DNA fragmentation was reported by Fitzgerald et al. [16]. These authors investigated the striking difference in stability between poly dT oligomers and oligomers of the other three deoxynucleotides. By incorporating a single G at different positions in a poly-T oligomer they showed unequivocally that fragmentation occurred dominantly at the G position, focusing attention on the apparent reactivity of the $-NH_2$ group in the purine bases and in cytidine. Nordhoff et al. have speculated that protonation of the amino group destabilizes the glycosidic bond and initiates fragmentation by base cleavage [17].

In early MALDI studies it was noted that RNA analytes appear to be much more stable than DNA, and a mass spectrum of a 461 nt RNA has been reported with quite good signal-to-noise and a mass resolution $\sim 1:30$ [18]. This observation has stimulated several efforts to define and synthesize more stable nucleic acid analytes. One obvious approach would be to use RNA transferase to synthesize RNA copies from a DNA template and develop a termination strategy that could produce RNA sequence ladders. This has been explored by Kirpekar et al. [18]. Indeed, mass spectra of RNA ladders

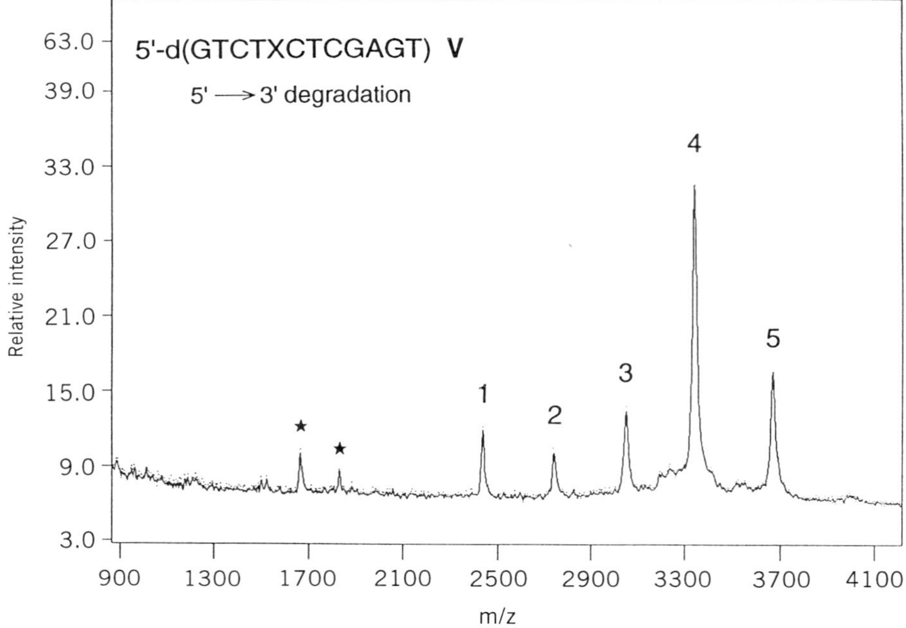

Peak	Sequence	Mass calc.	Mass found
1	5'-d(XCTCGAGT)	2438.7	2438.3
2	5'-d(TXCTCGAGT)	2742.9	2742.8
3	5'-d(TTXCTCGAGT)	3047.1	3046.9
4	5'-d(CTTXCTCGAGT)	3336.3	3336.6
5	5'-d(GCTTXCTCGAGT)	3665.5	3665.8

(a)

Fig. 11.10 Mass spectra of exonuclease digestion products of 12-mer oligonucleotide: 5'-d(GCTTXCTCGACT), where x = 2'-o-methyl adenosine. *Matrix:* 2,4,6-trihydroxy aceto-phenone, 337 nm laser. *Mass accuracy was sufficient to determine the entire sequence. (a) 5' ® 3' digestion using calf spleen phosphodiesterase; (b) 3' ® 5' digestion using snake venom phosphodiesterase. From Pieles et al. [14].*

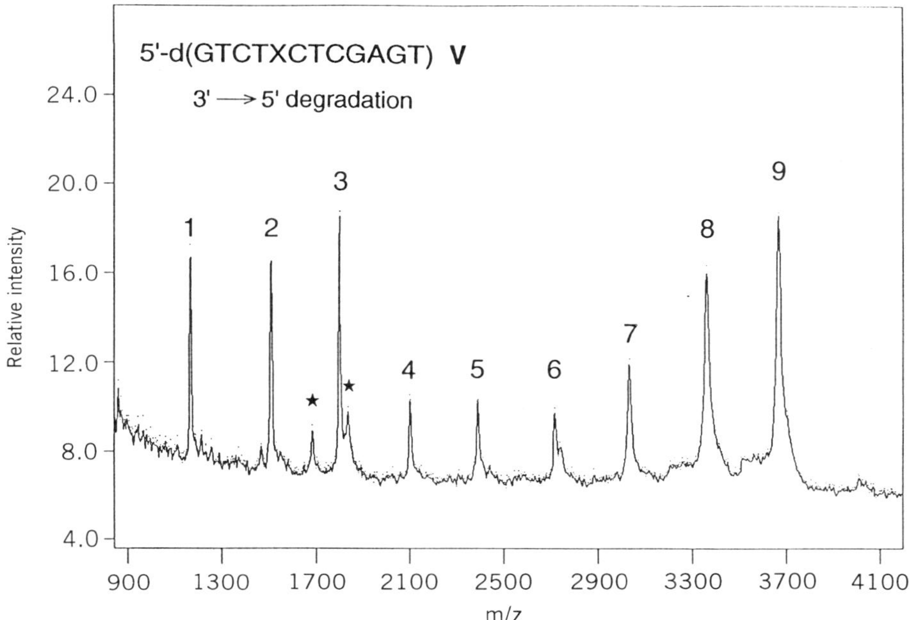

Peak	Sequence	Mass calc.	Mass *found*
1	5'-d(GCTT)	1163.8	*1163.9*
2	5'-d(GCTTX)	1507.1	*1506.9*
3	5'-d(GCTTXC)	1796.2	*1796.2*
4	5'-d(GCTTXCT)	2100.4	*2100.5*
5	5'-d(GCTTXCTC)	2389.6	*2389.8*
6	5'-d(GCTTXCTCG)	2718.9	*2719.3*
7	5'-d(GCTTXCTCGA)	3032.1	*3032.8*
8	5'-d(GCTTXCTCGAG)	3361.3	*3362.0*
9	5'-d(GCTTXCTCGAGT)	3665.5	*3666.0*

(*b*)

Fig. 11.10 *(Continued)*

(obtained in this case by exonuclease digestion) can be obtained well beyond what is possible for DNA (Fig. 11.11). Sanger-type sequence termination of RNA synthesis using 3′-deoxynucleotides might, in principle, generate RNA ladders comparable to DNA sequencing ladders. There are several problems to be overcome in working with RNA. A major difficulty is the much greater susceptibility of the RNA molecule to enzyme degradation. Also the mass similarity between uridine and cytidine nucleotides (1 Da different) would not allow full error recovery from RNA sequence ladders or exonuclease sequencing unless, as discussed, earlier the nucleotide masses were altered to improve this situation. Nevertheless, sequencing DNA by analysis of RNA ladders appears to be a promising approach. Also promising is the idea of using RNA analogues modified by introducing an unreactive electronegative substituent at the 2′ position of the ribose [19].

Schneider and Chait [20] and Kirpekar et al. [21] have suggested that fragmentation of purine-containing oligonucleotides in MALDI is initiated by protonation of the N7 nitrogen of the purine base. Using test compounds of the form $d(T_{10}XT_{14})$, Schneider and Chait showed that replacement of the N7 nitrogen by a CH in 7-deazaguanine and 7-deazaadenine gave significantly improved mass spectra with very little fragmentation. Figure 11.12 shows a comparison of the mass spectra for $d(T_{10}XT_{14})$, where X is adenine or guanine, and 7-deaza-adenine or guanine respectively. These spectra were obtained from a ferulic acid matrix, chosen to exaggerate any tendency of the analyte to fragment. If a comparably stabilizing modification of cytidine can be identified, a Sanger sequencing method utilizing deazapurines and modified cytidine might well allow longer sequence reads with better resolution than has been possible with the normal bases.

A significant improvement has also been made in time-of-flight instrument design with the introduction of delayed ion extraction. In this approach laser ablation occurs under electric field-free conditions, and the ion accelerating field is switched on after a submicrosecond delay. This technique leads to improvements in mass resolution and also to significantly improved ion stability, presumably because the analyte ions are not immediately accelerated through the high-density matrix vapor plume and thereby collisionally excited. Juhasz et al. [22] have demonstrated with this improved performance the ability to perform failure sequence analysis on a crude DNA 31-mer (Fig. 11.13). Exonuclease digestion techniques should similarly be improved. Delayed ion extraction is also being investigated as a means to improve the ionization efficiency in ice ablation, where it is speculated that the presence of a high electric field may suppress the ionization process.

ELECTROSPRAY IONIZATION MASS SPECTROMETRY APPLIED TO DNA CHARACTERIZATION

Research groups led by McLafferty at Cornell and by Smith at Pacific Northwest Laboratories (PNL) have made substantial progress in developing means to couple electrospray ionization of DNA analytes with high-resolution Fourier transform ion cyclotron resonance mass spectrometry (FTICRMS). These instruments are capable of extremely high mass resolving power (in excess of 1,000,000 for light ions), and resolving

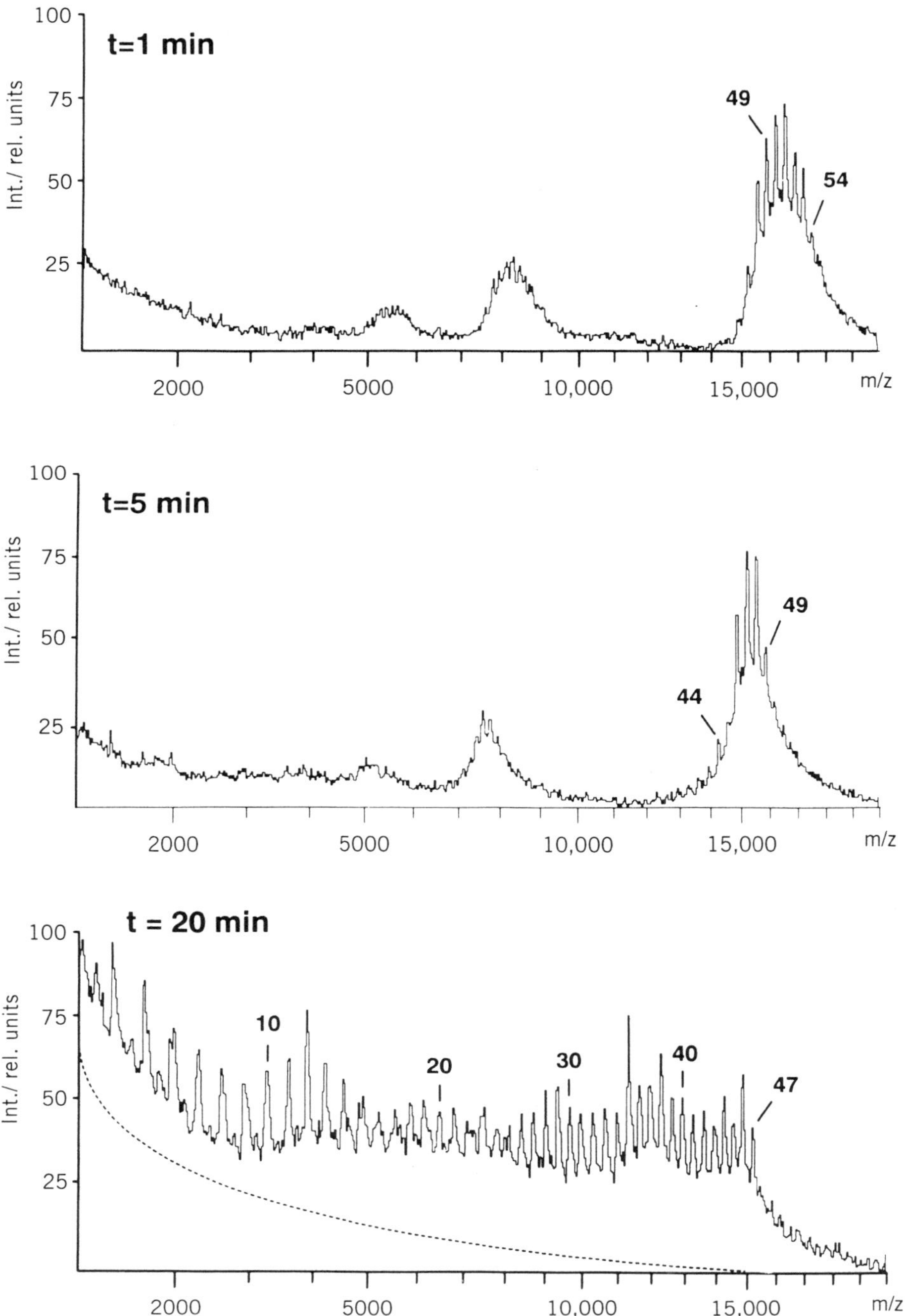

Fig. 11.11 Analysis of time course 3′ exonuclease digests of a 54/55 nt calf intestine phosphatase-treated *in vitro* RNA transcript. The numbers indicate the length of the digestion products in nucleotides. From Kirpekar et al. [18].

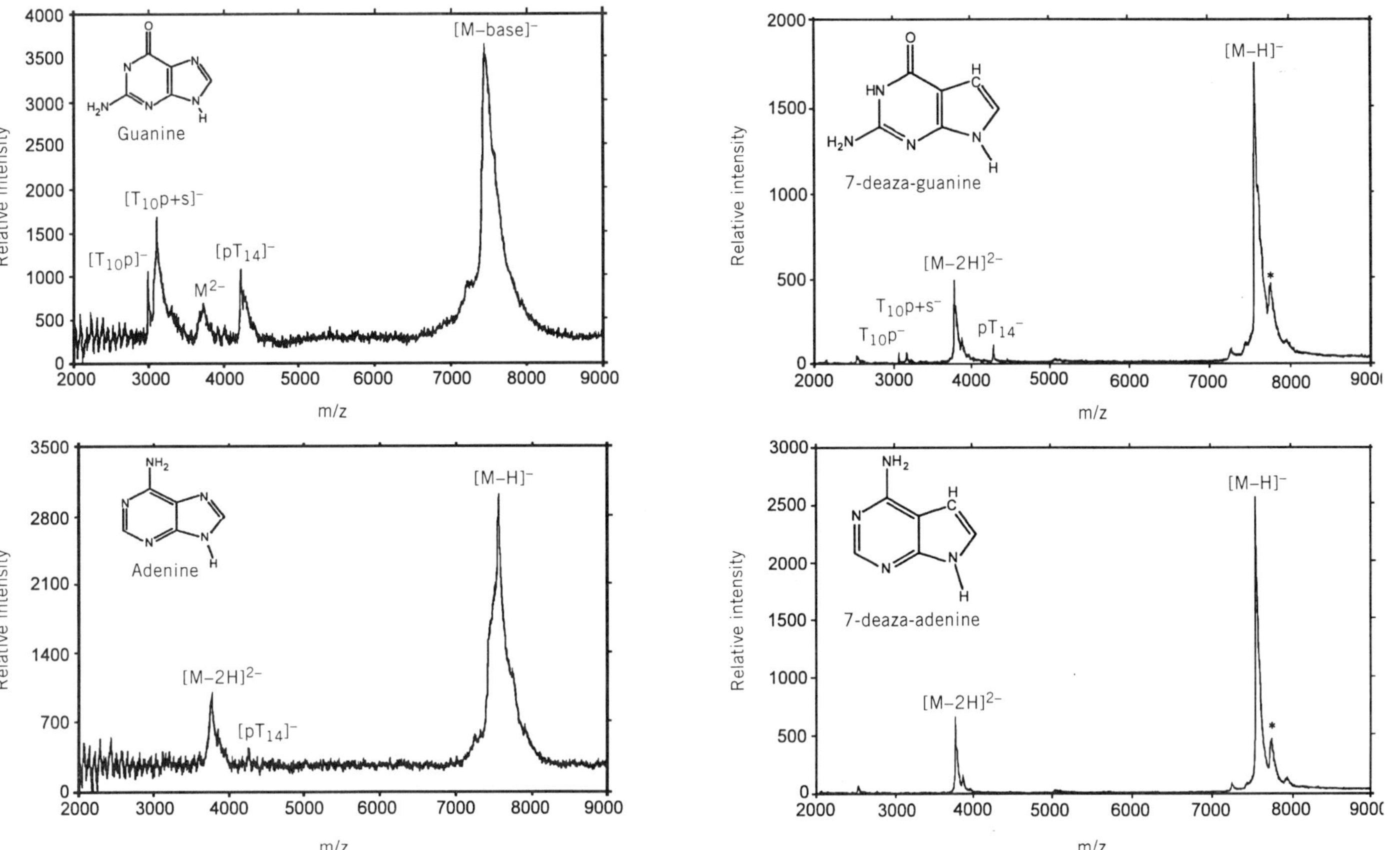

Fig. 11.12 Negative ion MALDI mass spectra of d(T₁₀XT₁₄). (*a*) X = adenine or guanine; (*b*) X = 7-deaza-adenine or 7-deazaguaninc. *Matrix:* ferulic acid, 355 nm laser. From Schneider and Chait [20].

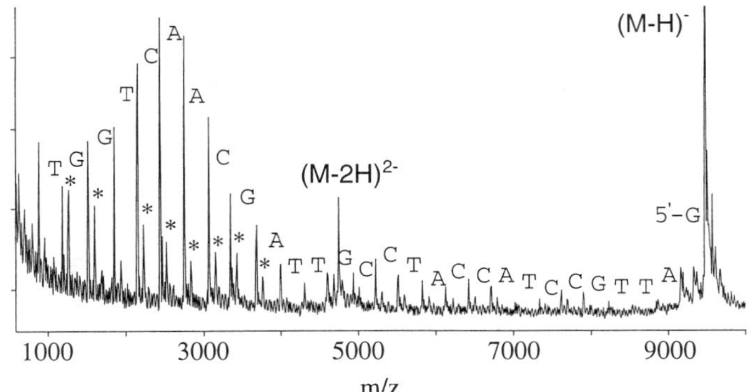

Fig. 11.13 Delayed ion extraction MALDI mass spectrum of a crude synthetic DNA 31-mer (M_r = 9486.2 calculated). Mass measurements on the failure products define the sequence up to the 3' trinucleotide. Asterisks indicate +80 Da satellite peaks. From Juhasz et al. [22].

powers ~50,000 or higher have now been achieved for oligonucleotide and protein molecular ions produced by electrospray ionization [23]. With such performance, individual isotopic peaks can be resolved, and the mass resolution limitation due to isotopic spread is removed. Because isotopic peaks are always separated by unit mass differences, they are useful to resolve the charge ambiguity issue that arises with the ESI technique; namely, if 12 peaks are detectable in a mass interval that would correspond to unit mass at unit charge, the ion must have 12 charges. A significant issue in precise mass measurement is to identify the isotopic species correctly. That is, for the case just cited, 12 possible exact masses are available, differing by 1 Da and each corresponding to a different isotopic composition of a single molecular species. It is important to correlate each mass with a correct isotopic composition for the ion in order to determine the precise elemental composition and identify the ion correctly. In essence, the mass of the pure ^{12}C (^{16}O, ^{14}N, ^{1}H) isotopic peak must correlate with that calculated from the empirical formula for the corresponding molecule. But, since the major isotope peak is usually not the most abundant and may not even be detectable at higher mass, identification is not trivial, and the danger exists that a mass measurement of exquisite accuracy (a few parts per milllion) may be attached to a peak whose identity is in error by one mass unit. McLafferty et al. have approached this problem by fitting a calculated isotopic distribution to their data and using the fit to identify the mass corresponding to the monoisotopic peak. The accuracy of this procedure is still limited by the noisiness of the isotopic distribution and by the fact that the isotopic distribution can change subtly with the natural variation in abundance of ^{13}C. The procedure adds yet another level of complexity to the data analysis, so that speedy analysis can only come with full computerization.

The first application of ESI-MS to oligonucleotide characterization by the McLafferty group was to verify the overall base composition of synthetic oligonucleotides by exact mass measurement [23]. An example is shown in Fig. 11.14. This figure shows a broad spectrum scan of the ESI FTICR mass spectrum of a synthetic 50-mer, $A_{10}T_{10}C_{17}G_{13}$. The broad range of charge states produced by ESI is evident. The left insert shows the resolved isotopic distribution for the most intense molecular ion peak which carries 20 negative charges. The dots indicate the calculated isotopic distribution, and clearly the fit is reasonable, allowing isotopic identification of each of the resolved peaks and a mass confirmation to within 7 ppm ($\sim$0.1 Da). This accuracy readily confirms the base composition of the material. Failure sequences corresponding to single nucleotide losses are visible in the right insert, together with an M + 329 Da peak indicating some addition of an extra G.

The charge spread produced by ESI is a major limitation to analysis of complex mixture such as sequence ladder mixtures, as is the fact that all ion species are compressed into a small m/z range. The PNL group has recently reported some progress in reducing the mean charge, and therefore the charge spread, in ESI [24]. If ESI mass spectra could be produced with ions carrying low charges, one to three, say, then the complexity might be reduced enough to allow sequence ladder readout. It should be noted that reducing charge increases m/z, and the resulting mass spectra will be moved beyond the mass range of all but time-of-flight mass analyzers. However, methods have now been developed to interface ESI ion sources with time-of-flight analyzers [25], so there is a real possibility that ESI-MS may play a role in the future in sequence ladder readout.

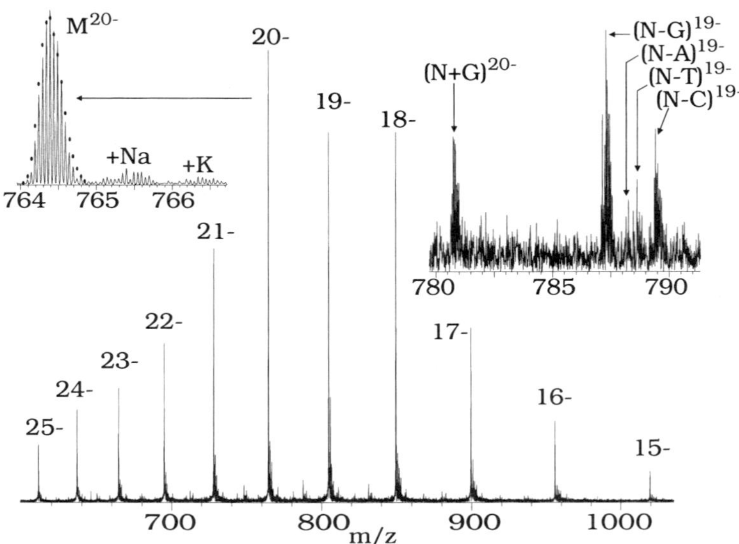

Fig. 11.14   ESI FTMS spectrum of the 50-mer $A_{10}T_{10}C_{17}G_{13}$. *Left inset:* Isotopic cluster for $(M\text{-}20H)^{20-}$; dots indicate the calculated isotopic distribution for this molecule. *Right inset:* Failure sequences with deletions of G, A, T, and C and addition of G. From Little et al. [23].

SEQUENCING METHODS BASED ON GAS-PHASE FRAGMENTATION

Fragment ladders from which sequence can be deduced for small molecules can be produced by inducing fragmentation of molecular ions in the gas phase. Extensive research on molecules of all sizes has repeatedly confirmed that collisionally or photonically excited gas-phase molecular ions tend to decompose through defined pathways, rather than falling apart with complete randomness. Reconstruction of molecular structure by analysis of fragmentation patterns has long been a staple of mass spectrometry, and starting in the early 1980s when methods of producing molecular ions of peptides became available, many groups demonstrated that such fragment reconstruction could yield sequence information for small peptides. The ability of ESI and MALDI techniques to yield molecular ions of oligonucleotides has prompted investigation of similar methods to derive sequence from fragmentation patterns for these compounds.

Fragmentation techniques, and the associated mass spectra, do not possess the simplicity and self-evident nature of the failure sequence or exonuclease digestion spectra shown earlier. Even though bond cleavage is selective (assuming that the fragmentation conditions are not too violent) a number of comparably efficient fragmentation routes compete and give a mix of fragments that at first sight appears bewildering. First, and most important, the ion charge may be located on any of the fragments, so fragmentation sequence ladders are inevitably bidirectional and overlapping. Second, multiple fragmentation pathways exist. Common fragmentation pathways for oligonucleotides include cleavage of the phosphodiester link at either the 3' or 5' side, and base loss. This can occur at any position along the oligo, subject to some degree of base-directed fragmentation. Therefore the mass spectrum typically contains all possible fragments resulting from all of these cleavages, plus other, less common fragmentation routes. Even if the various fragment peaks can be resolved and their masses determined, reassembling the sequence from these data is a complex intellectual endeavor at present, and consequently quite slow, although data acquisition may be rapid. However, the hope is that suites of fragmentation pathways will be identified and that tractable machine intelligence programs can be written to make the sequence reassembly facile. If computer programs of sufficient power can be developed, a complex mass spectrum may no longer be a headache but instead a rich source of data, quite possibly overdetermining the sequence to a significant level of redundancy.

The first requirement for deconvolution of the complex fragmentation mass spectrum of a large molecule is good mass resolution and mass accuracy. It is not clear that time-of-flight mass spectrometry, even with the improved resolution provided by delayed ion extraction, will have the requisite mass resolution for large DNA strands. Nordhoff et al. have reported a sequence analysis of a 19-mer using prompt and metastable fragmentation in a MALDI-TOF instrument [26]. Analysis of an odd-n oligomer is easier because the doubly charged parent ion does not overlap any of the fragments. The sequence of the chosen 19-mer is also conducive to sequencing because the bidirectional sequence ladders do not overlap strongly.

The high-resolution capabilities of FTICR mass spectrometry seem better suited to gas-phase fragmentation sequencing, and the Cornell group has begun to report results from this approach. Fragmentation ESI-FTICR mass spectra of a 42-mer obtained

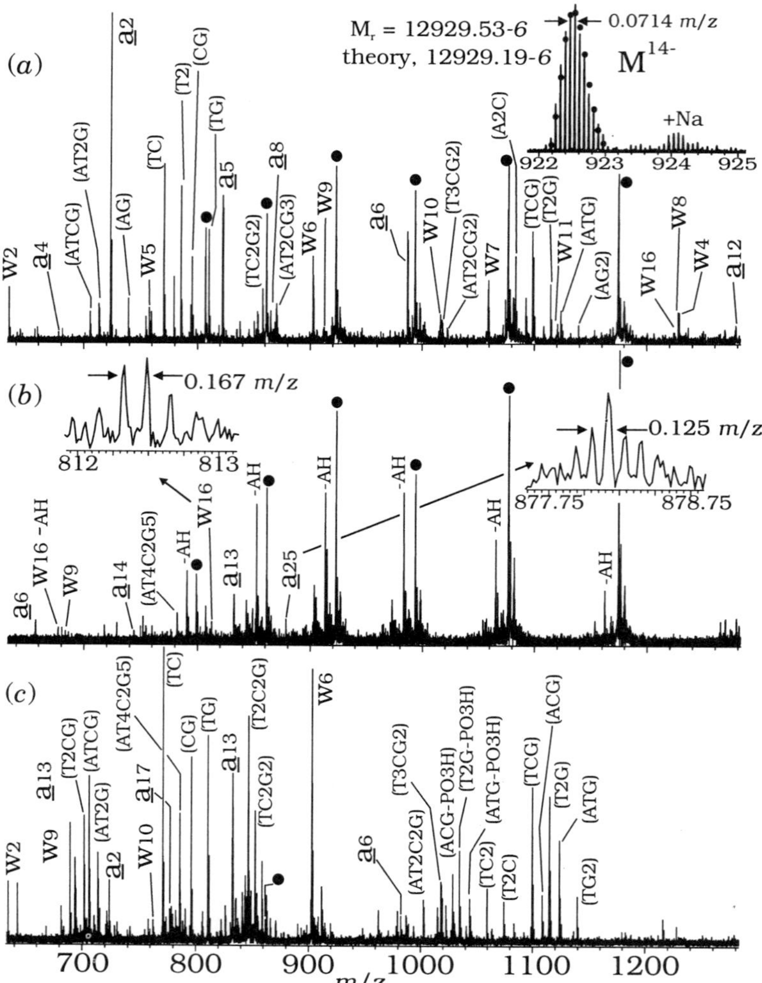

Fig. 11.15   ESI-FTMS spectra of fragments of a DNA 42-mer, excited (a) by acceleration between the nozzle and the skimmer of the ion source, (b) by infrared multiphoton excitation, and (c) by multiple collisions with N₂ gas. *Top inset:* Expanded (M-14H⁺)¹⁴⁻ region; small dots are best fit of theoretical isotope abundances. From Little et al. [27].

using different excitation techniques (by accelerating the ions in the high-pressure gas in the nozzle-skimmer region of the ion source, by infrared multiphoton excitation, and by multiple collisions with N_2 gas at low pressure in the mass spectrometer) are shown in Fig. 11.14. The wealth, and complexity, of the information available is readily apparent. The inset shows how the isotopic distribution in one of the parent molecular ion peaks may be used to determine the charge state (-14 in the example shown) from which the mass may be calculated from the measured *m/z* value. The authors [27] describe in detail how the information from such ion jigsaw puzzles may be assembled into a reasonable complete sequence. It remains to be seen whether this powerful but complex technique, requiring an expensive and complex instrument and substantial resources of human or machine intelligence, can be routinely applied to problems of DNA characterization.

THE FUTURE OF DNA MASS SPECTROMETRY

The use of MALDI mass spectrometry to characterize short DNA strands is already a reality. Either the failure sequence approach [13, 22] or exonuclease digestion [14] can generate short sequence ladders that can be read using existing MALDI-TOF technology. Accuracy of $\pm$ 1.5 Da distinguishes A from T with ~99.5% certainty. With present time-of-flight mass spectrometer technology and unmodified DNA, such accuracy is attainable for DNA analytes up to about a 30-mer [22] for DNA containing both normal and modified nucleotides. Fragmentation sequencing of significantly longer DNA strands has been reported by McLafferty's group using ESI-FTMS techniques [27]. However, it remains to be seen whether this much more complex and expensive approach will become widely available.

A possible application of MALDI TOFMS is detection of the products of PCR reactions. In contrast to sequence ladder mixtures, PCR products typically contain only a limited number of components that may differ in size by several nucleotide units. Also it is often possible to prepare relatively large quantities of PCR produces, so the required detection limits are not prohibitively low. Hurst et al. have demonstrated detection of 108- and 168-base PCR products using MALDI TOFMS [28]. They discuss the importance of using cleanup procedures to remove salts and unreacted primers that otherwise severely degrade the mass spectra. However, with current technology, mass spectrometric detection still falls short of the detection limits achievable using conventional gel electrophoresis technology.

Mass spectrometer performance that would allow readout of Sanger ladders of useful length is not yet available. The only technique that is potentially viable here is MALDI-TOFMS. There is debate about the read length at which mass spectrometry would become competitive with gels. Gel read lengths today are routinely at least 300 nucleotides, and read lengths of 500 or more are becoming common. Does mass spectrometry have to achieve similar read lengths before it will be competitive? There are several points to consider. Higher speed could make shorter mass spectrometric readouts competitive with gels. It has been suggested that even read lengths as short as 100 nucleotides would start to be useful [29]. Such read lengths may well become accessible in the near future if DNA analogues of greater stability can be identified and readily incorporated into the Sanger methodology, or if the reproducibility of the ice ablation approach can be improved. On the other hand, in shotgun sequencing, any time gained by having a short error-free read would be offset by having to assemble a larger number of short sequences, and in particular it might not be possible to span lengthy repeats.

In large-scale shotgun sequencing, the multiplexing advantage of gels raises the competitive barrier to mass spectrometry, which is inherently a single-channel technique. The raw speed advantage of mass spectrometry over highly multiplexed capillary gel techniques may only be about an order of magnitude. However, the potential for producing error-free sequence in a single pass may offer a clear speed advantage in eliminating the multiple resequencing required to reduce gel error rates to an acceptable level. It is important to realize that multiple re-sequencing using electrophoretic sequence ladder readout can only reduce, but not eliminate, errors. In research utilization of the finished human genome sequence, the costs associated with residual errors at the 0.1–0.01% level have yet to be fully assessed.

Recently Studier and others have advocated a new primer walking strategy as an alternative to the shotgun technique [30]. The key to the new approach is the rapid assembly of unligated primers from hexamer libraries. The rate-limiting step in primer walking is the wait for the end sequence that defines the sequence of the next primer. Here a rapid mass spectrometric readout could be extremely useful. Because the mass spectrometric readout should be much faster than the primer assembly and polymerase extension, a single mass spectrometer could conceivably keep pace with a large number of primer-walking robots operated asynchronously, while the multiplex advantage of gels would not be as obvious.

MALDI time-of-flight mass spectrometry is becoming widely available in biochemistry core facilities and is beginning to penetrate individual laboratories due to the demonstrated power of the technique for protein characterization. Sequence determination of short DNA oligomers can be accomplished today on the same instruments. Similarly for laboratory-scale Sanger sequencing, if long enough read lengths become possible, mass spectrometric readout should be much faster and more accurate than gels. It is not yet clear whether DNA mass spectrometry will be developed in time to make an impact on the Human Genome Project, where scale-up of established slab-gel technology will clearly carry the initial load. However, the completed human genome will stimulate a need for a great deal more DNA sequencing, for interspecies comparisons, disease studies, and diagnostics, and these will generally be on a smaller scale requiring rapid readout of one or a few sequence ladders. Mass spectrometry of DNA has the potential to play a key role in research utilization of the human genome database.

ACKNOWLEDGMENTS

We acknowledge support by the DOE Human Genome Project under Grant DE-FG02-91ER61127.

REFERENCES

1. M. Mann, C.K. Meng, J.B. Fenn (1989). "Interpreting the Mass-Spectra of Multiply Charged Ions," *Anal. Chem.* 61:1702–1707.

2. F. Hillenkamp, M. Karas, A. Ingendoh, B. Stahl (1990). Matrix-assisted UV-laser desorption/ionization: a new approach to mass spectrometry of large biomolecules. In A.L. Burlingame, J.A. McCloskey (Eds) *Biological Mass Spectrometry*. Elsevier, Amsterdam, pp 49–60.

3. R. Chen, X. Cheng, D.W. Mitchell, S.A. Hofstadler, Q. Wu, A.L. Rockwood, M.G. Sherman, R.D. Smith (1995). "Trapping, Detection, and Mass Determination of Coliphage T4 DNA Ions of 10(8) Da By Electrospray-Ionization Fourier-Transform Ion-Cyclotron Resonance Mass-Spectometry," *Anal. Chem.* 34:1159–1163.

4. R.W. Nelson, D. Dogruel, P. Williams (1994). "Mass Determination of Human-Immunoglobulin IgM Using Matrix-Assisted Laser-Desorption Ionization Time-Of-Flight Mass-Spectrometry," *Rapid Commun. Mass Spectrometry* 8:627–631.

5. C-W. Chou (1996). "Matrix-Assisted Laser Desorption Ionization Mass Spectrometry of Biomolecules," Ph.D. Thesis, Arizona State University.

6. K.J. Wu, A. Steding, C.H. Becker (1993). "Matrix-Assisted Laser Desorption Time-Of-Flight Mass-Spectrometry Of Oligonucleotides Using 3-Hydroxypicolinic Acid As An Ultraviolet-Sensitive Matrix," *Rapid Commun. Mass Spectrom.* 7:142–146.

7. C-W. Chou, S. Bingham, P. Williams. "Affinity Methods for Purification of DNA-Sequencing Reaction-Products for Mass-Spectrometric Analysis," *Rapid Commun. Mass Spectrometry*, 10(11), 1410–1414.

8. T.A. Shaler, Y. Tan, J.N. Wickham, K.J. Wu, C.H. Becker (1995). "Analysis of Enzymatic DNA-Sequencing Reactions By Matrix-Assisted Laser-Desorption Ionization Time-Of-Flight Mass-S[ectrometry]," *Rapid Comm. Mass Spectrometry* 9:942–947.

9. R.W. Nelson, M.J. Rainbow, D.E. Lohr, P. Williams (1989). "Volatilization of High Molecular Weight DNA By Pulsed Laser Ablation of Frozen Aqueous Solution," *Science* 246:1585–1597.

10. D.M. Schieltz, C-W. Chou, C-W. Luo, R. M. Thomas, P. Williams (1992). "Mass-Spectrometry of DNA Mixtures By Laser-Ablation From Frozen Aqueous Solution," *Rapid Commun. Mass Spectrometry,* 6:631–636.

11. P. Williams, R.W. Nelson (1991). "On the Mechanism of Volatilization of Large Biomolecules by Pulsed Laser Ablation of Frozen Aqueous Solutions," in K.G. Standing and W. Ens (eds) *Methods and Mechanisms for Producing Ions from Large Molecules* Plenum, pp. 265–273.

12. D.M. Schieltz (1994). "Advances in Nucleic Acid Mass Spectrometry With Applications to Sequencing," Ph.D. Thesis, Arizona State University.

13. T. Keough, T.R. Baker, R.L.M. Dobson, M. Lacey, T.A. Riley, J.A. Hasselfield, P.E. Hesselberth (1993). "Antisense DNA Oligonucleotides. 2. The Use of Matrix-Assisted Laser Desorption Ionization Mass-Spectrometry for the Sequence Verification of Methylphosphonate Oliogodeoxyribonucleotides," *Rapid Commun. Mass Spectrometry,* 7:195–200.

14. U. Pieles, W. Zürcher, M. Schär, H.E. Moser (1993). "Matrix-Assisted Laser Desorption Ionization Time-Of-Flight Mass Spectrometry: A Powerful Tool for the Mass and Sequence Analysis of Natural and Modified Oligonucleotides," *Nucleic Acids Res.* 21:3191–3196.

15. K. Tang, N.I. Taranenko, S.L. Allman, L.Y. Chang, C.H. Chen (1994). "Detection of 500-Nucleotide DNA By Laser Desorption Mass Spectrometry," *Rapid Commun. Mass Spectrometry,* 8:727–730.

16. L. Zhu, G.R. Parr, M.C. Fitgerald, C.M. Nelson, L.M. Smith (1995). "Oligodeoxynucleotide Fragmentation in MALDI/TOF Mass Spectrometry Using 355-nm Radiation," *J. Am. Chem. Soc.* 117:6048–6056.

17. E. Nordhoff, R. Cramer, M. Karas, F. Hillenkamp, K. Kristiansen, P. Roepstorff (1993). "Ion Stability of Nucleic Acids in Infrared Matrix-Assisted Laser Desorption Ionization Mass Spectrometry," *Nucleic Acids Res.* 21:3347–3357.

18. F. Kirpekar, E. Nordhoff, K. Kristiansen, P. Roepstorff, A. Lezius, S. Hahner, M. Karas, F. Hillenkamp (1994). "Matrix-Assisted Laser Desorption Ionization Mass Spectrometry of Enzymatically Synthesized RNA up to 150 kDa," *Nucleic Acids Res.* 22:3866–3870.

19. L.M. Smith (1996) Paper presented at 44th Annual Conference of American Society for Mass Spectrometry, Portland, OR.

20. K. Schneider, B.T. Chait (1995). "Increased Stability of Nucleic Acids Containing 7-Deaza-guanosine and 7-Deaza-adenisone May Enable Rapid DNA Sequencing by Matrix-Assisted Laser Desorption Mass Spectrometry," *Nucleic Acids Res* 23:1570–1575.

21. F. Kirpekar, E. Nordhoff, K. Kristiansen, P. Roepstorff, S. Hahner, F. Hillenkamp (1995). "7-Deaza-purine Bases Offer a Higher Ion Stability In the Analysis of DNA by Matrix-Assisted Laser Desorption Ionization Mass Spectrometry," *Rapid Commun. Mass Spectrometry* 9:525–531

22. P. Juhasz, M.T. Roskey, I.P. Smirnov, L.A. Haff, M.L. Vestal, S.A. Martin (1996). "Applications of Delayed Extraction Matrix-Assisted Laser Desorption Ionization Time-Of-Flight Mass Spectrometry to Oligonucleotide Analysis," *Anal. Chem.* 68:941–946

23. D.P. Little, T.W. Thannhauser, F. McLafferty (1995). "Verification of 50-mer DNA and RNA Sequences With High Resolution Mass Spectrometry," *Proc. Natl. Acad. Sci. USA* 92:2318–2322.

24. X. Cheng, D.C. Gale, H.R. Udseth, R.D. Smith (1995). "Charge-State Reduction of Oligonucleotide Negative Ions from Electrospray Ionization," *Anal. Chem.* 34:586–592.

25. A. Verentchikov, W. Ens, K. Standing (1994). "Reflecting Time-Of-Flight Mass Spectrometer with an Electrospray Ion Source and Orthogonal Extraction," *Anal. Chem.* 66:126–133.

26. E. Nordhoff, M. Karas, R. Cramer, S. Hahner, F. Hillenkamp, F. Kirpekar, A. Lezius, J. Muth, C. Meier, J.W. Engels (1995). "Direct Mass Spectrometric Sequencing of Low-Picomole Amounts of Oligodeoxynucleotides with up to 21 Bases by Matrix-Assisted Laser Desorption Ionization Mass Spectrometry," *J. Mass Spectrom.* 30:99–112.

27. D.P. Little, D.J. Aaserud, G.A. Valakovic, F.W. McLafferty, "Sequence Information from 42-108-mer DNAs (Complete for a 50-mer) By Tandem Mass Spectrometry," *J. Am. Chem. Soc.* 118:9352–9359.

28. G.B. Hurst, M.J. Doktycz, A.A. Vass, M.V. Buchanan (1996). "Detection of Bacterial-DNA Polymerase Chain Reaction Products by by Matrix-Assisted Laser Desorption Ionization Mass Spectrometry," *Rapid Commun. Mass Spectrometry* 10:377–382.

29. L.M. Smith, Private communication.

30. J. Kieleczawa, J.J. Dunn, F.W. Studier (1992). "DNA Sequencing by Primer Walking with Strings of Contiguous Hexamers," *Science* 258:1787–1790.

IV

ANALYSIS AND SYNTHESIS

12

Petri Net Modeling and Simulation for Automated Systems

DEIRDRE MELDRUM AND EMERSON TONGCO

CONTENTS

Introduction
Fundamental Structure of Petri Nets
Petri Net Definitions and Analysis
An Automated PCR System
The Genome Factory
Simulation
Control and Performance Evaluation
Conclusion
References

INTRODUCTION

The development of new and improved technology in the area of DNA sequencing is vital to the Human Genome Project if the desired goal of sequencing the entire human genome is to be achieved within the next 10 to 15 years. Clearly the introduction of automation is needed to dramatically lower the amount of manual labor needed to sequence DNA and thereby cut the costs of the Human Genome Project [1].

Automation Technologies for Genome Characterization, Edited by Tony J. Beugelsdijk.
ISBN 0-471-12806-6 © 1997 John Wiley & Sons, Inc.

One direction that is being taken is the concept of the Genome Factory. Olson [2] envisions a fully automated Genome Factory that is capable of processing one sample per second. This factory will contain labor-saving devices such as automatic pipettors, gel-loading machines, and other systems that remove the individual worker from the monotonous manual tasks inherent in current biotechnology laboratories. In addition, and more important, is the fact that these machines will be equipped with the capability to make real-time adaptive decisions during the process. Adaptive decision making is vital for the factory to function properly and independently. Although labor-saving devices as mentioned above have been built and tested, they still require the human factor in the loop to make decisions as to whether a process is done properly and whether results are acceptable.

The Genome Factory is very similar to a manufacturing plant factory except that the Genome Factory requires uniqueness of the information content associated with each item (sample or part) that traverses the automated process [2]. The productivity of a manufacturing plant is generally measured by the amount of throughput that it is capable of (in this case a factory throughput of a sample per second is desired). Even before any hardware is introduced, various schemes have been used to model automated/flexible manufacturing systems and to analyze their potential throughput. A Petri net is one method that has been studied and used extensively in manufacturing systems for modeling, control and performance analysis.

This chapter will introduce the reader to the concept of Petri nets and how they can be applied to an automated sample-handling system. It is envisioned that Petri nets will be used to model the Genome Factory, to develop the control for the entire system, and to evaluate the entire performance of the overall system. This chapter will not present an in-depth discussion of Petri nets but an overview of how they work, their uses, and properties. For those interested in learning more about Petri nets, the work by Murata [3] and by Desrochers and Al-Jaar [4] provide plenty of information.

FUNDAMENTAL STRUCTURE OF PETRI NETS

Petri nets were developed by German mathematician Carl Adam Petri in 1962 to model and analyze communication systems. They graphically describe relations between conditions and events of discrete event systems. The graphical nature of Petri nets allows for a good visualization of the flow of the system.

Figure 12.1 shows a simple Petri net model of one part of an automated sample-handling system. This model, which represents a capillary gripper (i.e., a robot) grabbing a capillary (i.e., a part), aspirating an amount of DNA sample (i.e., operating on a part), and storing the capillary in a chamber (i.e., placing a part in a bin or buffer), will be used to define the components in a Petri net. *Circles* are called *places*, and they represent a condition, be it resource availability or process status. For example, place p_1 represents the condition regarding the availability of the capillary gripper. A double circle (p_5) indicates a continuous resource. Table 12.1 gives the definitions for each place. The *dots* or *tokens* in the places give a dynamic representation of the system. The single token in p_1 indicates that the capillary gripper is available and ready while

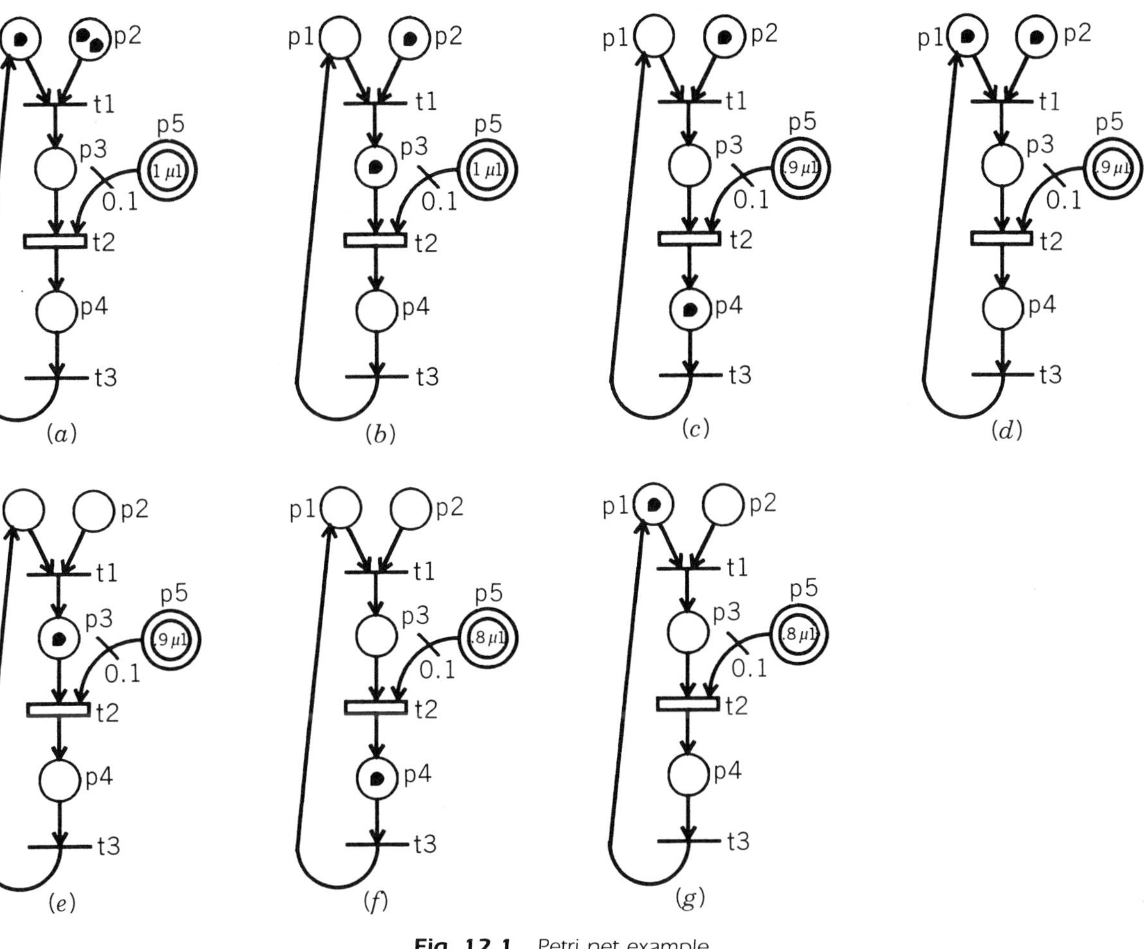

Fig. 12.1 Petri net example.

TABLE 12.1   Place and Bar Definitions for Fig. 12.1

p_1: Capillary gripper ready	t_1: Capillary gripper grabs a capillary
p_2: Capillaries available	t_2: Capillary aspirates 0.1 ml of DNA sample
p_3: Capillary in gripper; ready for aspirating	t_3: Finished storing capillary with sample in
p_4: Store capillary in a buffer chamber	buffer chamber
p_5: DNA sample ready/available	

the two tokens in p_2 indicate that there are currently two capillaries available to be used. The value in p_5 refers to the amount that is available in that resource. The number of tokens in a place are the *markings* of the net. The state of a Petri net is then indicated by its current markings at each setting.

The *bars* (both thinned and boxed) are called *transitions*, indicating that an associated event has occurred. In the example, t_1 is associated with the event that the capillary gripper has finished grabbing a capillary. Arcs are arrows that connect a transition to a place or a place to a transition. In general, arcs have a weight of one unless they have a cross bar indicating the weight value as in the arc directed from p_1. In Fig. 12.1*a* transition t_1 is enabled, since there are tokens available from both p_1 and p_2. When a transition fires, it takes away one token from each place pointing to it and drops one token on the place it is pointing to. The new markings are shown in Fig. 12.1*b*. The boxed transition t_2 indicates that it involves a continuous input. Transition t_2 is enabled when p_3 has a token and p_5 has enough resource to provide 0.1 ml (the specified weight of the arc from p_5). Each time transition t_2 is fired, the value in p_5 is decremented by a value of 0.1. This combination of discrete and continuous places and discrete and continuous bars defines a class of Petri nets called hybrid Petri nets. After t_2's firing, a new state is acquired as shown in Fig. 12.1*c*. Transition t_3 is now enabled and Fig. 12.1*d* shows the markings when the transition fires. Figures 12.1*e, f,* and *g* follow through the same steps and reasonings as before. Table 12.1 gives the definitions for each place and transition in Fig. 12.1.

PETRI NET DEFINITIONS AND ANALYSIS*

A Petri net is described by a five-tuple:

$$N = \{P, T, I, O, M_0\}$$

where

$P = \{p_1, p_2, \ldots , p_m\}$ is a finite set of places.
$T = \{t_1, t_2, \ldots , t_n\}$ is a finite set of transitions.
$P \cup T \neq \varnothing, \quad P \cap T \neq \varnothing$

---

* This section is based on DiCesare and Desrochers [8].

$I : P \cup T \Rightarrow \{0, 1, 2, . . .\}$ is an input function that defines directed arcs from places to transitions.

$O : P \cup T \Rightarrow \{0, 1, 2, . . .\}$ is an output function that defines directed arcs from transitions to places.

$M_0 : P \Rightarrow \{0, 1, 2, . . .\}$ is the initial marking where M indicates the number of tokens in each place.

Taking the Fig. 12.1 example,

$$P = \{p_1, p_2, p_3, p_4, p_5\}$$

$$T = \{t_1, t_2, t_3\}$$

$$I(p_1, t_1) = I(p_2, t_1) = I(p_3, t_2) = I(p_4, t_3) = 1$$

$$I(p_5, t_2) = 0.1$$

$$I(p_1, t_2) = I(p_1, t_3) = I(p_2, t_3) = I(p_2, t_3) = 0$$

$$I(p_3, t_1) = I(p_3, t_3) = I(p_4, t_1) = I(p_4, t_2) = 0$$

$$I(p_5, t_1) = I(p_5, t_3) = 0$$

Writing this in matrix form,

$$
I(p, t) = \begin{array}{c} \\ p_1 \\ p_2 \\ p_3 \\ p_4 \\ p_5 \end{array}
\begin{array}{ccc} t_1 & t_2 & t_3 \end{array}
\left[\begin{array}{ccc}
1 & 0 & 0 \\
1 & 0 & 0 \\
0 & 1 & 0 \\
0 & 0 & 1 \\
0 & 0.1 & 0
\end{array} \right]
$$

Matrix entry $I(p_5, t_2) = 0.1$ represents the fact that p_5 is a continuous place and an amount of 0.1 ml is taken each time transition t_2 triggers. Similarly

$$
O(p, t) = \left[\begin{array}{ccc}
0 & 0 & 1 \\
0 & 0 & 0 \\
1 & 0 & 0 \\
0 & 1 & 0 \\
0 & 0 & 0
\end{array} \right]
$$

And for the initial marking M_0,

$$
M_0 = \begin{array}{c} p_1 \\ p_2 \\ p_3 \\ p_4 \\ p_5 \end{array}
\left[\begin{array}{c}
1 \\
2 \\
0 \\
0 \\
1
\end{array} \right]
$$

Note that the marking $p_5 = 1$ states that the continuous place p_5 has an initial amount of 1 μl. Alternatively, one can convert this hybrid net to an ordinary Petri net (totally discrete) by letting $I(p_5, t_2) = 1$, and p_5 be represented with a discrete place containing ten tokens. The choice of generating either a hybrid type or ordinary type of Petri net is a matter of preference to the designer, although the type of Petri net chosen might later have an effect when the model is extended to control and simulation purposes.

The enabling of a transition t_j occurs for a marking M if and only if

$$M(p_i) \geq I(p_i, t_j)$$

for all members p_i of the set of input places for that transition t_j. The enabled transition can fire any time instantly (although there are classes of Petri nets that have a deterministic timed transition and a stochastically timed transition). The new marking attained after a transition firing is given by

$$M'\,(p_i) = M\,(p_i) + O\,(p_i,\, t_j) - I\,(p_i,\, t_j) \qquad \forall p_i \in P \qquad (1)$$

A marking M' is then said to be *reachable* from a marking M. Taking the example shown in Fig. 12.1, from the enabling rule for M_0,

$$M_0 = \begin{bmatrix} 1 \\ 2 \\ 0 \\ 0 \\ 1 \end{bmatrix} \geq \begin{bmatrix} 1 \\ 1 \\ 0 \\ 0 \\ 0 \end{bmatrix} = I(p, t_1) \qquad \forall p \in P$$

Transition t_1 is said to be enabled, since all members of M_0 are greater than or equal to all members of I. In a similar argument, t_2 and t_3 are not enabled, since some members of M_0 are less than some elements of I as shown by

$$M_0 = \begin{bmatrix} 1 \\ 2 \\ 0 \\ 0 \\ 1 \end{bmatrix} < \begin{bmatrix} 1 \\ 1 \\ 0 \\ 0 \\ 0.1 \end{bmatrix} = I(p, t_2) \qquad \forall p \in P$$

and

$$M_0 = \begin{bmatrix} 1 \\ 2 \\ 0 \\ 0 \\ 1 \end{bmatrix} < \begin{bmatrix} 0 \\ 0 \\ 0 \\ 1 \\ 0 \end{bmatrix} = I(p, t_3) \qquad \forall p \in P$$

The new marking attained after transition t_1 is fired, as shown in Fig. 12.1b, is

$$M_1 = M_0 + O(p, t_1) - I(p, t_1)$$

With the new marking M_1 attained, a comparison is made between this new marking and the transition columns of I to determine which transition is enabled next. This process is repeated continuously to generate the new markings.

To simplify the marking expression (Eq. 1), the following equation is defined:

$$C = C(p_i, t_j) = O(p_i, t_j) - I(p_i, t_j) \tag{2}$$

This equation (Eq. 2) is referred to as the *incidence* matrix C. For example, the incidence matrix for the Petri net in Fig. 12.1 is

$$C(p_i, t_j) = \begin{bmatrix} 0 & 0 & 1 \\ 0 & 0 & 0 \\ 1 & 0 & 0 \\ 0 & 1 & 0 \\ 0 & 0 & 0 \end{bmatrix} - \begin{bmatrix} 1 & 0 & 0 \\ 1 & 0 & 0 \\ 0 & 1 & 0 \\ 0 & 0 & 1 \\ 0 & 0.1 & 0 \end{bmatrix} = \begin{bmatrix} -1 & 0 & 1 \\ -1 & 0 & 0 \\ 1 & -1 & 0 \\ 0 & 1 & -1 \\ 0 & -0.1 & 0 \end{bmatrix}$$

Furthermore, by combining the marking expression (Eq. 1) and the *incidence* matrix expression (Eq. 2), a *firing* or *control* vector μ_k is defined. The firing or control vector μ_k designates the kth firing of the Petri net and has elements that are all zero except for the ith element which is one. This indicates that transition i fires at the kth firing. So the marking expression becomes

$$M_k = M_{k-1} + C\mu_k \tag{3}$$

In applying this to our example, $M_1 = M_0 + C\mu_1$, where μ_1 is $\mu_1 = [1 \quad 0 \quad 0]^T$, since transition t_1 is enabled. Then

$$M_1 = \begin{bmatrix} 1 \\ 2 \\ 0 \\ 0 \\ 1 \end{bmatrix} + \begin{bmatrix} -1 & 0 & 1 \\ -1 & 0 & 0 \\ 1 & -1 & 0 \\ 0 & 1 & -1 \\ 0 & -0.1 & 0 \end{bmatrix} \begin{bmatrix} 1 \\ 0 \\ 0 \end{bmatrix} = \begin{bmatrix} 1 \\ 2 \\ 0 \\ 0 \\ 1 \end{bmatrix} + \begin{bmatrix} -1 \\ -1 \\ 1 \\ 0 \\ 0 \end{bmatrix} = \begin{bmatrix} 0 \\ 1 \\ 1 \\ 0 \\ 1 \end{bmatrix}$$

and for M_2, where t_2 is the only transition enabled,

$$M_2 = \begin{bmatrix} 0 \\ 1 \\ 1 \\ 0 \\ 1 \end{bmatrix} + \begin{bmatrix} -1 & 0 & 1 \\ -1 & 0 & 0 \\ 1 & -1 & 0 \\ 0 & 1 & -1 \\ 0 & -0.1 & 0 \end{bmatrix} \begin{bmatrix} 0 \\ 1 \\ 0 \end{bmatrix} = \begin{bmatrix} 0 \\ 1 \\ 1 \\ 0 \\ 1 \end{bmatrix} + \begin{bmatrix} 0 \\ 0 \\ -1 \\ 1 \\ -0.1 \end{bmatrix} = \begin{bmatrix} 0 \\ 1 \\ 0 \\ 0 \\ 0.9 \end{bmatrix}$$

The process continues in the same manner to attain the next set of markings.

If the *control* vector μ_k is summed up as in

$$\sum_{k=1}^{d} \mu_k = f_d$$

where f_d is called the *firing count vector*, then the marking expression is written as

$$M_d = M_0 + Cf_d$$

This expression provides information only about the marking values after the number of transition firings d and gives no information about the actual firing sequence.

A useful analysis tool in Petri nets is the *reachability tree*. The tree represents the set of all reachable markings originating from the initial marking M_0. The tree shows the net's behavior, enumerates the possible firing sequences, and itemizes the reachable states for each firing sequence. The reachability tree for the simple Petri net in Fig. 12.1 is illustrated in Fig. 12.2. The reachability tree may be both finite and infinite but is usually infinite. In Fig. 12.2 a finite tree is shown representing a deadlock situation, since no transitions can be fired once the marking M_6 is attained. Although

$$M_0 = [1, 2, 0, 0, 1]^T$$
$$\downarrow t1$$
$$M_1 = [0, 1, 1, 0, 1]^T$$
$$\downarrow t2$$
$$M_2 = [0, 1, 0, 1, 0.9]^T$$
$$\downarrow t3$$
$$M_3 = [1, 1, 0, 0, 0.9]^T$$
$$\downarrow t1$$
$$M_4 = [0, 0, 1, 0, 0.9]^T$$
$$\downarrow t2$$
$$M_5 = [0, 0, 0, 1, 0.8]^T$$
$$\downarrow t3$$
$$M_6 = [1, 0, 0, 0, 0.8]^T$$

Fig. 12.2   Reachability tree of Petri net in Fig. 12.1.

the reachability tree for this simple Petri net example produces an unwanted deadlock, it also gives information regarding the firing sequences of the transitions and the various reachable states that are attained. The structure of a reachability tree commonly resembles that of binary trees. In this binary tree structure, information associated with dead-end nodes (terminal nodes), redundant nodes, and infinite loops can be concluded. By simulating the reachability tree in software, one can make decisions as to the efficiency and productivity of the system as modeled via a Petri net. A primary disadvantage of the reachability tree is the fact that when the system being modeled grows and becomes more complex, generating the tree can be nearly impossible and a computational nightmare.

AN AUTOMATED PCR SYSTEM

To demonstrate the application of Petri nets to automated sample handling, an example is presented below. Figure 12.3 shows the Petri net model of an automated sample-handling system (ASHS) performing a Polymerase Chain Reaction (PCR) with one primer pair. The ASHS is a hardware system being developed to automatically perform PCR, Restriction Enzyme Digest, and Sequencing Reactions [7]. PCR protocol calls for combining a typical sample mix of one part each of a DNA sample, primer A, primer B, 10XTaq buffer, Taq enzyme, dNTPs, and four parts water. This is a hybrid Petri net because of the combination of discrete and continuous places and bars. As an example, p_1 and p_2 are discrete *places*, while p_{15} and p_{16} are continuous places. An overview of the process is as follows: First, a capillary gripper acquires a capillary. The gripper then moves to microtiter well stations to aspirate desired quantities of a DNA sample and its corresponding primer pair. Once this is accomplished, the gripper moves to reagent dispensing stations where it receives the desired quantities of Taq enzyme, 10XTaq buffer, dNTPs, and water. The solution is mixed inside the capillary, and the gripper then maneuvers to a docking station where it loads the fluid-filled capillary into a special housing chamber for thermal cycling. Table 12.2 gives the various descriptions of the different places and transitions. Typically in a Petri net, the output arc of a *place* is directed towards a transition *bar*. From the transition bar another arc is then directed to a *place*.

A new structure has been defined specifically for this genome application because of the need to uniquely identify each sample traversing the automated process. This structure, which we will call the *unique choice generator,* is denoted by the label DNA Sample and Primer Pool. The *unique choice generator* shown in Fig. 12.3 is created to demonstrate that the samples flowing through a system have a unique identification associated with them. This is very different when compared to a manufacturing process where the unique identity of the raw parts or materials are irrelevant. In a mass production manufacturing plant, the final products bear no uniqueness to each other. This is not true, however, for a Genome Factory. Each DNA sample flowing through the system is uniquely identified and tagged. This is necessary especially when it comes time to analyze the final product. Each final product generated is unique relative to the other final product.

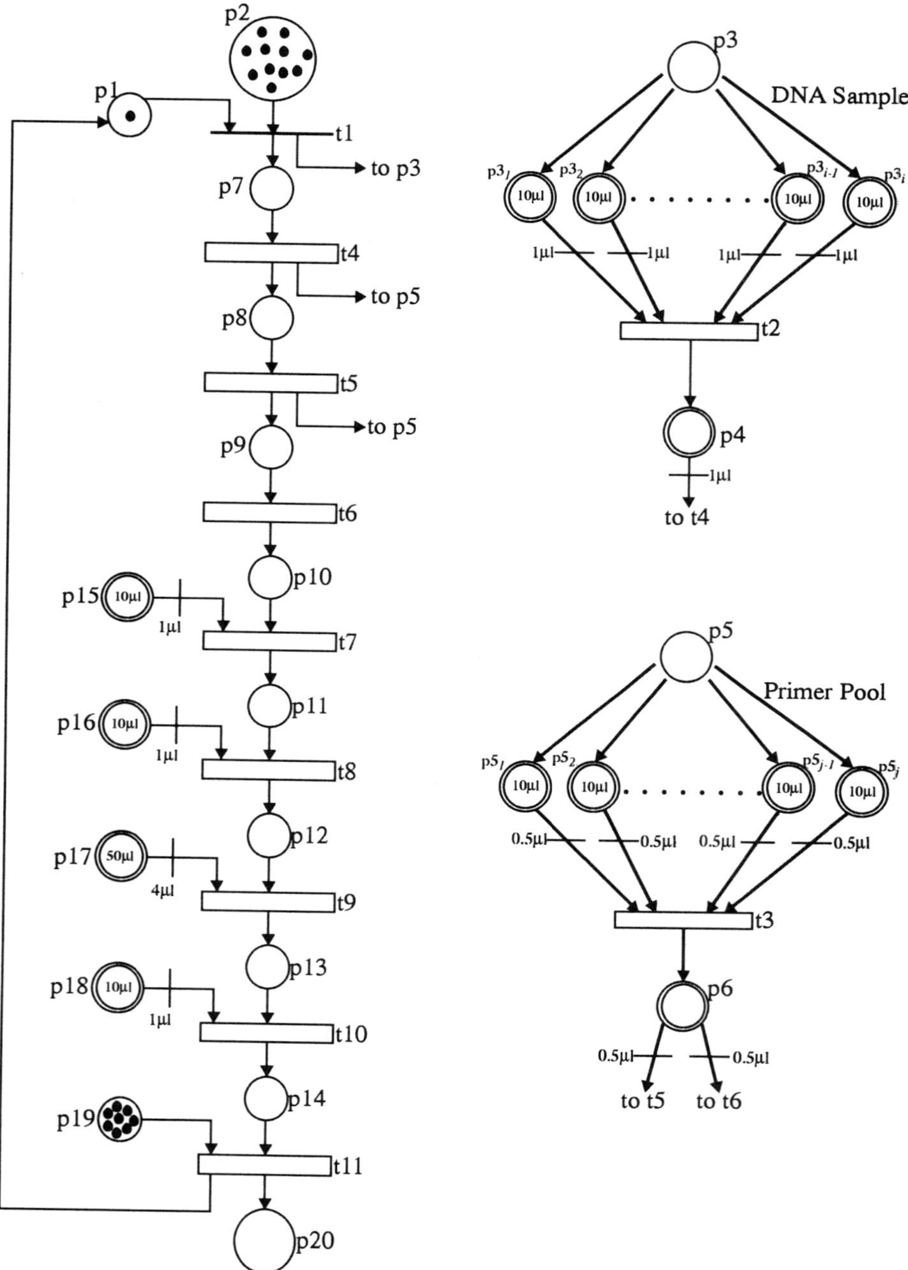

Fig. 12.3 Petri Net model for automated PCR.

TABLE 12.2 Place and Transition Definitions: PCR Gel Analysis

p_1: Capillary gripper ready	t_1: Capillary gripper grabs and loads a capillary
p_2: Capillary dispenser station ready	t_2: Particular DNA sample chosen from sample pool
p_3: DNA sample ready/choose	
$p_{3_{1-i}}$: Particular DNA sample chosen	t_3: One part of primer pair chosen from primer pool
p_4: Chosen DNA sample ready	
p_5: Primers ready/choose	t_4: Capillary aspirates 1 ml of DNA sample
$p_{5_{1-j}}$: Particular primer chosen	t_5: Capillary receives 0.5 ml of 1st part of primer pair
p_6: Chosen primer ready ($\frac{1}{2}$ of pair)	
p_7: Aliquot sample	t_6: Capillary receives 0.5 ml of 2nd part of primer pair
p_8: Aliquot first primer	
p_9: Aliquot second primer	t_7: Capillary receives 1 ml of Taq buffer
p_{10}: 10X Taq buffer dispenser	t_8: Capillary receives 1 ml of dNTP mix
p_{11}: dNTP dispenser	t_9: Capillary receives 4 ml of H$_2$0
p_{12}: H$_2$0 dispenser	t_{10}: Capillary receives 1 ml of Taq enzyme
p_{13}: Taq enzyme dispenser	t_{11}: Finished storing capillary with sample in docking chamber
p_{14}: Capillary ready to be docked	
p_{15}: Enzyme station ready	
p_{16}: dNTP station ready	
p_{17}: H$_2$0 station ready	
p_{18}: Taq buffer station ready	
p_{19}: Docking station ready/slots available	
p_{20}: Load to docking station	

This *unique choice generator* structure can be described as follows: Consider the structure labeled DNA Sample. When a token reaches the *place* p_3 (a discrete *place*), the token signifies the condition that the system is now ready to aspirate or retrieve a particular DNA sample for analysis. The numerous arcs coming out of p_3 imply that the net has to decide which sample it must choose. Remember that p_3 has only one token, and it has to allocate that one token to one of the DNA samples available. Imagine the *place* p_3 as the buffer bin (in the manufacturing setting sense) where a robot arm can come and pick up a raw part or material. But now each raw part has a unique identification tag associated with it, and this robot arm has to decide which part to pick from the original pool. The buffer bin in this system could be a 96-well microtiter plate, for instance. Parts are picked deterministically and tracked in the system software. The decision-making process is performed and also tracked systematically and can be easily implemented in software. As an example of how the net flows, say, that p_3 chose the place p_{3_1} (a continuous place). The arc from p_{3_1} has an associated weight of 1 μl, implying that the transition t_2 is enabled when an amount of 1 μl has been aliquotted from the well containing the DNA sample associated with p_{3_1}. The structure labeled Primer Pool operates under the same premises as described above. The flow of the main Petri net follows as before.

THE GENOME FACTORY

The section above demonstrated how Petri nets can be used to model a system performing PCR. Petri nets may also be used to analyze the process of larger systems such as the Genome Factory. An aggressive time schedule has been set in the scientific community to sequence the human genome within our lifetime, and this can only be achieved by scaling up the throughput rate by several orders of magnitude. Suggestions have been made to increase the throughput (from 10 Mbp/yr to 100 Mbp/yr) of current state-of-the-art sequencing groups by merely scaling-up current technology. This approach, however, exposes flaws, since current mapping procedures do not necessarily provide sufficiently high-quality templates for sequencing, and that proportion of the genome budget would become too great [6]. Olson suggests that to complete the sequencing in our lifetime, an unprecedented sequencing throughput of one clone per second is required. This throughput increase can be achieved by creating an automated sequencing factory.

What conceptually is this Genome Factory? The word *factory* generally describes a place where goods and products are manufactured or an industrial setting where operations are clearly defined and are highly repetitive. The Genome Factory resembles a manufacturing plant scenario. In this special factory, genome sequencing is broken down into cardinal operations that are carried out autonomously and repetitively by instruments that are linked in a pipe-lined process [6]. These modules will be single function modules that are highly robust and can be placed in series on a conveyor track with other modules. To increase the throughput considerably, parallel lines of these modules may be running concurrently.

As in every assembly or manufacturing plant, a model of the factory that describes the overall interaction of each machine is needed. The model also describes the flow of materials throughout its assembly. A model of the Genome Factory is needed to show the interaction between each module and to show how each of the unique samples are flowing throughout the system. Due to the nature of the interaction of these modules with each other, it is natural to ask about the occurrence of bottlenecks, deadlocks, and conflicts for resources available. Furthermore one can argue about the optimal number of modules needed for a particular process or step and/or the optimal number of parallel processes needed given the number of modules and resources available. Petri net modeling and analysis is a viable approach to provide answers to these questions.

The overall processes of the Genome Factory are divided into three main stages: cosmid library production, cosmid map generation, and cosmid sequencing. Figures 12.4a–d show the Petri net model of the entire process from the moment chromosome cells are sorted by the flow cytometer to the last step where the sequence data are being generated. Note that all four models are integrated to each other to form a huge model. They are separated here to show the different main stages in the process. Table 12.3 gives the place and transition definitions for this Petri net.

Note in Table 12.3 that place p_4 represents the readiness and availability of the sample transport system. The sample transport system is a conveyor belt-type system carrying the sample to the different modules or stations for servicing. For this setup, one transport system serves all of the different modules. Clearly a conflict in resource

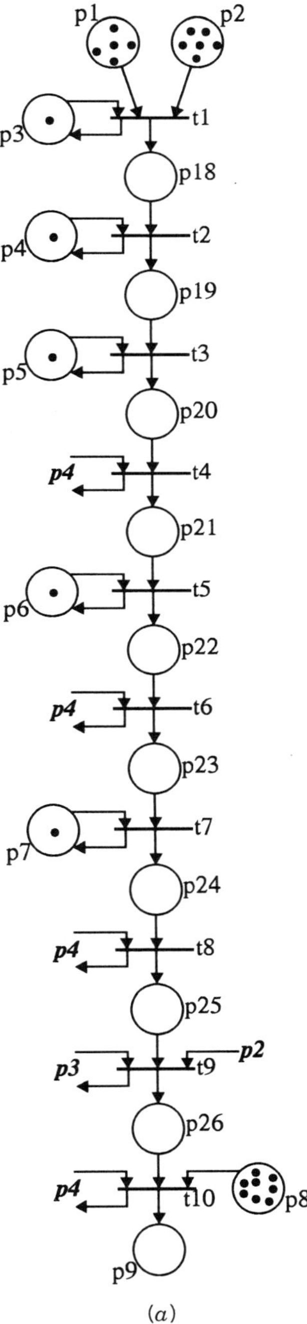

(a)

Fig. 12.4a Cosmid library production.

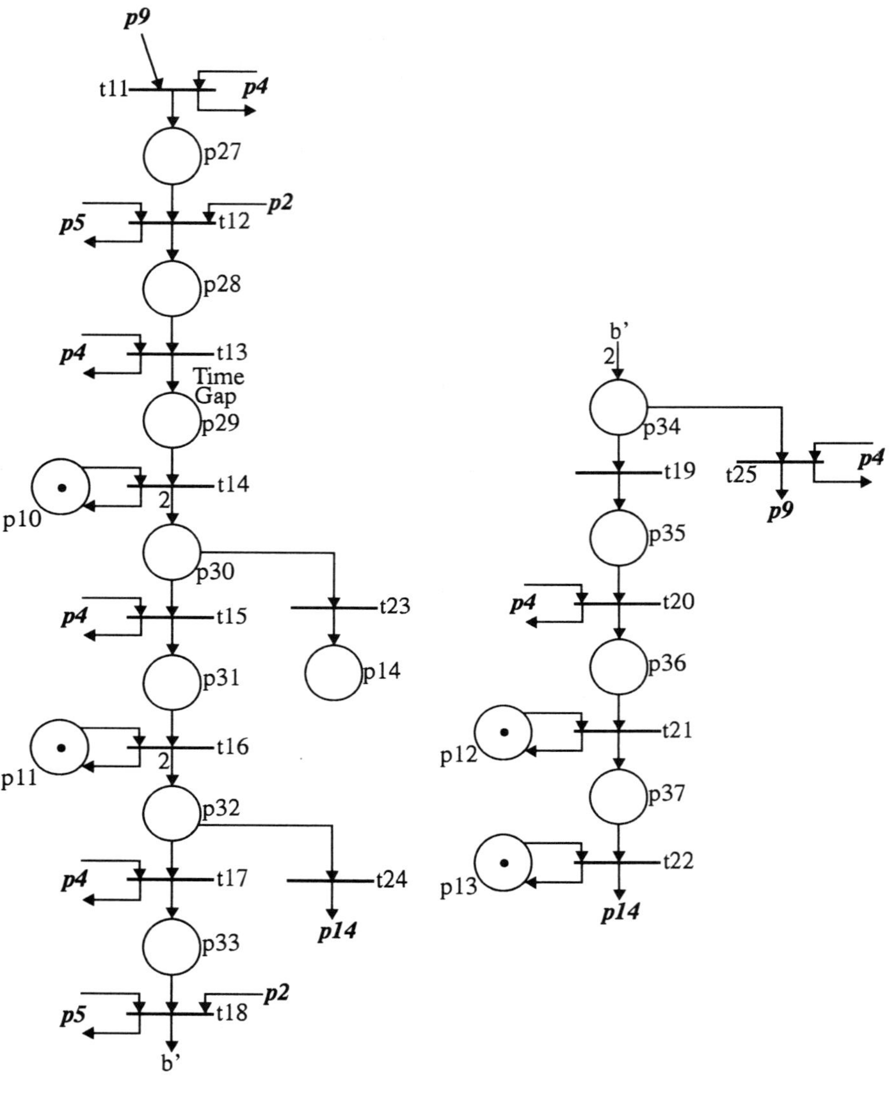

(b)

Fig. 12.4b First-level mapping.

availability will be present throughout, since the different modules will be "fighting" each other for the transport system to service them. Similarly some conflict will also be present in the storage chamber (p_9), since it serves three different clients. The automated sample-handling system (ASHS) described previously is also used numerous times throughout. In this example, samples (in containers) are continually fed and

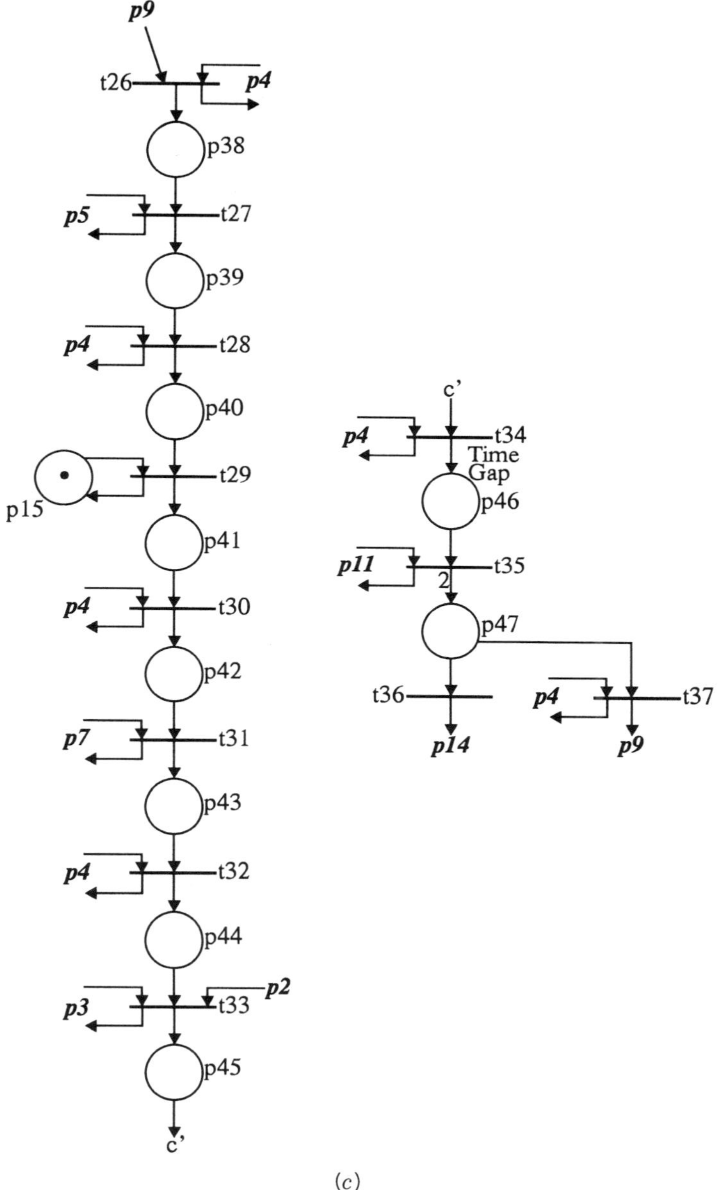

(c)

Fig. 12.4c Second-level mapping.

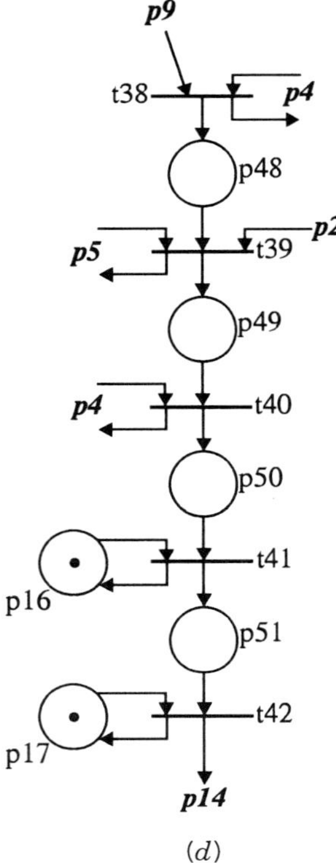

(*d*)

Fig. 12.4*d* Sequencing process.

transported throughout the system, and different modules act on the contents of these containers. The time each sample spends in a particular module varies throughout so careful planning must be taken to avoid bottlenecks. Petri net analysis and simulation will be able to show the occurrence of bottlenecks. The *time gap* (in places p_{29} and p_{46}) gives the time length for bacteria culture growth. In this example this time is spent in the conveyor transport system. As one can see, the flow in the model follows just like the PCR Petri net model described earlier, but now the model is a lot more complicated due to the numerous interactions in the entire system. Although the details of the modules are not given, this Petri model provides an understanding as to how the process flow of a sample occurs in the system. Further analysis with Petri nets could eventually provide insight to the designer as to how the factory should be designed and operated.

SIMULATION

Numerous commercial and noncommercial simulation packages have been written for Petri nets and are widely available. Due to the uniqueness and nature of the Genome

Factory, it is believed that an in-house custom discrete event simulation package might better serve this purpose. Figure 12.5 (on page 276) illustrates a doubly linked list structure very common in the computer sciences. This structure represents the necessary engine that drives the simulation. A specific data structure will have to be defined that describes and integrates the Petri net model of the Genome Factory. Each column of the list represents the time stamp of when an event occurs. In each column one or more events can have the same time stamp. This represents different machines (events) occurring at the same instance. The *ev_head* always points to the next event that occurs, and this event gets pulled off the list. An event with the same time stamp, or if this is not present, an event with the next time stamp, will now be pointed to by *ev_head*. New events with a time stamp defined get inserted into the list as needed. In the long term, a simulation of the Genome Factory will be performed once a complete Petri net model of the entire Genome Factory has been established. The model is envisioned to include concurrent assembly lines performing different tasks at the same instant. It will also include numerous shared resources. The intention is to use a simulation of the Petri net model to provide a clear representation of the flow of the process and to lead to a better design of the factory.

CONTROL AND PERFORMANCE EVALUATION

An advantage of Petri nets is that it is straightforward to go from a model to a simulation or control design. A computer-generated simulation of the Petri net shows the flow of the tokens in the system providing information on what might occur in the real system. For example, a simulation may show where a bottleneck or starvation might occur in the system, which would require a system redesign to eliminate this problem. The ability to simulate a large system such as the Genome Factory before it is even built would provide huge cost savings down the road.

A controller for a system is also easily generated from a Petri net model. The controller's goal is to establish and maintain a desired sequence of events and to handle asynchronous events. In a large system where there are concurrent events, the controller should be able to synchronize these concurrent events and resolve conflicting machine activities. Finally, the controller should be able to satisfy any constraints imposed on the system such as production rates and precedence of jobs.

The use of a Petri net based supervisory controller for a flexible manufacturing system has been designed, implemented, and illustrated in Desrochers and Al-Jaar [4]. Their system is a scaled model of an automated factory shop floor consisting of machining workstations, automated storage and retrieval systems, a robotic assembly workstation, and the like. The hierarchical control structure implemented is shown in Fig. 12.6 (on page 277) [4]. It is envisioned that the same hierarchical structure may be used for the Genome Factory. In considering the performance of any system, criteria need to be specified. For manufacturing systems, criteria such as makespan (total time to manufacture a product), machine idle time, and mean queue time serve in determining the overall performance of the system. A rigid criterion based on efficiency and productivity is the throughput rate (parts/unit time) of the system. As stated before, Olson projects the Genome Factory to be capable of a throughput of one sample per second.

TABLE 12.3 Place and Transition Definitions: Genome Factory

p_1: Chromosome cells available

p_2: Sample container ready/available

p_3: Flow cytometer ready

p_4: Sample transport system ready/ available

p_5: Automated sample-handling system (ASHS) ready/available

p_6: Cosmid packager

p_7: Ecoli transfecter ready/available

p_8: Storage space available

p_9: Storage chamber

p_{10}: Growth monitor (detector) ready

p_{11}: In-line centrifuge ready/available

p_{12}: Gel loader I ready/available

p_{13}: Gel scanner ready/available

p_{14}: Database

p_{15}: M13 packager ready/available

p_{16}: Gel loader II ready/available

p_{17}: Megasequencer

p_{18}: Sorting chromosomes by flow cytometer

p_{19}: Transporting sample (for partial restriction digest)

p_{20}: Partial restriction digest

p_{21}: Transporting sample (for packaging into cosmid)

p_{22}: Package into cosmid

p_{23}: Transporting sample (for transfecting Ecoli)

p_{24}: Transfect Ecoli with cosmid for library

p_{25}: Transporting sample (for flow cytometer)

p_{26}: Determining successfully transfected Ecoli

p_{27}: Transporting sample (for retrieving and aliquotting daily sample)

p_{28}: Retrieving and aliquotting daily sample

p_{29}: Transporting sample (time $\rightarrow$ bacterium cloning/clone expansion)

p_{30}: Detector determining properly grown cultures

p_{31}: Transporting sample (for DNA extraction)

p_{32}: In-line centrifuging/DNA extraction

p_{33}: Transporting sample (for restriction digest)

t_1: Start sorting chromosomes

t_2: Move to ASHS system

t_3: Perform partial restriction digest

t_4: Move to cosmid packager

t_5: Start packaging DNA into particular cosmid

t_6: Move to Ecoli transfecter

t_7: Start transfecting Ecoli with particular cosmid

t_8: Move to flow cytometer

t_9: Check successfully transfected Ecoli

t_{10}: Move sample to storage chamber

t_{11}: Retrieve microtiter plate from storage chamber

t_{12}: Aliquot daily sample to new sample container

t_{13}: Move new sample container to bacteria detector

t_{14}: Determine properly grown cultures

t_{15}: Move sample to in-line centrifuge

t_{16}: Centrifuge and extract DNA

t_{17}: Move sample to ASHS system

t_{18}: Aliquot DNA sample for storage

t_{19}: Start restriction digest

t_{20}: Move restriction enzyme digested plate to gel loader I

t_{21}: Load samples into gel

t_{22}: Scan electrophoresed gel/results to database

t_{23}: Results to database (from clone expansion)

t_{24}: Results to database (from DNA extraction)

t_{25}: Move other sample (from p_{34}) to storage

t_{26}: Retrieve sample (with clone DNA) from storage chamber; move to ASHS system

t_{27}: Start partial restriction digest

t_{28}: Move sample to M13 packager

t_{29}: Package digested DNA into M13 bacteria

t_{30}: Move sample to Ecoli transfecter

t_{31}: Start transfecting Ecoli with M13

t_{32}: Move sample to flow cytometer

t_{33}: Sort Ecoli with M13

t_{34}: Move sample to in-line centrifuge

t_{35}: Centrifuge and extract DNA

t_{36}: Results to database (from DNA extraction)

t_{37}: Move sample to storage

t_{38}: Retrieve sample (with sequence ready DNA) from storage chamber; move to ASHS system

t_{39}: Aliquot 4 samples for each DNA to new sample containers; Run PCR

t_{40}: Move polymerase chain reactioned plate to gel loader II

TABLE 12.3 *(Continued)*

p_{34}: Aliquotting DNA sample (have 2 separate microtiter wells)

p_{35}: Restriction digest

p_{36}: Transporting sample (for gel loader I)

p_{37}: Gel separating/loading

p_{38}: Transporting sample (with clone DNA)

p_{39}: Partial restriction digest (2nd-level map)

p_{40}: Transporting sample (for packaging into M13)

p_{41}: Package into M13

p_{42}: Transporting sample (to Ecoli transfecter)

p_{43}: Transfect Ecoli with M13

p_{44}: Transporting sample (to flow cytometer)

p_{45}: Sorting Ecoli with M13

p_{46}: Transporting sample (time $\rightarrow$ expanding M13 infected Ecoli)

p_{47}: DNA extraction (from M13 infected Ecoli)

p_{48}: Transporting sample (containing sequence ready DNA; to ASHS)

p_{49}: Aliquotting 4 samples for each DNA; Running PCR

p_{50}: Transporting sample (for gel loader II)

p_{51}: Gel loading of Megasequencers

t_{41}: Load samples into gel

t_{42}: Run Megasequencer/Send results to database

Ideally, performance analysis of a Petri net based model is best suited for stochastic Petri nets (SPN). SPN is another class of Petri nets where a random time is associated with the firing of the transitions. The transition times are commonly exponentially distributed, making the SPN identical to a homogeneous Markov process. A homogeneous Markov chain can then be associated with each SPN [5]. The performance measures are just functions of the steady state probabilities [4].

Performance analysis for a deterministic Petri net model is still viable. The criteria of the throughput rate is still in effect, but the approach is now different. Due to the large quantities of samples and reagents, it is natural to compare the effectiveness of either a batch approach or a sequential approach. A Petri net model using a batch approach and another model employing a sequential approach are easily simulated in a computer. These simulations can provide information on the throughput in each case.

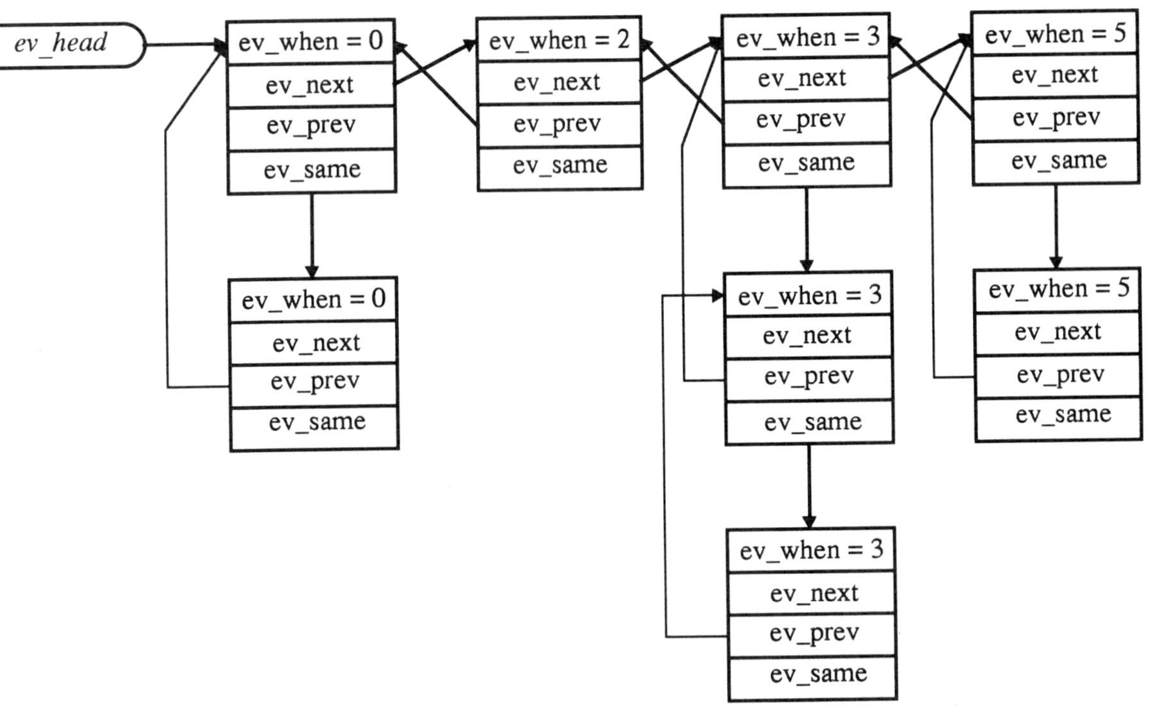

Fig. 12.5 Doubly linked list structure.

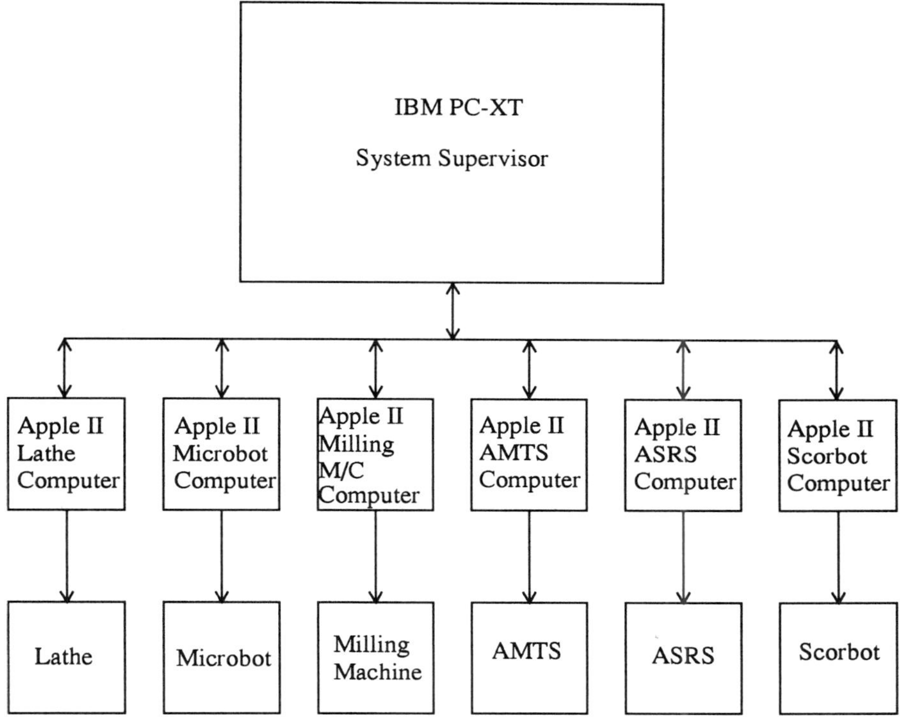

Fig. 12.6 Control architecture [4].

CONCLUSION

It has been shown that Petri nets can be used to model an automated sample-handling system. Their graphical representation allows one to visualize conceptually the relationship between conditions and events of discrete event systems and to analyze the performance of a given system. The graphical representation of the Petri net model makes it relatively easy to simulate on a computer and also to translate the model to control code.

We are presently in the process of using Petri nets for the modeling, control, and performance analysis of a Genome Factory. Work is underway for the design of this proposed factory that is capable of an ambitious throughput of one sample per second [6]. A Petri net based supervisory controller oversees the different phases and process in this factory, but more important, it provides a means to keep track of the unique identity of each sample that flows through the factory.

REFERENCES

1. Taking stock of the Genome Project (1993). *Science* 262:22–23.

2. M.V. Olson, G. van den Engh (1995). Why is it so hard to sequence a few billion pairs of DNA? Univ. of Washington, unpublished.

3. T. Murata (1989). Petri nets: Properties, analysis and applications. *Proc. IEEE* 77:541–580.

4. A.A. Desrochers, R.Y. Al-Jaar (1995). *Applications of Petri Nets in Manufacturing Systems.* IEEE Press, New York.

5. R. David, H. Alla (1994). Petri nets for modeling of dynamic systems—A survey. *Automatica* 30:175–202.

6. M.V. Olson, proposal to the Department of Energy for a UW Genome Center, May 1995.

7. D.R. Meldrum (1995). The interdisciplinary nature of genomics. *IEEE Eng. Med. Bio.* 14:443–447.

8. F. DiCesare, A.A. Desrochers (1991). Modeling, control, and performance analysis of automated manufacturing systems using Petri nets. In *Control and Dynamic Systems, Advances in Theory and Applications*, C.T. Leondes, ed. Vol. 47: *Manufacturing and Automation Systems: Techniques and Technologies.* Academic Press, San Diego, CA, 121–172.

13

Integrating Data Acquisition, Analysis, and Management

CHRIS FIELDS

CONTENTS

INTRODUCTION

The need for integrated information systems to support data acquisition, analysis, and management is one of the primary lessons learned in the spin-up phase of the Human Genome Project. In retrospect, the advantages of an integrated, highly automated approach to data handling are obvious: Genome projects are information-production activities, and they have the same basic requirements for efficiency and quality control as similar activities in other sectors. Most genome laboratories evolved, however, from

Automation Technologies for Genome Characterization, Edited by Tony J. Beugelsdijk.
ISBN 0-471-12806-6   © 1997 John Wiley & Sons, Inc.

traditional small-scale biology laboratories, where data were recorded in individual investigators' notebooks, data analysis programs were developed or purchased as needs became overwhelming, and no laboratorywide methodology was set up for local data or information management. Information was accumulated from multiple experiments, integrated at the stage of preparing a paper for publication, and presented in integrated form only in the printed literature. Example data, usually with some interpretive annotation, were deposited in archival community data banks such as GenBank [1, 2], the Human Genome Mapping Library [3, 4], or the Protein Data Bank [5]. Each published result was considered a "completed" piece of scientific work; indeed, this is what most scientific journals still insist on in their Instructions to Authors.

The traditional style of information management typical of small laboratories is entirely inappropriate to production laboratories in which high throughput, efficient division of labor, and cost effectiveness are central concerns. Today in leading genome laboratories, data and interpretive information are produced using uniform, extensively automated procedures and minimal staff. Process uniformity and replicability are key to efficient, high-quality operation; process breakdowns must be detected and resolved rapidly. The processes used in such laboratories evolve through identification and opening of bottlenecks, or by modification to satisfy new input or output requirements or accommodate new equipment or methods. Information systems are integral components of the infrastructure of high-throughput laboratories, and their development and maintenance account for a significant (20%–30% in leading production laboratories) fraction of the operating budget.

The evolution of laboratories has been accompanied by an evolution of both the type of data produced and the expectations with which data and interpretive information are made public. Genome data are widely recognized as foundational data for molecular, cellular, developmental, and eventually organismic and community biology. Map and sequence data are by their nature meant to be built upon by further investigations. These further investigations are moreover generally recognized as the duty of multiple investigators over an extended period of time. Published results from genome laboratories, even complete genome maps or sequences, are not "completed"; they are starting points for further work, which in all likelihood will be carried out by others. While essentially all scientific results have this provisional or "working" character—"last words" rarely survive as such—the genome project has raised the recognition of this point to a matter of funding-agency policy with the implementation of requirements for rapid data release by publicly supported laboratories.

Both the procedures employed by high-throughput laboratories and the expectations with which their output is greeted by the scientific and applications communities impose demands for information systems well beyond those of traditional, small-laboratory biology. This chapter will focus on DNA sequencing laboratories and the data they produce as examples, but the requirements and characteristics of successful systems apply to mapping, gene expression analysis, functional characterization, and downstream applications as well. As biology and biotechnology become progressively more automated, the types of information systems developed for genome project laboratories will become both ubiquitous and essential for effective research.

ANATOMY OF A HIGH-THROUGHPUT SEQUENCING LABORATORY

The development of DNA sequencing methods and tools has been reviewed by [6]. From an engineering perspective, DNA sequencing is a production activity that uses biological samples as a raw material and produces interpreted DNA sequence data as output. The major steps of this process are shown in Fig. 13.1. At each stage ancillary input materials and a variety of tools, ranging from clone-handling robots to software for predicting gene structure, are required. Automation of the process has involved the development of novel ancillary materials, such as fluorescent dye-labeled DNA primers [7] and dideoxy-termination nucleotides, and novel instruments, such as automated DNA sequencers [8] and sequencing-reaction robots [9]. Over the past five years the sustainable rate of automated DNA sequencing has increased by over a factor of six, from about 8400 raw bases per day (24 lanes, 350 bases per lane, 1 run per day on an Applied Biosystems 373) to 54,000 raw bases per day (36 lanes, 500 bases per lane, 3 runs per day on an Applied Biosystems 377) for a single instrument under typical conditions. This increase in sequencing throughput has driven corresponding increases in the throughput of upstream template production and downstream data analysis and interpretation. The latter is especially significant: In previous years data analysis was viewed as an activity separate from sequencing, to be carried out after the experiments were complete. In a production laboratory the data generated in a day must be analyzed in a day to avoid falling behind. The efficiency and cost-effectiveness constraints of production apply to the science of sequence data analysis and interpretation, not just to the technical components of sequencing.

The increase in per-instrument throughput, and the fact that single laboratories may operate ten or more instruments simultaneously, has placed personnel, process management, and quality control requirements on sequencing laboratories analogous to those faced by industrial production operations generally. These include organization of personnel into work teams, often with specific shifts and responsibilities, institution of procedures for process-load balancing and bottleneck elimination, process control, inventory control, and development of quality control standards [10]. These innovations, in turn, have further increased the requirements for laboratory information systems: Such systems must now support work-flow management and quality control as well as sequence data and analysis. Different laboratories have taken different approaches to these expanding requirements [11]. A few have attempted to maintain paper laboratory notebooks and flat data files, often paying a price in flexibility, efficiency, and data quality. Most successful high-throughput laboratories have developed specialized laboratory databases, generally based on commercial relational management systems [12–15].

Integrated laboratory information systems developed on robust platforms provide laboratories striving for efficiency and cost effectiveness with the flexibility needed to take advantage of new protocols and procedures without systemwide alterations in methodology or expensive special-purpose software development. The basic components of a database system to support the sequencing process reflect the components of the process itself, and hence correspond to the data and protocol types shown in Fig. 13.1.

Sequencing Materials Flow

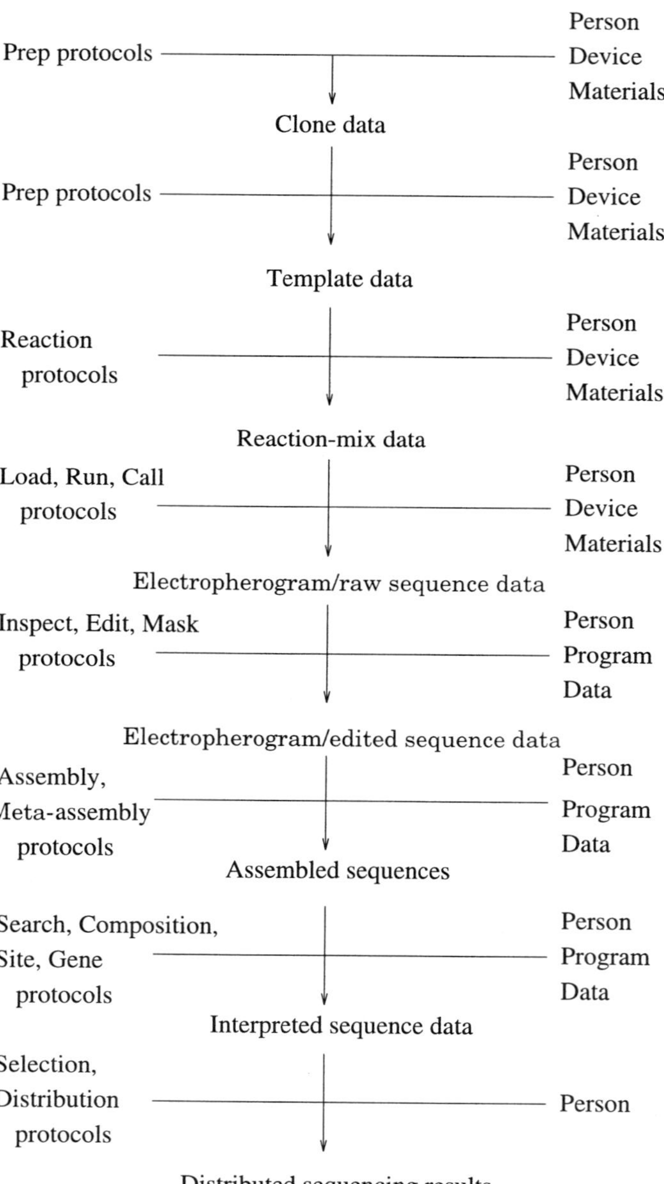

The database design can be approached from two complementary perspectives: that of the materials flowing through the process and that of the protocols that compose the process. The materials-flow perspective is typically that of the technician, who wants to know what has been done, and what needs to be done next, to a given sample as it flows through the process. The protocol perspective is more likely to be that of the instrument that needs precise instructions on what to do to each sample that it receives as input, or of the protocol designer. In a setting in which protocols change slowly, the database may be able to represent protocols simply by name, brief description, and expected duration; in a setting in which the protocols are regularly modified, a more detailed representation may be necessary. Schema fragments illustrating these two perspectives within a relational model are shown in Fig. 13.2. Databases may support these perspectives to different extents; for example, a sequencing laboratory that employing a fixed sequencing chemistry (e.g., a particular vendor's kit), but running similarity searches with a variety of mismatch and gap parameter settings might require a database supporting the materials-flow perspective throughout as well as the protocol perspective for sequence analysis protocols.

DESIGN DECISIONS AND THEIR CONSEQUENCES

In general, the perspective taken by the database design, the semantics used to describe the domain, and the architecture and platform chosen for implementation determine the types of queries that a database can answer. The choices of perspective and semantics are tightly interdependent; a greater number of entites and process components must be distinguished if a protocol perspective is to be supported as well as a materials-flow perspective. The choice of architecture and implementation platform is dependent on the level of use and complexity of query capability required. Relatively simple reports describing the current status of a process can be generated with a very simple relational or object database; retrospective or other cross-instance comparative queries generally require a relational architecture and a development platform supporting a general query

Fig. 13.1 Materials and data flow in a typical sequencing laboratory, with types of data needed to track the sequencing and analysis process indicated. Clones for sequencing may be prepared locally or obtained from another site. Sequencing templates are typically prepared in-house, either by PCR or subcloning. A variety of protocols and some commercial kits are available for 96-well format bulk template preparation. Sequencing reactions are generally automated, and at the least involve use of thermal cyclers or similar instruments. Automated sequencers yield electropherograms (traces) that are generally carried through the editing and assembly process. A variety of analysis programs, typically including similarity-searching, compositional analysis, and site-searching programs, are generally run on all data not rejected for failing quality-control standards. Distribution may be to public databases or in-house R&D groups.

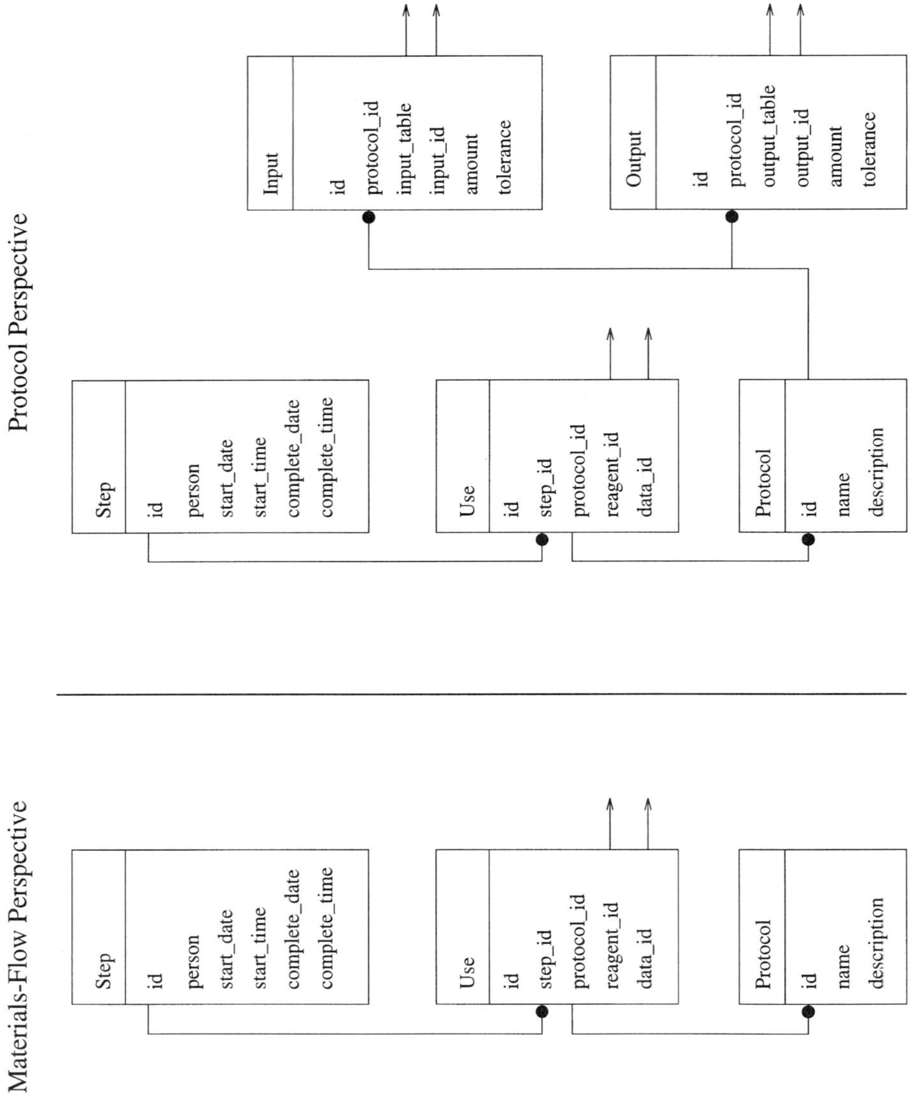

Protocol Perspective

Input
id
protocol_id
input_table
input_id
amount
tolerance

Output
id
protocol_id
output_table
output_id
amount
tolerance

Step
id
person
start_date
start_time
complete_date
complete_time

Use
id
step_id
protocol_id
reagent_id
data_id

Protocol
id
name
description

Materials-Flow Perspective

Step
id
person
start_date
start_time
complete_date
complete_time

Use
id
step_id
protocol_id
reagent_id
data_id

Protocol
id
name
description

language such as SQL. Support for multiple simultaneous users performing read and write operations generally also requires a robust relational database management system.

The naming of entities flowing through a complex process is a universal challenge in database design. Names must be unique if the named entities are to be identified precisely. This requirement often clashes with the tradition in biology laboratories of making names describe the named entities. For example, clones are often named by their well locations in standard 96- or 384-well microtiter plates. This convention makes it easy to find a clone by its name, and simplifies the often manual task of moving clones from one plate or plate format to another. Sequences are similarly often named for the clones from which they are derived. Naming conventions along these lines can lead to inconsistencies, however, whenever what is to count as "the same thing" is context dependent or subject to change if a protocol changes. For example, naming clones for plate locations can cause havoc if clones from several plates are re-arrayed in multiple ways to create custom grids, requiring open-ended thesaurus files that relate the multiple names for a single clone, and relate alternative names to different uses. Similarly naming sequences after the clones from which they are derived presents difficulties if sequences are run from different primers, with different chemistries, or even multiple times to compensate for gel or equipment failures. Naming sequences after genes that they contain, or after other sequences to which they are similar, leads to analogous problems. For example, if a new sequence is reported to which a previously named sequence has an even greater similarity, either the name must be changed or an explicit exception created. As exceptions accumulate, the naming convention is effectively abandoned. The assignment of arbitrary, unique, semantics-free identifiers that serve as rigid names for an entity throughout the sequencing and analysis process is essential to avoid such problems. These identifiers are used as primary, unchanging names; descriptive names can be used alongside such rigid identifiers to make queries easier. Most commercial database systems support rigid identifiers as well as multiple descriptors.

Sequencing strategies vary in the complexity of the representations and types of names required. Some laboratories engaged in high-redundancy shotgun sequencing, for example, name templates for the cosmid, or other large-scale clone from which they are derived, sequence the templates from a single cosmid simultaneously, manage

---

Fig. 13.2 *Partial entity-relationship diagrams showing some of the fields that may be employed in tables describing steps in a laboratory or analysis procedure and the protocols that they employ. Outgoing arrows indicate links to additional tables that are not shown. From a materials-flow perspective, it may be sufficient to simply name the protocols, and provide a human-readable, but otherwise unqueryable, description. A protocol perspective requires describing the protocol in queryable form, minimally by explicitly representing its inputs and outputs. Since many different protocols may use the same inputs and outputs, this is most efficiently done by breaking these out into separate tables. A similar approach may be taken to specifying durations, instrument usage, or internal structures of protocols, the latter using a "step" construct such as used here for the overall procedure.*

the raw data from the templates from a single cosmid as a "project" within a sequence assembly tool, maintaining records on reaction protocols, additional sequence walks, or special chemistries in paper logbooks. This style of data management effectively treats "sequencing" as a single protocol with a cosmid or similar clone as the input and its assembled sequence as the output, and is highly dependent on the memories and note-taking discipline of the personnel implementing it. Access to what goes on "within" this protocol by anyone other than the implementing personnel is assumed to be necessary only rarely, and hence not to justify development and maintenance of a database to represent the internals of the protocol. The sequence derived from any individual template in a shotgun project, for example, may not be of sufficient interest to be tracked individually by a database; once an assembly has been completed, the component sequences are unlikely to be revisited. In an expressed sequence tag (EST) project, in contrast, each individual sequence may represent a unique gene [16], and the ability to easily trace back from a sequence to its clone and library is essential for downstream analysis. Similarly sampling and semidirected sequencing methods such as ordered shotgun sequencing [17], sequence-map gaps[18], or end-sample sequencing [19] require careful tracking of sequences from individual templates or other clones.

The level of process management and quality control adopted by a laboratory also has a profound impact on data management requirements. Some quality control steps, such as the inspection of sequence electropherograms as they are produced or the analysis of sequences to remove vector fragments, are typically carried out in real time. For such methods the database needs only to be able to record the fact that a procedure was carried out and its result, with details of the procedure used in some cases. Retrospective quality control analyses that require comparing data obtained over an extended period, in contrast, generally require an ability to construct complex queries over both experimental and process data. Assessing the closure efficiencies of different sampling strategies or the quality of libraries are problems of this type. Retrospective quality control procedures are often very difficult to implement if data describing different steps in a process are maintained in separate databases or by different methods; they become progressively more tractable as the laboratory database is extended to cover all aspects of the clone preparation, sequencing, and analysis process.

Databases supporting complex, ad hoc queries can also greatly increase the flexibility and power of the data analysis process by providing a mechanism for automatically relating different analysis results. Sequence (or physical map) assembly provides a case in point. By saving intermediate assembly results in a queryable database, they are made available for additional processing using alternative overlap detection or layout algorithms, heuristics to take template length, end-pair relations, or restriction map data into effect, or direct user manipulation. Such capabilities are not provided by present-day assemblers but are very useful in genome sampling and partially directed sequencing strategies in which, for example, assemblies of low-pass shotgun or other partial data sets are used to efficiently choose which sequences to determine next. Maintaining analysis results in the laboratory database also allows them to serve as inputs to retrospective quality-control analyses.

DESIGN METHODOLOGIES AND PLATFORMS

Genome informatics has developed during a period of rapid development in computer science and engineering, and various artificial intelligence, neural network, and especially object-oriented design methods have been applied to genome-related problems. Some highly successful systems, such as the GRAIL3 gene-structure prediction system, which combines neural network and expert system methods [20], have been developed using relatively nontraditional tools and methods. Especially in the area of databases, however, system functionality has often been conflated with design methodology, and claims that certain desireable functionality can only be achieved with particular design methodologies or tools have been common. Such claims contradict the fundamental computing principle that any function can be implemented with many algorithms, and any algorithm with arbitrarily many distinct programs. As applications have matured and been put to the test of laboratory use, much of the idealism of early design arguments has been replaced with more pragmatic thinking. Heterogeneity of both design methodologies and platforms has emerged as a dominant theme of successful systems.

A number of groups have experimented extensively with either commercial object-oriented database systems or *de novo*, bottom-up implementations of object management systems in C or C++. The most widely used noncommercial database system is ACeDB, originally developed to support the *Caenorhabditis elegans* genome project [21] and since adopted by many other genome projects [22]. ACeDB is an object-management system written in C that includes a sophisticated graphic interface and (in later versions) a modest query language. Although additional features are continually being developed by laboratories employing ACeDB, its main strengths are ease of data entry and graphically guided browsing. ACeDB does not currently support the multiple-user, multiple-task management mechanisms provided by robust commercial systems, and it is not designed for query distribution or other forms of automatic communication between distributed databases. In general, task-specific *de novo* database implementations tend to lack sophisticated query languages and management functions, which often become essential as the data throughput, complexity of operations, and user base grow [23]. Several projects have abandoned commercial object-oriented managers in favor of relational managers for the same reasons.

An alternative methodology for developing an object-oriented system is to build an object layer on top of a relational database. Many of the relational databases developed for genome laboratories are in fact object-oriented at the user-interface level (unless the user is interacting with the data directly in SQL). Implementation of an object-oriented "buffer" between users and external tools and an underlying relational manager allows the system to take advantage of both the conceptual and software-engineering advantages of an object-oriented methodology and the relative maturity of commercial relational database management systems [24].

Intense debate has surrounded choices of hardware platforms in many genome laboratories. The consequences of selecting a single platform often include users dissatisfied with an unfamiliar operating-system interface, unacceptably poor performance of the database server, or inability to implement a robust system at all due to limitations in the available software for a given hardware platform. As with software methodologies and platforms, adopting a heterogeneous hardware solution and a client-server architecture offers distinct advantages. The availability of World-Wide Web interface standards, and of a variety of commercial cross-platform interface development tools, allows the development of user interfaces that run on the personal machines with which users are already familiar. The core database management software can then run on a platform with the processor, operating system, and memory resources required for acceptable performance. While such solutions may have higher initial costs, the benefits of appropriate functionality and relative scalability can greatly outweigh them in even the medium term.

FROM ARCHIVES TO SHARED RESOURCES: THE NEW ROLE OF COMMUNITY DATABASES

The development of high-throughput laboratories is having a profound impact on the division of labor in biology. Large-scale mapping efforts have increased the efficiency of postional cloning by covering almost every gene with a variety of clones [25]. Sequences that researchers have spent considerable effort trying unsuccessfully to clone by traditional means often turn out to be already available from EST projects [26]. High-throughput genome sequencing efforts perform routine analysis to predict locations of new genes but leave detailed functional characterization of most if not all of them to other researchers [21, 27–29]. This division of labor benefits all concerned. Laboratories with interest and expertise in particular genes or proteins are saved the effort of cloning and sequencing them, while high-throughput genomics laboratories are saved the diversion of energy and resources needed for long-term studies of single genes. Genes of medical or technological interest are delivered to biotechnology companies and the public more rapidly and cost-effectively. Open-access community databases are key intermediaries in making this division of labor possible, and such databases will play an increasingly critical role as high-throughput characterization of genomes progresses.

High-throughput laboratories are major contributors to and users of community databases. Consequently these resources are driven to make a transition from their traditional role as static archives of completed results to dynamic communication media that support data exchange and progressive, continuous data analysis and integration [30]. Fundamental to this transition is the realization that most of the work of characterizing a new gene will be done by scientists other than the gene's discoverer. This has always been the case, but the follow-on, third-party work on gene structure confirmation, expression analysis, and functional characterization has traditionally been presented only in the paper literature; sequence archives such as GenBank, for example, store only the original sequencing results. Results published in journals are difficult to locate and access; they would be far more usable if presented in computer-searchable and readable form, linked directly to the underlying sequence data. All interested researchers could then participate in the analysis of a given gene, communicating their intermediate results in a common database. Systems to support this type of collaborative effort have been pursued for some time by a variety of businesses, under rubrics such as "computer-supported cooperative work." The Worm Community System, a collaborative database developed for the *Caenorhabditis* research community, was one of the first experiments of this type in molecular biology [31].

As the genome projects progress, and as methods for expression analysis, reverse genetics, and other follow-on biology become progressively more automated, the distinction between laboratory databases, and community databases will become progressively more blurred. The old static data archives may survive in a data-maintenance role but will be replaced in day-to-day use by databases that span communities, but allow the same kinds of interactive analysis, multiple-user updating, and collaborative construction of results that laboratory databases permit [32]. Data in such databases will be subject to constant manipulation as old analyses are revised, new relevance links are made, and the data themselves are joined to other data obtained by other investigators. The standard laboratory database issues of account protection, input attribution, multiple-user contention, and collision avoidance, with their attendant requirements for robust management systems that support multiple users executing reads and writes, will become dominant issues for community databases. Support for complex queries spanning different types of data will be essential.

Initial steps in the direction of developing interactive community databases are being taken by both the Genome Data Base[4] and the new Genome Sequence DataBase (GSDB; [24]). Both of these databases are implemented on a robust relational management system (Sybase, Emeryville, CA) and offer users anonymous access for ad hoc queries formulated in SQL. Both have developed schema and user-interface support for arbitrary third-party annotation of existing data, and are designed to allow distributed joins that access data in other relational databases.

INTEGRATING HETEROGENEOUS DATA: SEMANTICS AND SCALABILITY

Genome projects have tended to focus on single organisms or chromosomes and on the technology required to achieve the immediate goals of mapping and sequencing. While the number of distinct types of data required by such projects in the laboratory is large, they are highly related. The information accompanying genome data into the community databases has, thus far, typically been restricted to standard feature annotations and some source and phenotype information. The information relevant to characterizing the functions of a gene and its products, or assessing variability in gene structure or function across a population, in contrast, spans much of biology. Some of this information is illustrated schematically in Fig. 13.3. As genome projects move into the application phase, progressively more of these additional types of information will need to be integrated with genome data to provide the background information needed by working biologists.

Integration of data of multiple types from multiple sources can be pursued using two fundamentally different strategies. The more straightforward strategy is to locally integrate all data in a single, centralized system. This generally involves obtaining data from multiple sources, transforming it into a common representation, and linking data

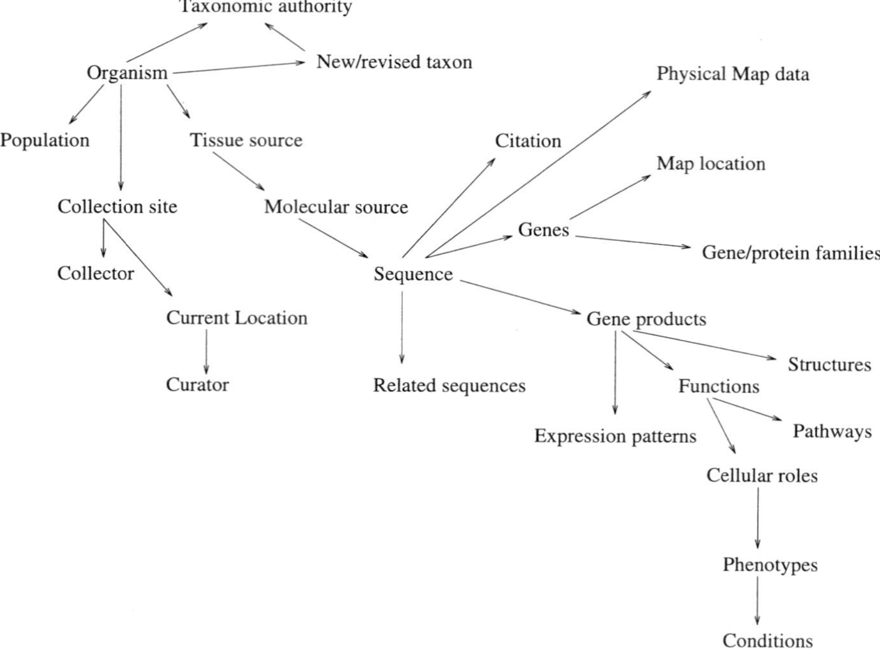

Fig. 13.3 Some of the types of data relevant to characterizing sequences, from population variation and biogeography through product structure, function, and expression. Support for ad hoc queries spanning all of these types of data will be needed by multidisciplinary programs for high-throughput functional characterization of genes and genomes.

items that are relevant to each other. The representation into which data are transformed must be sufficiently expressive to capture the semantic distinctions made by all data sources; otherwise, some information will be lost, and the resulting integrated database may be wrong or inconsistent. Construction and curation of such systems is generally carried out by a centralized database staff, which acts authoritatively to resolve semantic conflicts; such systems are therefore essentially archival. The Entrez system developed by the National Center for Biotechnology Information (NCBI), which integrates DNA and protein sequence data with citations using a precomputed similarity measure as an index of relevance between entries [2] is a centralized archival system. The ACeDB-based databases developed by several genome projects, which typically integrate strain, phenotype, map, and sequence data with literature citations and informal comment [22], are also systems of this type.

Centrally integrated archival systems work well in domains in which the types of information that are relevant to a given data item can be strictly limited and are not subject to change. As the scope and semantic complexity of the domain increase, such "circumscription" becomes progressively more difficult. Scientific domains are notoriously "open"—uncircumscribable—due both the incompleteness of scientific knowledge and to the tendency of scientists in different disciplines to employ distinct and often incommensurable concepts and vocabularies. For example, the concept of "gene" used in classical genetics has no clear definition at the level of resolution provided by DNA sequencing; hence links between genetic maps and physical maps or their underlying sequences are often problematic. This situation is typical whenever scientific data obtained at radically different spatial or temporal resolutions are integrated. The difficulties posed by open domains have been investigated for some time by the database and artificial intelligence communities, and it is widely accepted that no general solution to the circumscription problem exists [33–36]. Without such a general circumscription method, centralized approaches to data integration cannot scale as the size and conceptual diversity of the problem domain increase.

Decentralized, loosely coupled database systems provide an alternative to centralized archival systems. Such systems typically comprise multiple, independently developed and curated databases related by conventions that link unique identifiers with minimal assumptions about semantics. Because such federated systems do not impose a common representation on the data, they can support multiple interpretations of a datum and multiple descriptive languages for subdomains. The task of circumscription is devolved to the user, who effectively constructs a task-specific circumscription of the domain with each query and must assume responsibility for that circumscription being coherent. Decentralized systems are essentially arbitrarily scalable to open domains. The World-Wide Web, with its multiple cross-domain links, is a prominent example of such a system.

Development of a federated network of autonomous, distributed community databases capable of the type of anonymous interoperability supported by the Web is essential for the multidisciplinary application of genome data. Such a federation will need to support complex, ad hoc queries joining data items from multiple databases in addition to providing browsing capabilities like those of the WWW [32]. Steps toward the development of such a federation are currently being taken by GDB, GSDB, the Mouse Genome Database (MGD) under development at the Jackson Laboratory, and the Sequences, Sources, and Taxa (SST) database being developed at The Institute for Genomic Research.

AFTER THE GENOME: FUNCTIONAL BIOLOGY AND BIOTECHNOLOGY

The genome projects have transformed bioinformatics from a somewhat obscure theoretical discipline into a robust engineering discipline. As the drive for greater automation, higher throughput, and improved cost effectiveness has intensified, the demands placed on information systems have become progressively more stringent. The level of sophistication of design and software engineering has increased, and the robustness and scalability of genome information systems has benefited. It is no longer appropriate for informatics to be an afterthought in designing a high-throughput laboratory or to design experimental strategies without considering the feasibility and cost of the information systems required to support them.

An evolutionary process at least as far-reaching as that instigated by genomics will be required as biology moves from a focus on genomes to large-scale, high-throughput, functional biology and biochemistry. Computational chemists already view high-resolution, manipulable three-dimensional graphics as standard; advanced molecular design laboratories now incorporate walk-around virtual reality systems with high-speed computers to simulate molecular docking interactively. Systems of this sort, expanded to the scale of the cell, will be the standards of tomorrow's developmental biology and cellular biochemistry. Programmable simulation systems analogous to the Genesis system for neuron and synapse modeling [37] will be applied to pathway modeling problems. Robust geographical information systems will be used for correlating genetic with environmental variation in population studies. As these capabilities mature, bioinformatics will become an integral component of biology and biotechnology.

REFERENCES

1. C. Burks, et al. (1989). GenBank: Current status and future directions. *Meth. Enzymol.* 183:1–22.

2. D. Benson, M. Boguski, D. Lipman, J. Ostell (1994). GenBank. *Nuc. Acid. Res.* 22:3441–3444.

3. J.C. Stephens, I. Cohen, K. Kidd (1990). The human gene mapping library: Present status and future directions. In: *Computers and DNA,* G. Bell T. Marr, eds. Addison-Wesley. Reading, MA, 69–81.

4. K. Fasman, A. J. Cuticchia, D. Kingsbury (1994). The GDB human genome database anno 1994. *Nuc. Acid Res.* 22:3462–3469.

5. D. George, W. Barker, H.-W. Mewes, F. Pfeiffer, A. Tsugita (1994). The PIR-International protein sequence database. *Nuc. Acid. Res.* 22:3569–3573.

6. E. Chen E (1994). The efficiency of automated DNA sequencing. In *Automated DNA Sequencing and Analysis,* M. Adams, C. Fields, and J. C. Venter, eds. Academic Press, San Diego, 3–10.

7. L. Smith, S. Fung, M. Hunkapillar, T. Hunkapillar, L. Hood (1985). The synthesis of oligonucleotides containing an aliphatic amino group at the 5′ terminus: Synthesis of fluorescent primers for use in DNA sequence analysis. *Nuc. Acid. Res.* 13:2399–2412.

8. L. Smith, J. Sanders, R. Kaiser, P. Hughes, C. Dodd, C. Connell, C. Heiner, S. Kent, L. Hood (1986). Fluorescence detection in automated DNA sequence analysis. *Nature* 321:674–679.

9. R. Cathcart (1990). Advances in automated DNA sequencing. *Nature* 347:310.

10. M. Adams, A. Kerlavage, J. Kelley, J. Gocayne, C. Fields, C. Fraser, J. C. Venter (1994). A model for high-throughput DNA sequencing and analysis core facilities. *Nature* 368:474–475.

11. S. Lewis (1994). Design issues in developing laboratory information management systems. In *Automated DNA Sequencing and Analysis,* M. Adams, C. Fields, J. C. Venter eds. Academic Press, 329–338.

12. A. Kerlavage, M. Adams, J. Kelley, M. Dubnick, J. Powell, P. Shanmugam, J.C. Venter, C. Fields (1993). Analysis and management of data from high-throughput expressed sequence tag projects. *Proc. Twenty-Sixth Annual Hawaii International Conference on System Sciences.* Los Alamitos, CA. IEEE Computer Society Press, 585–594.

13. S. Clark, G. Evans, H. Garner (1994). Informatics and automation used in physical mapping of the genome. In: *Biocomputing: Informatics and Genome Projects,* D. Smith, ed. Academic Press, San Diego, CA, 13–49.

14. T. Slezak, M. Wagner, M. Yeh, L. Ashworth, D. Nelson, D. Ow, E. Branscomb, A. Carrano (1995). A database system for constructing, integrating, and displaying maps of chromosome 19. *Proc. Twenty-eighth Annual Hawaii International Conference on System Sciences,* in press.

15. A.R. Kerlavage, W. FitzHugh, A. Glodek, J. Kelley, J. Scott, R. Shirley, G. Sutton, M. Wai-Chiu, O. White, M. Adams (1995). Data management and analysis for high-throughput DNA sequencing projects. *IEEE Engineering in Medicine and Biology* 14:710–717.

16. M. Adams, et al. (1991). Complementary DNA sequencing: Expressed sequence tags and the human genome project. *Science* 252:1651–1656.

17. E. Chen, D. Schlessinger, J. Kere (1993). Ordered shotgun sequencing: A strategy for integrating mapping and sequencing of YAC clones. *Genomics* 17:651–656.

18. S. Richards, D. Muzny, A. Civitello, F. Lu, R. Gibbs (1994). Sequence map gaps and directed reverse sequencing for the completion of large sequencing projects. In *Automated Sequencing and Analysis,* M. Adams, C. Fields, J. C. Venter, eds. Academic, Press, San Diego, CA, 191–198.

19. M. Smith, A. Holmsen, Y. Wei, M. Peterson, G. Evans (1994). Genomic sequence sampling: A strategy for high-resolution sequence-based physical mapping of genomes. *Nature Genet.* 7:40–47.

20. E. Uberbacher, J. Einstein, X. Guan, R. Mural (1993). Gene recognition and assembly in the GRAIL system: Progress and challenges. In *Bioinformatics, Supercomputing, and Complex Genome Analysis,* H. Lim, J. Fickett, C. Cantor, and R. Robbins, eds. World Scientific, Singapore, 465–476.

21. J. Sulston, J. et al. (1992). The *C. elegans* genome sequencing project: A beginning. *Nature* 356:37–41.

22. J. Cherry, S. Cartinhour (1994). ACEDB: A tool for biological information. In *Automated DNA Sequencing and Analysis,* M. Adams, C. Fields, and J. C. Venter, eds. Academic Press, San Diego, CA, 347–356.

23. N. Goodman, (1994). An object-oriented DBMS war story: Developing a genome mapping database in C++. In *Modern Database Systems,* W. Kim, ed. ACM Press, New York, 216–237.

24. G. Keen, et al. (1996). The genome sequence database (GSDB): meeting the challenge of genomic sequencing. *Nucleic Acids Research,* Vol. 24, No. 1:13–16.

25. F. Collins, (1995). Positional cloning moves from perditional to traditional. *Nature Genet.* 9:347–350.

26. N. Papadopolous, et al. (1994). Mutation of a *mutL* homolog in heriditary colon cancer. *Science* 263:1625–1629.

27. S. Oliver, et al. (1992). The complete DNA sequence of yeast chromosome III. *Nature* 357:38–46.

28. R. Wilson, et al. (1994). 2.2 Mb of contiguous nucleotide sequence from chromosome III of *C. elegans. Nature* 368:32–38.

29. R. Fleischmann, et al. (1995). Whole-genome random shotgun sequencing and assembly of *Haemophilus influenae* Rd. *Science* 269:496–508.

30. C. Fields (1992). Data exchange and inter-database communication in genome projects. *Trends Biotech.* 10:58–61.

31. B. Schatz (1992). Building an electronic scientific community. *J. Managem. Info. Sci.* 8(3):87–107.

32. M. Waterman, et al. (1994). Genome informatics I: Community databases. *J. Comput. Bio.* 1:173–190.

33. P. Hayes (1973). The frame problem and related problems in artificial intelligence. In *Artificial and Human Thinking,* A. Elithorne and D. Jones, eds. Elsevier, Amsterdam, 45–59.

34. M. Coombs, J. Alty (1984). Expert systems: An alternative paradigm. *Int. J. Man-Machine Studies* 20:21–44.

35. C. Hewitt (1985). The challenge of open systems. *Byte* 10:223–242.

36. C. Fields, E. Dietrich (1991). Engineering artificial intelligence applications in unstructured task environments: Some methodological issues. In *Artificial Intelligence and Software Engineering,* D. Partridge, ed. Ablex, Norwood, NJ, 369–381.

37. M. Wilson, J. Bower (1989). The simulation of large-scale neural networks. In *Methods in Neuronal Modeling,* C. Koch, I. Segev, eds. MIT Press, Cambridge, MA, 291–333.

Index